Algorithm-Structured
Computer Arrays and Networks

Architectures and Processes
for Images, Percepts, Models, Information

This is a volume in
COMPUTER SCIENCE AND APPLIED MATHEMATICS
A Series of Monographs and Textbooks

Editor: WERNER RHEINBOLDT

A complete list of titles in this series appears at the end of this volume.

Algorithm-Structured Computer Arrays and Networks

*Architectures and Processes
for Images, Percepts, Models, Information*

LEONARD UHR

*Department of Computer Sciences
University of Wisconsin
Madison, Wisconsin*

1984

ACADEMIC PRESS
(Harcourt Brace Jovanovich, Publishers)

Orlando   San Diego   San Francisco   New York   London
Toronto   Montreal   Sydney   Tokyo   São Paulo

ACADEMIC PRESS, INC.
Orlando, Florida 32887

United Kingdom Edition published by
ACADEMIC PRESS, INC. (LONDON) LTD.
24/28 Oval Road, London NW1 7DX

Library of Congress Cataloging in Publication Data

Uhr, Leonard Merrick, Date.
   Algorithm-Structured Computer Arrays and Networks.

   (Computer science and applied mathematics)
   Bibliography: p.
   Includes index.
   1. Parallel processing (Electronic computers)
2. Computer networks.   3. Computer architecture.
I. Title.   II. Series.
QA76.6.U37 1984   001.64   82–8878
ISBN 0–12–706960–7   AACR2

PRINTED IN THE UNITED STATES OF AMERICA

84 85 86 87   9 8 7 6 5 4 3 2 1

Contents

PART I   AN INTRODUCTION TO COMPUTERS

Introduction   Toward Algorithm-Structured Architectures

Chapter 1   Conventional Computers and Loosely Distributed Networks

v

PART II **ARRAYS AND NETWORKS BUILT OR DESIGNED**

Chapter 2 **First Attempts at Designing and Organizing Multicomputers**

Chapter 3 **Large and Powerful Arrays and Pipelines**

Chapter 4 **More or Less Tightly Coupled Networks**

Chapter 5 **Parallel Parallel Computers**

Chapter 6   **Converging Pyramids of Arrays**

Chapter 7   **Comparisons among Different Interconnection Strategies**

PART III   **DEVELOPING PARALLEL ALGORITHMS, LANGUAGES, DATA FLOW**

Chapter 8   **Formulating, Programming, and Executing Efficient Parallel Algorithms**

Chapter 9   **Development of Algorithm–Program– Architecture–Data Flow**

Chapter 10   **Some Short Examples Mapping Algorithms onto Networks**

Chapter 11 A Language for Parallel Processing, Embedded in Pascal

Chapter 12 A Programming Language for a Very Large Array

PART IV TOWARD GENERAL AND POWERFUL FUTURE NETWORK ARCHITECTURES

Chapter 13 A Quick Introduction to Future Possibilities

Chapter 14 Construction Principles for Efficient Parallel Structures

PART V   APPENDIXES: BACKGROUND MATTERS AND RELATED ISSUES

Appendix A   Applied Graphs, to Describe and Construct Computer Networks

Appendix B   Bits of Logical Design Embodied in Binary Switches

Appendix C   Component Cost and Chip Packing Estimates for VLSI

Appendix D   Design of Basic Components and Modules for Very Large Networks

Preface

The traditional single central processing unit (1-CPU) serial computer is only one of a potentially infinite number of possible computers (those with 1, 2, 3, . . . , n processors, configured in all possible ways). It is the simplest, and also the slowest. It will soon reach its absolute "speed-of-light-bound" limit. Yet there will still be many very important problems far too large for it to handle.

An amazing variety of new parallel array and network multicomputer structures can now be built. Very large scale integration (VLSI) of many thousands, or millions, of devices on a single chip that costs only a few dollars will soon make feasible many completely new computer architectures that promise to be extremely powerful and very cheap.

This book examines the new parallel-array, pipeline, and other network multicomputers that today are just beginning to be designed and built, but promise, during the next 10–50 years, to give enormous increases in processing power and to successively transform and broaden our understanding of what a computer is.

This book describes and explores these new arrays and networks, both those built and those now being designed or proposed. The problems of developing higher-level languages for such systems and of designing algorithm, program, data flow, and computer structure all together, in an integrated way, are considered. The book surveys suggestions people have made for future networks with many thousands, or millions, of computers. It also describes several sequences of successively more general attempts to combine the power of arrays with the flexibility of networks into structures that reflect and embody the flow of information through their processors.

The book thus gives a picture of what has been and is being done, along with interesting possible directions for the future.

The Necessity of Parallel Computers
for Large Real-Time Problems

There are many very large problems that no single-processor serial computer can possibly handle. This is especially true for those problems where the computer must solve some problem in "real time" (that is, fast enough to interact with the really changing outside world, rather than fast enough that somebody can pick up the results whenever they happen to get done). These problems include (among many others) the perception of moving objects, the modeling of the atmosphere and the earth's crust, and the modeling of the brain's large network of neurons.

Only very large networks of computers, all working together, have any hope of succeeding at such tasks.

I approach arrays and networks from a very definite perspective, one that has probably not been the dominant perspective among those who have been exploring networks (usually loosely distributed networks like ETHERNET). My basic premise is that very large networks (that is, ones with at least thousands of computers, and probably, when we learn how to use them—whether in 10 or in 50 years—millions of computers) are needed to handle a number of key problems, and that the network's structure should reflect the problem's structure.

I focus on large networks of relatively closely coupled computers whose overall architecture has been designed so that many processors can work together, efficiently, on the same problem.

In particular, my personal interests are the modeling of intelligence, with special emphasis on image processing, the perception of patterned objects, and perceptual learning (because I feel that only by having them learn through perceptual–behavioral interactions with an environment rich in information do we have any hope of achieving computers with real intelligence, and of developing powerful models of intelligent thinking).

These problem areas are emphasized throughout this book, because they are the most familiar to me and because they are of tremendous potential importance. But there are a number of other problems, some of greater practical and immediate interest, including the processing of a continuing stream of large images (such as those transmitted by satellites) and the modeling of three-dimensional masses of matter (as for meteorological, wind-tunnel, or oceanographic research) that will benefit tremendously from large networks.

Computers have been so successful, in the way they have blossomed almost everywhere in a bewildering variety of applications and in the phenomenal way in which they have increased in power (at the same time decreasing in price) because of a steady and continuing stream of technological advances, that most people are reasonably happy with computers in their present, single-processor–serial guise.

This may well mean that parallel multicomputers will have an especially difficult time getting started. A standard rejoinder to the suggestion that a parallel computer would better handle a particular type of problem has been that a suitably bigger serial computer would be better still. Marvin Minsky and Seymour Papert (1971), Eugene Amdahl (1967), and a number of other distinguished computer scientists have long argued against the value of parallel computing.

An ironic voice was that of John von Neumann, who, in his seminal (1945) paper sketching out the design of the modern serial single-CPU computer, argued that only one processor should be used, as the best interim compromise, given the current state of failure-prone vacuum-tube technology, between programming ease (in cumbersome machine languages), design of these brand-new computer architectures, and mean time to failure needed to handle the war-motivated numerical problems contemplated. But von Neumann's ultimate goal (1959) was highly parallel nerve-network-like structures. The single-CPU "von Neumann machine" (with what Backus, 1978, named the von Neumann bottleneck) was suggested simply as a beginner's first step.

Only time, and much long-avoided hard work, will show whether we are able to uncover the secrets of very fast, very powerful parallel processing that natural evolution achieved millions of years ago and that our largely unconscious thought processes handle with mastery!

Computer Architecture := {Becomes} Algorithm Structure

The purpose of computers is to solve problems. Computers may well be the most powerful tool human beings have ever invented. In terms of their profound importance, they rank with fire, wheels, numbers, alphabets, and motors. Books can be written on this issue, even though we are barely beginning to appreciate the importance of computers. But their basic purpose is to serve as a tool; our basic problem is to discover how to use and exploit this unboundedly general, versatile, and powerful tool.

The first computer, built in 1945, was a marvel of technology. It used roughly 50,000 vacuum tubes, cost millions of dollars, filled a large room with far more electronic equipment than had ever before been combined into one system, and accurately computed the equations people programmed for it many thousands of times faster than any human being. Today, substantially more powerful computers are built on a single 4-mm-square silicon chip that costs a few dollars. This means we can design computers many thousands of times faster and more powerful, by structuring large numbers of chips into networks.

Today's large computers are still basically like the first. They have clearly found their niche. But there are many kinds of problems for which they are quite poorly structured, and intolerably slow. To solve these problems, a variety of parallel architectures are needed. This book explores the following approach to the development of multicomputer architectures:

(1) Consider a problem domain and try to find or to develop techniques for solving the problems it poses.

(2a) Try to develop precise computational algorithms.

(2A) At the same time, think in terms of the structures of physical computer processors that would best carry out those algorithms.

(3) Design, simulate, and build as general and flexible as possible a version of the architecture that seems indicated.

(4) Similarly build good programming languages and operating and support systems.

(5) Code and run programs.

(6) Evaluate, benchmark, and compare.

(7) Try to develop tools and theory for conceptualizing, analyzing, and examining these systems.

(8) Continue to cycle through all these steps, to improve algorithms and architectures (toward more powerful specialized systems and also more generally usable systems) and our understanding of how to design, build, and use these systems.

The Flow of This Book

This book should be useful as a textbook or auxiliary textbook in courses on computer architecture, parallel computers, arrays and networks, and image processing and pattern recognition; but it should be especially appropriate for independent reading. It is the only book of which I am aware that examines:

the present arrays and networks (Chapters 2–7, 14);

the problems of developing basic algorithms and complex programs, and of programming and using these architectures (Chapters 9–12, 16–18);

comparisons among different networks using a variety of different formal, empirical, and structural criteria (Chapters 4, 7, 14);

new graph theory results that suggest good new interconnection patterns for networks (Chapters 4, 7, 14, Appendix A);

network flow modeling techniques for evaluating architectures, programs, and the way programs are mapped onto and executed by hardware (Chapters 2, 7, 8, Appendix A); and

architectures that combine arrays and networks into very large, potentially very powerful, three-dimensional pipelines that transform information flowing through them (Chapters 15–19).

It attempts to do this in an integrated way, from the perspective that a network's structure should embody the structure of processes effected by the algorithms being executed, and that algorithm, program, language, operating system, and physical network should all be designed in a continuing cycle that attempts to maximize the correspondence between program structure and multicomputer structure.

Background and review chapters and appendixes cover much of the material on computers, graphs, logic design, VLSI components and their costs, and other topics needed to follow the book's exploration of arrays and networks, languages and techniques for using them, and some of their future possibilities.

The introduction examines some of the problems that can be handled only with large networks and gives a brief idea of the astonishing progress we can expect to see during the next 20 or so years in the fabrication of more and more sophisticated chips with millions of devices packed on each.

Chapter 1 describes traditional single-CPU computers, the general-purpose Turing machine that they embody in much more powerful form, several formalisms for describing and analyzing computers, and some of the supercomputers and distributed networks of today. Appendix A briefly looks at graph theory as applied to computer networks, and Appendix B describes how computers are built from switches that execute logical operations.

Chapter 2 explores network flow models for representing and evaluating the flow of information through a multicomputer network as that network executes programs mapped onto it. It surveys key arrays, networks, logic-in-memory computers, and supercomputer systems designed or built in the pre-LSI era that ended in the early 1970s. And it describes how computers are coordinated by clocks, controllers, and operating systems.

Chapter 3 examines the very large arrays of literally thousands of small computers that have already been built or designed, and also several powerful pipelines and more or less specialized and special-purpose systems.

Chapter 4 surveys the great variety of network architectures being explored today (mostly proposed, a few small ones actually built). Chapters 5 and 6 examine several designs that begin to combine promising array and network structures.

Chapter 5 suggests the possibility of combining several different kinds of structures, including arrays, pipelines, and other types of networks, into a system where each can be used, as appropriate, and from which we can learn how to use each, and how useful each is—this as a step toward designing better integrated and more useful systems.

Chapter 6 begins to examine some possibilities for combining arrays and networks into multilayered converging systems, moving toward a pyramidal shape.

Chapter 7 explores some of the similarities among the very large number of different network structures that have been proposed, and are now beginning to be built.

The next several chapters focus on issues of software development in conjunction with hardware architecture. These include the development of languages, interactive tools, and more parallel ways of thinking about the flow of data through processes, to encourage the integrated design of parallel algorithms, programs, data flow, and architectures.

Chapter 8 explores some of the issues involved in integrating algorithm, program, data flow, and network structure. It examines issues of formulating parallel algorithms, several alternative possibilities for future programming languages, and a variety of specific algorithms.

Chapter 9 examines some of the ways in which we might design, program, and use networks. It focuses on the development of highly structured architecture most appropriate for a particular type of complex algorithm. Chapter 10 examines a wider variety of problem areas and possible algorithms.

Chapter 11 surveys the "higher-level" languages that have been developed for networks—almost all for arrays—and describes one example language in some detail. Chapter 12 gives a taste of how to program an array, both in a somewhat lower-level language and in machine language, along with more detail about how one array (CLIP) works. Appendix E gives examples of code and programs in several different languages.

We then shift perspective back to architecture, looking into the more distant future (but only 5, 10, or 20 years away), starting, in Chapter 13, with interesting speculations by a number of people who are actually building arrays and networks today.

Chapter 14 examines the various possible sources of parallelism in a multicomputer, and of speedups from parallelism and from hardware and software design. Then it continues the comparison of different network architectures.

Chapter 15 explores a variety of issues related to the future design possibilities of VLSI, when the design of formulation, program, and specific algorithms may well become the design of silicon chips.

Chapter 16 begins the exploration of possible future designs that is the major theme of the rest of the book.

Chapters 16 and 17 attempt to develop a variety of array, pyramid, lattice, torus, sphere, and "whorl" designs for networks, with Appendixes C and D examining the types of components and modules such networks might use.

Chapter 18 suggests several especially interesting designs, ones that have the kind of regular micromodular structure suggested by VLSI technologies, and offer the possibility of pipeline flow of information through local network structures designed to execute programs efficiently. Chapter 19 recapitulates some of the major possible architectures.

Also included: Appendix F on getting groups to work together (whether groups of computers, or groups of neurons, animals, or human beings); Appendix G on explorations of what we should mean by "messages"; and Appendix H on what "real time" "really" means. The glossary defines and describes a number of terms, and the suggested readings and bibliography complete the book and point to more.

Acknowledgments

This book has benefited immeasurably from continuing contacts and discussions (both in person and by phone and mail) with many of the people involved in designing, building, developing programming languages for, and using the very large parallel arrays, and also the closely coupled networks being developed to use many processors on a single program (with special emphasis on image processing, pattern recognition, and scene description).

I am especially grateful to Michael Duff, the architect of CLIP, University College London, for the opportunity to work in his laboratory during my sabbatical in 1978. There I got my first taste of programming parallel arrays (in CLIP3 assembly language), and first began to think seriously about parallel architectures and higher-level languages for parallel computers. (I have since formulated and coded a specific language that compiles to either CLIP3 or CLIP4 code, and two versions of a more general parallel language, one that currently compiles to PASCAL code for a simulation of a parallel machine.) Frequent discussions with Dr. Duff, and also with Terry Fountain, Alan Wood, and other memebers of his group, were, and continue to be, a great source of knowledge and insight.

During my period in London, and by correspondence since that time, I have been fortunate to have had a number of very interesting interactions with Stewart Reddaway, the architect of the distributed array processor (DAP), and with Peter Flanders, David Hunt, Robin Gostick, and Colin Aldredge of ICL, and also with Philip Marks, of Queen Mary College, and

Paul Redstone, of Logica, the two persons programming and evaluating the DAP for image processing and pattern recognition. (While in London I wrote some simple code for the DAP, and discussed this code and programming techniques with Phil Marks. I also sent Jack Stansbury, a graduate student working with me, to spend the summer of 1979 programming the DAP.)

I visited Bjorn Kruse's laboratory in Linkoping, and examined and wrote simple code for PICAP-1, in addition to having many interesting discussions with members of his group, and with Goesta Granlund, who was then designing his GOP.

I have visited Stanley Sternberg several times in Ann Arbor, and worked through some simple code for his Cytocomputer. I have also profited tremendously from a number of stimulating discussions with him over the phone and, occasionally, at meetings in person.

And I have been in continuing contact with Jerry Potter, who is in charge of software development for Goodyear-Aerospace and has kept me up to date about developments of the MPP, ASPRO, and STARAN. (I have also benefited from interesting conversations with Kenneth Batcher, and with David Schaeffer and James Strong, of NASA-Goddard, who have been monitoring the MPP development.)

My firsthand experience with networks other than arrays has been less extensive (partly reflecting the fact that only a few networks have been developed with user problems, and especially with perceptual tasks, in mind). But here I have been in continuing, albeit sporadic, contact with Larry Wittie, and have benefited from a number of interesting conversations and interchanges. I have also been fortunate to have examined mutual problems with H. J. Siegel and with Faye Briggs, K. S. Fu, and Kai Hwang, all of Purdue University.

And I have learned from interactions with people at the University of Wisconsin working with networks: Raphael Finkel, Jack Fishburn, and Marvin Solomon working on Arachne; Andy Frank, Mike Redmont, and Vincent Rideout working on the WISPAC three-dimensional lattice; James Goodman (who had worked with the X-tree group at Berkeley); and Diane Smith (who had worked with H. J. Siegel at Purdue). New graph theory results achieved along with Will Leland and Li Qiao were helped by discussions with John Halton and Pat Hanrahan, and by insightful suggestions from E. F. Moore.

In terms of the development of higher-level languages, Robert Douglass (Los Alamos), Stefano Levialdi (University of Rome), Andrea Magiolli-Schettini (University of Pisa) and I have all been working closely together (considering our physical distance) on the problems of developing a higher-level (for the moment PASCAL-based) language. I have also been fortunate to have had a number of extensive discussions of these languages, and of array architectures, with Anthony Reeves, Cornell, and, during the period of two four-day workshops that we both attended, with Adolfo Guzman, University of Mexico. I mention this group together since it is, so far as I am aware, virtually the entire group of people who have been working on this problem (along with Jerry Potter, Stanley Sternberg, and Alan Wood, whom I have already mentioned).

I have especially profited from a large number of detailed and often heated discussions with the people at Wisconsin with whom I worked in developing ideas for and designing our Parallel Pyramidal Array/Network, including Murray Thompson, Larry Schmitt, Colin Cryer, Heinz Kuettner, Alwyn Scott, Larry Travis, and, above all, Joseph Lackey.

I make these acknowledgments in this form to emphasize that I have been fortunate in firsthand contact with many of the systems that I consider to be the most important in this field, and especially with the researchers who designed and built them.

PART I

An Introduction to Computers

Toward Algorithm-Structured
Architectures

Traditional single-processor serial computers will soon reach their limit in speed and power, for they are rapidly approaching the "speed-of-light barrier," where the time needed for current to flow between the computer's devices will be greater than the time needed by these devices to compute and transform information.

Yet traditional computers will still not be nearly fast enough to handle many extremely important problems, for example, the perceptual recognition of and interaction with objects in motion, the modeling of three-dimensional masses of matter, to predict weather, earthquakes, or other large-scale phenomena, and the handling of inventories, records, and very large bodies of data.

Arrays and Networks of Large Numbers
of Closely Coupled Computers

The only hope of handling such problems is with the new kinds of "parallel arrays" and other kinds of "networks" (including systems where information is "pipelined" through structures of processors) that are just beginning to be designed and constructed.

Such systems today use 4−80 computers working together, or even (when specialized into arrays optimized for certain classes of problems) 4000−16,000 computers, all working together on the same problem in a relatively tightly coupled, lockstep organized way.

3

By 1985 or 1990 we shall be able to build networks of many thousands of computers. By 1995 or 2001 there is a good chance that we shall be able to achieve millions of computers. (See, e.g., Mead and Conway, 1980; and browse through issues of *Electronics* magazine for continuing updates on technological advances and prospects.)

New kinds of arrays and networks of large numbers of computers, working together at the same time, in parallel, offer great promise of continuing, and even accelerating, the rapid progress that we have seen since 1945 (when the modern computer was invented) in increasing the computer's speed and power.

New "very large-scale integration" (VLSI) technologies are about to allow us to fabricate millions of transistor-equivalent devices on a single 4-mm-square silicon chip. We shall then be in a position to use several thousand of these chips to build enormously powerful computer networks with billions of devices, systems many thousands of times more powerful than the computers of today.

This book explores the great promise of these multicomputers, and also the very difficult—and very interesting—research problems that they pose. We can think roughly of three impending eras:

(1) From 1975, continuing to around 1985, technology has allowed and will continue to allow the building of networks that link together a few dozen traditional computers, or a few thousand computers when each is made as simple as possible and all are run and interconnected in a simple manner (e.g., as an array with every processor executing the same instruction).

(2) From 1983 to 1990, people will be able to use the new computers on VLSI chips, each chip containing up to one million or more transistors, with a few thousand chips in the total system.

(3) By the 1990s a greater variety of chips, each containing one or more computers, with a variety of mixes and interconnection topologies, should be available as the basic building blocks for a much greater variety of much larger networks.

A traditional processor has $2000-50,000$ devices; each 1-bit processor in today's arrays has $50-800$ devices; a logic gate (e.g., to AND, OR, threshold) has $1-20$ devices. The range of basic building blocks is already large, and many more different kinds of processors can be expected. Therefore the computer architect will soon be able to design in terms of $1-10,000$ or more chips each with $1-10,000,000$ or more devices. This will be the landscape on which networks of processor and memories, each built from a handful or many thousands of gates, will be configured.

Toward Architectures That Mirror
Algorithms' Information Flow

These new kinds of networks will make possible far more integrated and coherent approaches to the design and development of large programmed systems, and of the computer systems (networks) on which they run.

Today one codes one's program in a specific programming language, and then executes that program on any computer for which a compiler for that language has been written. But soon we should be able to design algorithm, program, language, and computer in a single integrated architectural exercise. We can then strive toward the following:

The "algorithm" (that is, the set of procedures used to solve the problem) will determine the flow of information through a network of processors.

This flow will determine the structure that a set of processors should have to efficiently effect the algorithm.

This structure will determine the architecture of the network of processors that executes the program.

The operating system will then either choose the appropriate architecture, or it will actually carve and reconfigure that architecture out of the large general-purpose network of the thousands or even millions of computers or individual devices at its disposal.

We thus have a problem of mapping the graph that represents the algorithm's flow of information, through the structure of processes that transform the input information into the solution to be output, into the physical processor-memory graph structure that will actually execute those processes. Graph matching is a very difficult problem. At best we shall achieve less-than-perfect mappings, after much hard work. But there are a variety of ways in which algorithm and network can be brought closer together in structure. To the extent that this is achieved, a program is executed simply by letting information flow through the network, with a minimum number of delays or inefficiencies.

To the extent that we fall short of a good mapping, the program is forced to pass data and send other kinds of information at run time, to achieve, in effect, the proper structure in software. And to do this it must use what (at least with today's operating systems) are excessively slow "message-passing protocols" that take thousands of times longer to send a piece of data to a processor than it takes for that processor to execute an instruction on and process that data.

The Traditional Single "Central Processing Unit"
Serial Computer

From 1945 to 1980 the word "computer" has meant *one single* "central processing unit" (CPU), plus a memory from which it fetches all instructions and data, plus input-output connections to the external world. Such a "serial" computer, if given a rather simple basic set of instructions, *is capable of executing any conceivable program;* it is therefore called "general purpose." But such a serial computer must execute a program *one instruction after the other, in strict serial fashion.*

Computers are amazingly fast. By 1980 ordinary computers took roughly 1 μsec (microsecond, 1 millionth of a second) to fetch and then execute a single instruction. The fastest took 10 nsec (10 billionths of a second). By 1990 even faster "cryogenic" computers, cooled to almost absolute zero, should be $100-1000$ times faster still, executing instructions in $10-100$ psec (trillionths of a second). But then there will no longer be any possibility of still faster speeds; the absolute limit of the speed-of-light barrier will have been reached.

These incredible speeds, and the equally incredible speed at which technology has advanced, so that every $3-6$ years computers grow $5-10$ times faster and more powerful, have made most people feel that the one-processor general-purpose serial computer is more than sufficient (or, if not, soon will be).

Problems for Which the One-CPU
Serial Computer Is Inadequate

But there are many problems where even the largest and most powerful one-processor computer is hopelessly slow. These include not only image processing, perception, and the modeling of masses of matter, but also the handling and accessing of large bodies of information (e.g., tax records, inventories, books, libraries), and the development and modeling of intelligent thinking systems.

Almost certainly, 50 years from now people will look back and wonder why we thought computers were so powerful when we used them only for such a restricted set of simple problems. Many of the problems for which the computers of the 21st century will be used have not even been considered or formulated for computers today, because they are far too difficult to contemplate.

The Enormous Computing Burden in Perceiving Pictorial Images

Consider the (seemingly) relatively simple problems of enhancing and recognizing the objects in a picture input from a television camera, or of modeling a tornado or other section of the atmosphere. To handle these (and many other) problems, a one-processor computer must process each portion of the picture in turn, serially.

A television picture contains from 250×250 to 4000×4000 tiny pieces (often called picture elements, or "pixels"). Typically, each pixel is examined with respect to its immediately surrounding pixels, in order to begin to uncover simple gradients, edges, and regions.

For example, a program might compute a weighted average of the pixel and its 8 nearest neighbors. It would have to fetch the pixel's value from memory, multiply that value by the appropriate weight, store it temporarily; fetch the next neighbor's value, multiply it by the appropriate weight, add these weights; and so on for each neighbor; and finally divide the summed weights by a normalizing constant and store the results.

This will (typically) need 9 fetch instructions, 9 multiply instructions, 9 add instructions, 9 store instructions, 1 divide instruction, and 1 (final) store instruction. That totals 38 instructions. (Computers vary in the specific instructions that have been wired into their hardware. So some will need more, some less. And this has assumed that each pixel contains one number, signifying the intensity of grayscale light at that point. Pictures in color will need 3 numbers, one for each of the primary colors, and at least 3 times the number of instructions to process them.)

For convenience, let us assume the picture image is 1000×1000 giving 1,000,000 pixels. (This is the size of the relatively high-resolution television pictures that are being beamed 30 times a second from Landsat and other satellites in orbit around the earth.) The simple weighted averaging operation just described will take about 50,000,000 instructions (including the additional instructions needed to change the addresses of the 9 cells being looked at as the program iterates from one pixel to the next).

A typical serial computer that executes one instruction each microsecond will therefore need 50 seconds for just this one very simple process! A very fast computer, one that takes only 10 nsec per instruction, will need 0.5 sec.

But this is only the first in a long sequence of transformations that a program must effect to do something useful. A next transformation might be one that looked at each pixel and its near neighbors to begin to get a gradient, e.g., by subtracting the (weighted) value of each of the neighbors from the (weighted) pixel value. This would take another 50 (or more) instructions per pixel.

Even a relatively simple program to eliminate and smooth out local specks of noise, sharpen edges a bit, and, in general, "enhance" the image will need 5, 20, 50, or several hundred such instructions. Often it is necessary to have the program look at a larger neighborhood of 25 or even several hundred nearby pixels; this will multiply the time needed by from 10 to thousands.

For simplicity, let us assume 100 8-neighbor instructions. This adds up to 5000 sec (1 hr 14 min), or 50 sec on the fastest of today's computers.

But the program must do a lot more. To determine what information of interest might be there, in the attempt to recognize objects in and describe that scene, will take far longer still. This is an extremely difficult problem, one that needs so much computing time that people are only beginning to get results (and these on carefully chosen and much-simplified pictures). Very roughly, whereas for image enhancement a program might need to execute $5-500$ transforms on each pixel, for pattern recognition and scene description a program might need to execute from 100 to 5000, or more.

A very simple and quite ubiquitous problem that, ultimately, would be of great importance if computers could handle it is the recognition and description of scenes of *moving objects*. Successful programs would make possible self-steering rockets, satellites, airplanes, buses, and autos, along with artificial visual systems for the blind, perception-controlled robots, surveillance devices, and systems for a variety of other applications.

But television pictures are taken in a continuing stream of $20-50$ per second; so *each must be processed in at most 0.05 sec!*

Modeling and Examining Masses of Matter

To model some mass of matter, whether a wing or a missile in a wind tunnel, or a windstorm, or an earthquake, or a larger portion of the atmosphere or the ground beneath us, a three-dimensional array can be stored and processed region by region in much the same way as just described for two-dimensional images. The design of airplanes and missiles would benefit greatly from increases in the ability to model their behavior. And meteorologists have had relatively limited success in predicting the weather from such models to a great extent because the size of the atmosphere that can be handled in the detail needed is still far too small with even the biggest of today's (serial) computers.

As computers grow more powerful, it will become possible to model larger and larger arrays, with (when needed) finer and finer resolution. But we will probably never reach the point where we can model as large a mass of interesting matter in as much detail as we would like to do.

Serial computers will *never* be able to handle such tasks. But large arrays and networks of many computers, all working together in a parallel-serial manner, offer great promise for these kinds of problems. And, by the 1990s, they will not only be powerful and fast enough, but they may also be quite small (1000-computer array/networks the size of a pocket calculator, weighing a pound or two), and cheap ($100 or less).

Modeling the Human Brain and Developing Intelligent Programs

The human brain has roughly 12 billion neurons. Each neuron is an extremely complex system, with hundreds or thousands of "dendrites" connecting to other neurons over "synaptic junctions." To model the brain at this level of detailed complexity, a serial computer would need to iterate 12 billion times to handle just one step of brain activity. The transformation it would effect at each neuron is itself far more complex than the simple near-neighbor transforms we examined for image processing; each would take hundreds or thousands of instructions.

A neuron conducts a train of impulses and fires another neuron across a synapse in $1-2$ msec. So a program would have to repeat this loop of 12 billion neurons $500-1000$ times just to model one second of thinking!

Modeling the mind/brain is an extremely important and exciting scientific problem in its own right. It also has great potential promise for valuable applications, as we develop a better understanding of intelligent processes and, finally, succeed in building more intelligent computer systems. But such systems will have to be highly parallel, to be both intelligent enough and fast enough to exercise that intelligence on real-world problems. For they will have to recognize the objects with which they must interact, decide what to do, and generate and effect appropriate behaviors—all within the period of a second or two, or less.

Toward Developing Powerfully Structured Arrays and Networks

There are many similarly large problems of real-time interaction and control of such physical systems as missiles, satellites, robots, and complex manufacturing processes.

We cannot continue to build faster and more powerful traditional serial computers for very much longer; we will be stopped by the speed-of-light barrier. But we can build ever-larger parallel-serial computers by adding more processors and having more and more of these processors work on the same problem, at the same time, in parallel.

For certain problems this may be relatively straightforward. For example, we might build 1000×1000 arrays, or even $10,000 \times 10,000$ arrays, with much the same design as the 64×64 or 128×128 arrays being built today, simply by adding more of the same. Then a 1024×1024 array will process a 1024×1024 image (or one that is an integer multiple larger) at least 256 times as fast as a 64×64 array.

But for most problems it will not be nearly so simple to map the program's set of procedures onto ever-larger numbers of processors, all executing at the same time, with efficiency. Our progress will depend upon a continuing attack on a set of difficult research problems:

We must reformulate our ways of solving problems, our "algorithms," into as parallel as possible a form. We must even reformulate our ways of thinking and problem solving, since our "conscious" mind is quite serial and our whole development and tradition of mathematical, logical, and rational thinking has been serial. (*But most of our thinking is actually done unconsciously, and at such a fast speed that it must be highly parallel*—so we know and have existence proofs that it can be done, and that *parallel algorithms do in fact exist*.)

We must develop computer architectures that put large numbers of computers into appropriately coordinated interaction with one another, so that they can all work efficiently together on the same problem. And we must develop programming languages and operating systems that allow for convenient and efficient development and execution of programs on such systems.

Very Large-Scale Integration (VLSI)
and Very Large Networks

In 1979 manufacturers succeeded in fabricating single 4-mm-square silicon "chips" with 100,000 transistor-equivalent devices on each. In 1981 450,000 devices were achieved in Hewlett-Packard's microcomputer chip. We are beginning to reap the fruits of roughly 30 years of continuing advances in miniaturizing this silicon-based technology (started with the invention of the "transistor" in 1948). Every $12-30$ months researchers succeed in packing $2-4$ times as many devices on a single chip. We can expect this rate of improvement to continue for at least 10, and most likely 20 or 30 more years.

Thus in the 1980s we can expect single chips with 1,000,000 devices. In the 1990s we may well see single chips with 10,000,000 or more devices.

The chip is the basic building block of a computer. A microcomputer might have 10 or 50, or even 1, chip. A minicomputer might have 200 or 2000 chips; a very large computer might have 10,000 chips, or even more. And the finished computer will cost roughly $5-100$ times as much (in dollars) as the chip count. When each chip contains 1,000,000 or 10,000,000 times as many "transistor-equivalent devices" as a 1950s transistor (which is much larger than today's chips!) the increase in power will be overwhelming.

Therefore the arrays and networks described in the next few chapters, and the suggestions for future possibilities of the last few chapters, should be placed on the following background:

Every $5-10$ years chip packing densities will increase by one order of magnitude. Sutherland, Mead, and Everhart in 1976 concluded: "the past 15 years appears to be only the first half of a potential eight-orders-of-magnitude development. There seem to be no fundamental obstacles to 10^7 to 10^8 device integrated circuits." Thus we shall move from 10^5 devices per chip to 10^6 around 1985-1990 to 10^7 or even 10^8 around 1990-2000.

The number of chips in a computer ranges from 10^0 to 10^5, so we should consider what kinds of computers should be built with 10^5, 10^6, ..., 10^{13} devices.

Certainly many of these computers will be relatively traditional one-CPU serial machines. But the potential muscle at the computer architect's disposal, with many millions, or even billions, of devices waiting to be structured as one desires, strongly suggests that a variety of new kinds of processors and parallel architectures should be considered.

Steps toward Efficient, Powerful
Algorithm-Structured Architectures

New and nonconventional multiprocessor, multicomputer array and network architectures are today being developed from several different perspectives:

Architects of conventional single-CPU serial computers are interested in adding more special-purpose processors to speed up the CPU, along with other specialized processors to handle input-output, memory management, and other peripheral operations.

People building distributed networks of many conventional computers linked via a coaxial cable or a similar kind of "bus" to peripheral input, output, mass memory, and other peripheral devices are beginning to think about the problems of having several, or many, of these computers work together on the same program.

Researchers interested in parallel processes, some at a theoretical level, some at an engineering architecture level, some at a programming software level, are beginning to explore the possibilities of networks:

(1) The theoretical people have traditionally been interested in developing (and occasionally applying) formal models and tools, like graph theory or Petri nets.

(2) The hardware people have been interested in the overall structure of the interconnection pattern of the network, and the details of structure and interconnection of the parts.

(3) The software people have been interested in the operating system and the programming languages.

Researchers developing VLSI technologies find the highly modular parallel structures attractive architectures to explore (although this work today takes place chiefly within the laboratories of chip manufacturers that cannot justify research on nonconventional computers with unknown markets).

Researchers interested in using parallel networks and arrays (that is, "users" with a variety of different applications that lend themselves to one or another parallel-serial structure) have tended to propose, design, and (occasionally) build structures specialized to their specific problem areas.

This book attempts to examine and to integrate work done from all these perspectives and points of entry:

It surveys array, pipeline, and other network architectures.

It explores architectures from the point of view of graph-theoretical results, and of network flow concepts.

It explores the problems of developing parallel-serial algorithms.

It examines several architectures, several languages, and several algorithms in some detail.

It then explores attractive possibilities for very large algorithm-structured networks, using the rapidly improving VLSI technologies.

It suggests and explores the following steps, ones that are being taken or might be taken to develop good architectures and programmed systems:

(a) *Start with local "clusters" of computers (e.g., arrays, or specialized configurations of gates) whose size and structure are most appropriate for the algorithms* to be programmed and executed. Compound many such clusters together into the overall architecture. Large arrays, pyramids, and lattices are good examples of clusters that have been found useful for a variety of image-processing and number-crunching tasks.

(b) *If the "appropriate structure" is not known, or the system must handle a wide variety of programs with unknown structure, add "reconfiguring" capabilities,* using switches, locally common memory, or other devices, so that the operating system or the user program can change the structure, as needed.

(c) *If reconfiguring hardware is too expensive, start with a tree* (which is a surprisingly good structure; for example, as we shall see, it is better than n-cubes and most other nontree structures that have been proposed).

(d) *But then try to improve upon the tree by "augmenting" it,* that is, by adding new links (especially to the buds, that is, the nodes with only one link) to build a good "regular" graph (this gives a graph with the same number of links to every node, and with many cycles and alternative paths and a diameter appreciably smaller than the diameter of the original tree). *The best that can possibly be achieved is to reduce the tree's diameter k to $\frac{1}{2}k + 1$ (except for three graphs of diameter 4, which can be reduced to $k=2$). Note that this is not such a significant reduction, in the context of and compared with computers and VLSI chips that double in size and power every year or two.*

(e) *Develop algorithms and programs whose flows of processes and data mirror the structure of the network as closely as possible.* At the same time develop network structures and processors that mirror the flows of processes of programs, and of important classes of programs. (This may well entail tearing apart the traditional processor, with its thousands of gates, and building chips with processor gates, memory gates, and switches intermingled. These might be special-purpose "systolic" processors, or more or less specialized structures through which information can be pipelined. This very interesting alternative is quite reminiscent of a network of neurons that successively transforms information that flows through it.) Here we need new languages and new interactive environments for efficient program building, debugging, and testing.

(f) *Develop operating systems that assign programs and procedures to local clusters of processors* so that they fit one another as well as possible (making use of the efforts of the architects and programmers who designed structures and algorithms to fit one another).

(g) *Have the operating system keep watch on performance, and reassign programs, reconfigure subnetwork clusters,* tell the programmer about inefficiencies and suggest possible alternatives, and/or use whatever other tools can be put at its disposal.

(h) *Continue to cycle through these steps, to give the gradual improvements in architectures, algorithms, languages, programs, and operating procedures that will come only with experience.*

(i) *Develop architectures, languages, and operating systems to fit one another, and have the language encourage good structuring in user programs,* and discourage or forbid practices that would lead to poor mappings of programs onto architectures and therefore inefficient execution. Ultimately, the architectures available should suggest what is "good structure" for a program, while the structures of information flow that underlie programs should determine the structures that are built into multicomputer architectures. Successive generations of algorithms, programs, languages, and architectures will cycle through continuing improvements within this ultimately circular situation.

(j) *Everybody may need to help at the very difficult task of building networks and programs that execute algorithms as efficiently as possible.* It may will be that, at least for a while, research programmers must be asked to use programming languages that prompt them to tell the operating system about the program's structure, and architect and programmer must work closely enough to understand each other's problems and possibilities.

We appear to be moving toward the day when an algorithm will be expressed directly in a graph that specifies a structure of gates/processors through which information flows. A physical network will then be chosen, reconfigured, or constructed, as most feasible. Ultimately, today's "programs" may well be replaced by flow diagrams that specify and are executed by isomorphic hardware networks.

CHAPTER 1

Conventional Computers
and Loosely Distributed Networks

For over 35 years the word "computer" has conventionally meant "serial" "single-central processing unit" (CPU) "digital" "computer." The computer's single processor accesses a single memory, and inputs and outputs information to and from the external world. The processor's "arithmetic logic unit" (ALU) (which is often a whole set of simple special-purpose processors) plus controller, high-speed registers, etc., make up the single CPU. Often there are several input devices and several output devices, and a hierarchy of successively slower and larger memories. But computing is done by the single processor, using data and program stored in its "main" memory.

This chapter examines this kind of traditional system, and some of the ways these computers have been analyzed and described. It then surveys several relatively straightforward first steps toward building computer systems that have more than one processor—computers with two or three processors sharing the main memory—and distributed networks of computers.

The rest of this book explores the great variety of computer networks (including arrays and pipelines) that have been built or designed, and could be built, *where many processors and many main memories are combined into a single structure, all working together on a single program.* This opens up a variety of possibilities, including systems where separate traditional computers, or processors, are linked together; where all processors share a common memory; where several processors communicate to one another via locally shared memories or registers; and where the processor—memory network mirrors a program's flow of information.

General-Purpose Computers
and Turing Machines

Emil Post (1936) and Alan Turing (1936) independently proposed a very simple kind of computer (a suitably powerful "finite-state automaton" with "potentially infinite" memory, often called the "Turing machine") that is capable of doing anything that any other computer might conceivably do—given enough time. A number of other independent formulations have also been proved identical, e.g., Alonzo Church's 1936 lambda calculus, Stephen Kleene's 1936 recursive functions, and Post's 1943 productions. Figure 1 gives Post's formulation of one example of such a computer.

(a) *Basic Structure* (A "read—write head" looks at an arbitrarily long ("potentially infinite") tape with symbols:

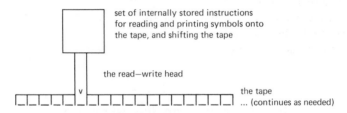

set of internally stored instructions for reading and printing symbols onto the tape, and shifting the tape

the read—write head

the tape
... (continues as needed)

(b) *Emil Post's 1936 formulation of a necessary and sufficient set of primitive instructions:*

"The worker [i.e., the computer] is assumed to be capable of performing the following primitive acts:

(a) Marking the box he is in (assumed empty),
(b) erasing the mark in the box he is in (assumed marked),
(c) moving to the box on his right,
(d) moving to the box on his left,
(e) determining whether the box he is in, is or is not marked."

Fig. 1 Emil Post's 1936 formulation of a "Turing Machine."

Thus a general-purpose computer can be amazingly simple. All that Post's version does is execute a sequence of instructions from the simple repertoire:

READ the current symbol;
SHIFT to the next symbol;
WRITE the current symbol onto the current memory/tape location;
IF the current symbol is x, THEN DO instruction i, ELSE DO j.

Information is read from and written onto a "tape"—which serves as the memory of the system. The tape contains all needed information, which means both the data on which the program must act, and also *the program itself* (whence the term "stored-program" computer). The processor that executes sequences of these instructions is a simple example of what is meant by a "finite-state automaton."

Note the extremely simple, weak, "low-level" instructions. A program of even a little bit of power will need a very long sequence of such instructions, and it will take a very long time to execute. For example, to add two six-digit numbers such a computer must read each digit, make a sequence of successive shifts to find free scratchpad space on the tape, add the digits, write the result, shift suitably and write the carry, shift back to get the next digits, and iterate this procedure through all the digits. Before each of these steps the head must shift to the region of the tape that contains the next instruction, which in turn tells it where to shift for the next step, and for the next instruction. [One of the great sources of power for stored-program computers lies in the fact that the program is input to and stored in the same tape/memory that contains all (other) kinds of data.]

The General-Purpose Single-Processor
Serial Computer Described

Words and Random Access Memories
to Speed Up the Turing Machine

Such a procedure is intolerably slow, and two straightforward modifications have given computers improvements of several orders of magnitude in simplifying the program to a much shorter sequence of much more powerful instructions, and in cutting down the number of instructions, and the time needed to execute a program. The key modifications have been the following:

(a) *The processor can read, write, and operate on a whole "word" (32 binary digits is the typical size of a word).*

(b) *The processor is given the "random access" ability to immediately fetch any of a rather large number of such words* (64,000−1,000,000 are typical numbers), rather than needing to execute a long, slow, serial sequence of shifts from where the processor is now looking to where the word is stored. Thus the processor is directly linked, as through the center of a star graph, to every memory location.

The Program (Which Is Just Data Stored in Memory) Runs the Computer

The "program" (which is simply a set of "instructions" stored in the computer's memory) tells the computer exactly what to do. The first instruction is loaded into the Instruction Register and "decoded" to get the operation to be performed, and the addresses where (a) operands are to be fetched, (b) results are to be stored, and (c) the next instruction will be found. The operands are then fetched and processed, the results stored, and the next instruction fetched and loaded into the Instruction Register.

This process continues until the "end" instruction is reached.

There is a very important type of instruction that allows programs to make decisions that change their flow of processes: A "conditional" instruction makes a test on the contents of a specified register. The computer branches to one instruction if the test succeeds, and to another instruction if the test fails.

The programmer almost always writes instructions in a language that is much "higher-level" than the "machine language" the computer is actually wired to decode. A "compiler" (which is just another program) translates from the higher-level-language program to the computer's "machine code." This entails replacing one instruction by a whole sequence of much smaller steps, each specified in complete detail, with the exact machine addresses.

The "General-Purpose" Capability of Serial (and Parallel) Computers

All modern computers are "general purpose" in the sense that each can compute anything that any other computer can compute—given enough time. This is achieved simply by giving the computer (a) a set of instructions that includes the basic Turing machine functions, and (b) a (potentially) unlimited amount of memory—that is, as much memory as is needed to handle the program being executed.

We sometimes say all computers are "universal," or "Turing-machine equivalent." But even a small "microcomputer" like the Apple or Pet is millions of times faster than a simple Turing machine. And a more powerful and more expensive computer like the DEC VAX or Cray-1 is hundreds or thousands of times faster still.

This kind of general-purpose computer can be built with any number of processors, all working at the same time, in parallel. But a computer with one single processor is the simplest, and has full generality. So from the very first modern computers (designed and built by Zuse, Atanasoff, and Eckert, Mauchley, Goldstine, and von Neumann; see Randell, 1973 for their original descriptions) people have concentrated on improving the single central processing unit (CPU) serial computer.

It is the fact that only one processor is working at any moment in time that makes such a system a single-processor serial computer.

The Very Simple Structure of the Basic General-Purpose Serial Computer

We usually describe this kind of computer as having the very simple structure of Figure 2.

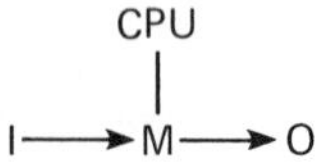

```
            CPU
             |
    I———►M———►O
```

Fig. 2   The (idealized) structure of a single-CPU serial computer. (Note: —, | are links; > gives direction; M = memory store; I = input device; O = output device.)

The CPU (the single processor) fetches information from the memory and stores results into that memory; input and output devices (e.g., tape readers, printers) input and output information.

This is an unusually simple and elegant structure when we consider everything that is being handled. For among the information input to the computer's memory store and fetched by the processor are all kinds of data, and also the actual program that tells the computer what to do.

Typical Additions for Power That Complicate a Computer's Structure

An actual computer will usually be a good bit more complex, with several kinds of input and output devices and several successively faster (and, because therefore more expensive, smaller) memories (e.g., tapes, disks, high-speed main memory, higher speed registers).

Often input and output are handled by special I-O processors working in parallel with the CPU. Similarly, the next program instruction is often fetched at the same time the present instruction is being executed. Several high-speed registers are usually used to store the current instruction, its operands, and its resulting output. A larger "cache" memory of high-speed registers is often used to contain instructions and data that will probably be needed in the near future.

Often the computer is given special-purpose hardware for several types of processes (e.g., addition, floating-point multiplication, division, string matching, input and output) computer manufacturers and computing centers expect will frequently occur. Such a computer's processor is really a whole set of specific processors. One might say that these processors "sit around in parallel"; but they actually work only one at a time, in series.

Figure 3 indicates how these additions are typically interconnected.

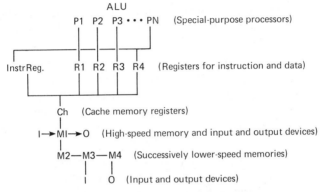

Fig. 3 A typical computer with a hierarchy of memories, I-O devices and special-purpose processors.

How Computers Actually Work

We must incorporate into the electronic circuits of a given technology the ability to execute all the detailed steps that must be taken to effect the basic operations, and to put them together into larger processes and programs. A particular computer can have many different memory stores (high-speed registers, caches, main memory, disks, tapes) and special-purpose processors for input-output, floating-point multiplication, and other special purposes. But its basic operations boil down to the copying of a string of binary digits from one register to another ("fetching" and "storing") and the pumping of bit strings through gates whose outputs are simple logical functions like "AND," "OR," and "NOT." Addition, multiplication, string matching, and all the more complex functions that make up a whole program are simply very large structures of such simple processes.

Consider the following simple conditional expression:

IF c + x > t THEN v := c + x

(this is a valid statement in PASCAL; it means: "If c plus x is greater than t, then assign the value of $c + x$ to be the value of v").

Even such a simple instruction in a "higher-level language" like PASCAL involves a whole set of "machine-language" instructions, and a long sequence of detailed actions by the machine:

First, the value of the constant c must be fetched; then the current value of the variable x must be fetched; then these must be multiplied together; then the results compared with t (probably by subtracting t); then if the result of the subtraction of t is positive, the result of $c + x$ must be stored in the memory register assigned to the variable v.

The PASCAL compiler must convert such statements into the machine-language representation of this detailed sequence, assigning locations for storage of the different variables and, possibly, building tables that can be used when the program runs to find those locations. That is a long story in itself. But, essentially, it is a matter of assigning all the detailed names of locations and operations, correctly sequenced.

First, to fetch c, the particular location that the symbol "c" names (which may have actually been put in place of c during the compilation phase, or may now be looked up in a table) is accessed. This entails an actual fanout into a bank of memory registers, e.g., into a set of 32 16K RAMs, finally reaching the specific bit location where c is stored across all RAMs (1 bit of the 32-bit number in each). Note that with today's technologies this often entails a search into (binary) trees. (With the magnetic core memories that predominated in the 1960s it entailed the specification of a row and a column, the memory being in the cell where they intersected.) Now these 32 bits are copied into one of the high-speed "accumulator" registers associated with the processors (the bits in the memory register remain unchanged).

The same process is repeated for the value stored in x, except that it is copied into a second accumulator.

Now both these bit strings are pumped into the structure of logic gates that effects a multiplication. (Additional signals are used to keep track of the loading of accumulators and to signal the controller that the two operands are fetched and ready.)

The appropriate processor is addressed and fanned out to much as memory fetches are handled (except that the fanout is only a few steps and the operands are copied to the processor, whereas the value of a memory register is copied to the accumulator into which it is fetched). Let us consider the processor that does the multiplication as a black box (as, fortunately, we can do at most points that would otherwise become terribly tedious to follow through in detail) designed along the lines described in Appendix B. Depending upon its speed and generality it will be a fairly large network of dozens, or hundreds, of gates. But finally (in a few microseconds or less) it will output the result of multiplying two 32-bit (roughly, ten decimal digit) numbers. This result will be output into the particular accumulator the compiler had designated for it (possibly one of the two used for the operands but no longer needed, or possibly a third).

Now t is similarly fetched, after a ready signal that the next operation can start.

Now the appropriate processor is used to subtract t from the previous result $(c + x)$, and this new result is stored in the designated accumulator.

Now an instruction that the compiler inserted during its translation process is effected, to check whether this new result is positive or negative (by testing the bit that is set to signal this aspect of the number). If positive, the program stores the result of $c + x$ (which still rests in one of the accumulators; the compiler should have taken care not to destroy it when loading t) in the memory location that has been assigned to the variable v. This storing process entails fanning out to the location where v is stored, then copying the accumulator into it (destroying what had previously been there).

This description is already tedious, but I have overlooked a number of details, ignored the big black processor boxes, and resisted descending to lower levels of machine cycles and register and RAM architectures, much less the electrical activities in individual components.

The major processes I have ignored were the loading of each instruction in its turn into the "instruction register" and the details of processes whereby that instruction is decoded and used to set the computer to execute the instructed operation. This is handled by a program counter that simply adds one to a count and then uses this number as the name—address of the program instruction to be executed. This instruction must then be fetched from memory, just as variables and constants are fetched. Then the instruction is decoded and used to initiate the operation that it instructs the computer to effect.

One important point: A conditional test like the IF...THEN... shown entails the possibility of executing either of several different instructions. This means that the instructions' names must be stored as part of the conditional, and when the branch is to an instruction that is not simply one greater than the previous instruction, that different instruction's name (number) must be loaded (just like a fetch) into the instruction counter.

Consider the following statement:

IF c(a+b+c+d+e+f+g)/z > t THEN v := true

and see whether you can spell out all its tedious details. (It's pretty much more of the same, except that the seven values should first be added, followed by a single division, and the Boolean variable "true" will be represented by a single bit set to one.)

This is about all there is to it! Different instructions will call different processors, and different operations are effected by different configurations of gates. But, essentially, all programs are simply large structures of such instructions that, on a serial computer, are loaded and executed in this sequential manner. Data (constants, variables, instructions) are pointed to, accessed, and copied from one memory register to another. Processors are enabled and the data waiting in their registers pumped through them, the results output to the appropriate register. Even when the data are on disk and tape, the operations are essentially the same, although the details of accessing and transferring these data will be entirely different.

This is an important point to generalize upon. At a conceptual level computers are amazingly simple (and amazingly repetitive and boring as a result). Actual hardware implementation descriptions quickly become amazingly detailed—still relatively simple, but because of the enormous and intricate detail, it is very easy to overlook something and to make trivial mistakes.

Graphs, Automata, Petri Nets, Information Flow, Network Flow

A large number of different more or less formal models have been proposed for computers, the details of their structure and behavior, and their basic components. I shall describe several here, side by side. They give interesting perspectives on the architectures we shall examine, and the way we picture them.

Computers, and also most of the notational systems that are used to describe, model, and study computers, are probably best thought of as "graphs." A graph is composed of "nodes" (often called "vertices" or "points") connected by "links" (often called "edges" or "lines"). "Directed" graphs, with one-way links, that have values associated with nodes or links are often called "networks." (See Appendix A for a fuller description of graphs, B for logic design, and D for basic components and building-blocks.)

McCulloch — Pitts Neurons and Finite-State Automata

The neuroscientist Warren McCulloch and the mathematician Walter Pitts in 1943 proposed an idealized neuron-net calculus for modeling the brain's computations. This consists of a directed graph whose "neuron" links are either +1 or inhibitory and whose "synapse" nodes are assigned an integer threshold. When enough neurons fire into it to exceed its threshold, and no inhibitory neurons fire into it, a synapse fires (see Fig. 4).

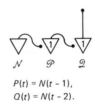

$$P(t) = N(t-1),$$
$$Q(t) = N(t-2).$$

Fig. 4 McCulloch — Pitts neuron nets for conjunction, disjunction. Solid dots excite; circles inhibit (see Fig. 6); threshold shown in triangle. [From McCulloch and Pitts (1943).]

This model had a great influence, as part of the more general cybernetic approach, for a number of years. It appeared to make possible much more precise, analyzable formulations of brain functions, at both biological and psychological levels. John von Neumann, who is credited with some of the key concepts in the development of the modern computer and who wrote the design specifications of the first stored-program computer, was deeply influenced by this model, and considered computers as embodying basic elements of this sort.

"Finite-state automata" (see Moore, 1956, 1964) are, essentially, Turing machines with finite, not potentially infinite, memories. They are directed graphs with integer inputs labeling each link and an integer output labeling each node. The automaton moves from one state to the next state as determined by the present input; it then outputs as determined by the new state (see Fig. 5).

Kleene (1956) showed that finite-state automata and McCulloch — Pitts neurons are equivalent. When augmented with a potentially infinite memory they become equivalent to Turing machines and general-purpose computers. Since any actual computer is finite, finite-state automata are often considered the more appropriate model.

Machine A

Previous state	Previous input		Present state	Present output
	0	1		
q_1	q_4	q_3	q_1	0
q_2	q_1	q_3	q_2	0
q_3	q_4	q_4	q_3	0
q_4	q_2	q_2	q_4	1

Fig. 5   A finite-state automaton (Moore's version). Tabled definitions of a machine and its transition diagram. [C. E. Shannon and J. McCarthy, eds., *Automata Studies*, Annals of Mathematics Studies No. 34. Copyright © 1956 by Princeton University Press. Selections, pp. 28−29, reprinted by permission of Princeton University Press.]

Petri Nets and Flow-Chart Schemata

C.A. Petri (1962, 1979; see Baer, 1973, Peterson, 1977, Brauer, 1980) formulated a widely studied "Petri-net" model. Noe (1971, 1979), among others, has developed several variants and used them to model the CDC 6400 and other computers. Karp and Miller (1969) formulated the very similar parallel program and flow-chart schemata.

A Petri net is a directed bipartite graph whose links connect "places" to "transitions." A "token" is put on several places, "filling" each. A transition "fires" if all its input places are full; the result is to remove one token from each of its input places, and to put one token on each of its output places. This can also be viewed as a directed graph whose links are labeled with nonnegative integers, where a node is fired when each of its input links' labels is greater than zero, and a node firing adds one to each of its output links' labels, and subtracts one from each of its input links' labels (see Fig. 6).

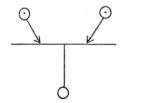

Fig. 6 A Petri net.   **Fig. 7** Equivalent McCulloch−Pitts net.

Figure 7 shows a similar McCulloch−Pitts net to indicate their close relation. The Petri net's basic construct may be more appropriate for an information-flow net, since a token (which represents a job, or a packet of information) should remain until it is processed. Concurrent processes can be represented conveniently, since tokens can be placed so several nodes fire in parallel. Data-flow computers (Dennis *et al.*, 1980) and data-flow languages (McGraw, 1980, Arvind *et al.*, 1978) are designed with this type of data-flow network in mind (although the actual hardware topologies proposed to date to execute data-flow programs do not preserve that data-flow net topology).

Network Flow Models and Data-Flow Programming Languages

Network flow models (Ford and Fulkerson, 1962; Lawler, 1976) have been used to study and help design highways, railroads, manufacturing plants, and a variety of complex systems that must carry and process material. A network is a (usually directed) graph whose links are assigned positive integer values (channel capacity or bandwidth), with two (usually disjoint) sets of nodes, designated as "source" and "sink." Networks without cycles where links are tasks and nodes represent their completion have been studied as PERT (project evaluation and review technique) and CPM (critical-path method) networks in job-shop scheduling (Conway *et al.*, 1967; Coffman, 1976; Gonzales, 1977). The shortest possible time for completing a project is the longest (critical) path. One can then attempt to improve throughput by shortening that path (see Fig. 8).

Shortest Paths

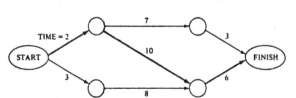

Fig. 8 A PERT network. [From Lawler (1976).]

A factory assembly line transforms raw materials into finished products, whether cars, computers, chips, perfume, or tapes. A computer transforms information (questions and problems) into information (answers, solutions, useful commentaries). Network flow models are used extensively in the design and evaluation of communications systems, including computer networks like the ARPANET described below. But the combinatorial problems are overwhelming, and they serve as aids rather than as complete tools. They are useful if only to describe a multicomputer network, and to make clear the close analogy with the flow of data through a set of processes that is especially emphasized by the data-flow languages.

Parallel Hardware Additions for Super (Traditional) Computers

Subsequent chapters of this book will examine the new kinds of parallel computers, in which a whole array or other kind of network of processors works at the same time on the same problem. But a number of parallel capabilities are often built into the traditional serial computer. This section briefly examines the parallel aspects of serial computers.

Parallel Hardware in the CPU's Special-Purpose Processors

From the beginning, computers have been given a good bit of parallel capability within their individual components. Thus a whole "word" is input, output, fetched, and stored by using parallel input–output lines. (The typical word size for a large computer is 32 bits; for a small computer it is 16 or 8 bits.)

The CPU is made of a number of individual processors, each using the most appropriate, fastest (and therefore often highly parallel) hardware design. Thus a number of parallel processors are used (but only one at a time) and each of these processors usually has a number of parallel components. The CPU can access any (one) of the words in memory and any one of the individual processors. Here too is a type of potential parallelism, although only serial access can be effected.

Often additional processors are added, to handle input and output from external devices, including disks and other mass memory stores. Sometimes hardware is added that fetches and decodes the next instruction while the present instruction is being executed.

Here we see the beginnings of parallelism, in several different directions. But (though this can become a fuzzy distinction) the components of such systems that work in parallel are all working under the skin of one single computer that executes instructions one after the other, serially.

Computers with Two, or Several, Central Processing Units

One type of multiprocessor system is probably best examined here—one with two, or a few, CPUs, each executing different programs, but both part of the same computer, using the same memory. The IBM 360/67, the Univac 1100 series, and the Burroughs B-7700 are good examples of such computers (see Enslow, 1977, for a survey). In each case, one or more CPUs are added to the system, in such a way that all CPUs are executing programs at the same time, in parallel, using the same common high-speed memory and auxiliary slower memory stores on drums, disks, tapes, etc. Each CPU executes a different program; the savings come in their shared use of memory, tape drives, and other expensive resources (see Fig. 9).

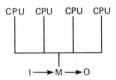

Fig. 9 Computers with several CPUs sharing other resources.

The Burroughs B-7700 and Univac 1110 were built to use up to six processors. But it appears that people have had little success with more than two to four. One of the few comparisons made, by Miller *et al.*, (1970), found that two processors increased throughput to 1.8, but three processors to only 2.1. Patton (1972a,b; see Thurber, 1976, pp. 308,309) found the throughput shown in Table I.

Table I

Examples of Throughput
for *N*-CPU Multiprocessors

Computer	Number of Processors			
	1	2	3	4
Univac 1108	1	1.85	2.40	2.00
CDC Cyber 74	1	0.75	2.20	2.50
Honeywell 6050/60	1	1.90	2.50	3.00

The Univac actually gave poorer throughput with four CPUs, because of memory access conflicts. Larger local memories would reduce these conflicts. But these CPUs are already very expensive additions to very expensive systems.

These computers were also designed to "time share"—that is, to execute a number of programs at the same time, the processor doing a bit of work on one, then going to the next, and so on, so that each program moves forward, although at the level of the processors only one program is being executed by each processor at each "slice" of time. This gives users the impression of parallel execution, but at the machine's level everything is still serial.

Pipelines and Parallelism to Speed Up Vector Arithmetic

Several of the most powerful "number-cruncher" "supercomputers" have been given "pipelines" to increase their speed. A pipeline is much like an assembly line of processors, each processor repeatedly executing the same instruction on successive pieces of data (see Chap. 3, Fig. 2). It is very useful when a procedure or an instruction can be broken down into small components, so that each component can be assigned to a different stage in the pipe; and when a sequence of such components is executed over and over again, as when working with vectors of numbers, or large images.

The Control Data 6600, STAR, and Cyber 203 and 205; the IBM S/360 model 91; the Burroughs BSP; and the Cray-1 are very powerful (and very expensive) examples of such computers (see Kozdrowicki and Theis, 1980; Lincoln, 1982). The Cray-1 (Russell, 1978) has twelve special-purpose processors plus a pipeline of up to fourteen processors that can all be used concurrently. These plus very fast logic (12.5 nsec clock cycle) and outstanding architectural design make the Cray-1 one of the world's fastest one-CPU computers. For example, in executing a DO loop through a vector (a one-dimensional array) of numbers, five overhead operations (for indexing and testing bounds) take place in parallel, so that a 64-bit floating-point multiplication is done every 12.5 nsec. The Cyber 205, using slower logic but two (or, optionally, four) pipelines achieves even faster 32-bit arithmetic. The BSP (Kuck and Stokes, 1982) combines a pipeline with 16 additional parallel processors.

Networks of Loosely Coupled Distributed Computers

The rest of this chapter examines "loosely" "distributed" networks (see Tanenbaum, 1981), in which a number of traditional single-processor computers are linked together, but in a relatively "loosely coupled" way. They are being examined separately because (in most cases) the coupling is too loose for our purposes: too much time must be taken in passing information (usually called "messages") for several processors to be able to work together efficiently on the same program and thus significantly increase the speed at which that program is executed.

The best way to give the flavor of distributed networks is to describe the two models that have had by far the most influence:

(a) the ARPANET of computers spanning the entire U.S. and parts of Europe, and

(b) the ETHERNET of small personal computers, one in each user's office, connected for purposes of electronic mail and other kinds of message sending, group use of large databases, and to share such expensive resources as large mass memories and input—output devices.

The ARPANET of Widely Distributed Large Computers

The ARPANET (McQuillan and Walden, 1977) was established by the Advanced Research Projects Agency of the U.S. Department of Defense to interconnect the large computers at a number of universities and research organizations around the United States, along with several in Europe. This was desirable for several reasons:

It allows researchers to interact with one another, sending mail, accessing and reading one another's working drafts of papers, engaging in on-going (typed) "conversations," etc.

It makes possible the transfer of large banks of data, and of programs, and the executing remotely from one's own terminal of a program that runs only on some special computer somewhere else on the network.

The development of such networks is an important research problem in itself, and one of great relevance to the military, which must be able to tie together and effectively use together a large number of computers scattered over the potential and actual battlefields, in the air and at sea, and at permanent installations in the home country.

The ARPANET connects different kinds of computers, with different operating systems. Computers are hundreds or thousands of miles apart. Some (an increasing number) of the connections are by satellite, which gives quite significant transmission delays. All these factors (plus the fact that this was not the purpose of the ARPANET) mean that a single program is never run on several of the computers at the same time, all processors working in parallel, in order to speed up execution (see Fig. 10).

Fig. 10 An example fragment of an ARPANET-type network.

The major problems addressed by the ARPANET development are of an entirely different sort: how can messages be routed over a network of 20, 40, or even more computers spread over thousands of miles? For example, in order to travel from Los Angeles to Boston the message might have to go through San Francisco, Salt Lake City, and Pittsburgh. How can computers be linked together at such distances to be most efficiently used?

Making decisions about and setting up an appropriate routing, and guaranteeing the accuracy of the received message, are major research issues in themselves. To this end each installation was given a special extra computer (an "IMP") to handle information transfer, and an elaborate system was worked out to collect, send, receive, and verify "packets" of messages, each packet headed by a string of "protocol" information describing it, its routing, its destination, etc.—with this very complex and general network in mind.

The physical distance between computers, the differences between computers in both hardware and software, and the packeting of information—all necessary and desirable for the purpose of connecting widely distributed computers together—all slow down the speed at which these connections can be traversed, to the point where message passing takes far longer (on the order of milliseconds or longer) than does processing of those messages by the computer (on the order of microseconds or nanoseconds). But in order to use a number of processors effectively on the same program, the time spent passing messages should be smaller than, or at worst not much larger than, the time actually computing.

The ETHERNET of Personal Computers

A relatively powerful and easy to use personal computer has been developed at Xerox Parc, where about 200 scientists are investigating the uses of computers in automating the office. [See Metcalfe and Boggs (1976) and Shoch and Hupp (1980); and see Thurber and Freeman (1979) for a survey of related systems.] Each scientist has her/his own personal computer, just as most of us have our own cars, typewriters, and television sets.

Note that until very recently the model for using computers has been that a number of people share the use of a "computer facility"—whether by submitting cards at the input—output desk of the computing center or by interacting with a time-shared multiprogrammed computer through a terminal. But "mini" and "micro" computers are now becoming cheap enough so that the convenience of having one completely at the disposal of a single person can seriously be considered. Large numbers of $500 to $5000 computers are now being sold by Radio Shack, IBM, Apple, and others for use at home and in small businesses. The Xerox Alto computer is a good bit larger, with a number of nice features in terms of language and graphics capabilities, and costs around $30,000−$60,000. But this is still cheap compared to the much more powerful computers, costing from $200,000 to $5,000,000 or more, that are used by the computing centers of large businesses, universities, and research organizations.

The Xerox Parc personal computers are all connected together over an ETHERNET bus—a coaxial cable that runs through the building to which each computer, along with a number of printers, plotters, tape readers, disks, and other input, output, and memory devices, are connected. This allows programmers to output their programs on the printer or plotter down the hall, to type their manuscripts on the higher-quality (but much slower) typewriter in the basement, or to print (more expensive) book-quality output using the typesetter in the publications office. One can also read-in large blocks of data from remote disks or other mass storage devices. And one can send messages and, with a little delay, manuscripts and longer bodies of information, to colleagues anywhere in the building (see Fig. 11).

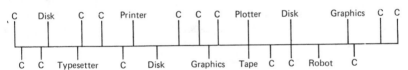

Fig. 11   Fragment of an ETHERNET-type network.

But once again, although there are a number of computers in this network, they have been interconnected in a way that makes message passing too slow for efficient use of several computers working in parallel on a single program. Communication among all processors, memories, and input−output devices takes place over the single common bus. This is usually handled by having each processor "broadcast" its messages to every node on the net, and that means every node must spend time listening to and deciding whether to bother with every message. As more nodes are added to the system the possibilities for contention and overload increase. The operating system can assume nothing in terms of topology; it therefore must expect the worst cases —which means a very high overhead of operating-system processes to handle all possible contingencies.

The ETHERNET has, in a local environment, much the same set of goals as the much more widely distributed ARPANET. The two complement each other very nicely. Indeed they are the two keystones in systems we will soon find in widespread use—very large global networks spanning several continents, connecting local networks of personal computers to one another.

Problems with Loosely Coupled Networks for Parallel Processing

ARPANET and ETHERNET, along with a number of newer networks that were greatly influenced by them and serve the same sets of purposes, are sometimes thought of as the only kind of network. Most research in computer science on networks in recent years has focused on this kind of network. Even many of those people who envision the computing center of the future as being a network of many computers think of an operating system that will assign each job to one computer.

Today's distributed networks pass information far too slowly for several processors to be able to work together effectively on the same program. But there are many different structures in which *more than one* processor can be coupled together so that they can effectively and efficiently work on a *single program, in a true parallel fashion.* There are varying degrees of "tightness" of coupling, and a variety of ways to tighten. Some of these architectures are relatively simple variants on the ARPANET and ETHERNET; others are intricately interconnected or reconfigurable; others are highly specialized. Chapter 3 begins to examine such systems.

Summary Discussion

A "general-purpose stored-program single-CPU serial computer" consists, essentially, of one processor [often called the central processing unit, or (CPU)]; a memory store that contains all needed information, including data and programs; and input and output devices, to communicate with the external world.

All modern computers are "general purpose"—which means that they are all capable of computing anything, of executing any program, that any other "general-purpose computer" can execute. This very powerful capability depends upon their having the elegant basic set of operating functions of a simple "Turing machine." (These are: read a symbol; write a symbol; shift; make a conditional test and act accordingly.)

Actual computers are made millions of times faster than Turing machines, basically by having their processors work with whole words (usually 32 bits of information at a time), and be able to "randomly access" any from a very large number of words in the computer's high-speed memory (usually 64,000 to 1,000,000 or more words).

The computer can be made even faster and more powerful by building into its processor a number of special processors for the different kinds of operations (e.g., addition, floating-point multiplication, matching of strings of symbols) that experience with programs shows are most frequently used. In addition, faster and faster technologies, and smaller and smaller, cheaper and cheaper technologies, add several more orders of magnitude in increases in speed and power.

Several large computers have been manufactured that can be bought with two CPUs on a bus, both linked to a common memory, each working on a different program. Such systems can almost double throughput. But systems with three, four, or more CPUs have given relatively smaller further increases (especially when judged with respect to their added costs).

A good deal of parallel hardware has always been built into conventional computers, to have them handle a whole word of information in parallel, and to use the most appropriate special-purpose processor to execute each instruction as quickly as possible. The most powerful supercomputers (and also the most expensive) have also made use of pipelines of assembly-line-like processors and other types of parallelism to further increase performance.

Two major loosely coupled distributed networks are today often considered the models for interconnecting several single-CPU computers: the ARPANET and the ETHERNET. The ARPANET connects a number of large computers that are distributed over several continents. The ETHERNET connects a number of usually identical, smaller, but relatively powerful, personal computers that are used by people working in the same building.

Both these networks serve very effectively to allow people to send a variety of types of messages (electronic mail, manuscripts, programs, bodies of data, etc.) to one another. The ETHERNET allows the individual programmer to use other resources (printers, plotters, disks, etc.) on the net, when needed, as well as her/his own personal computer. But these nets have been designed with other purposes in mind than the execution of a single program by several, or many, processors in the network at the same time. And the speed with which they send messages and transfer information is several orders of magnitude slower than the speed with which each computer can execute a process.

So more tightly coupled networks and arrays, of the sort that will be examined throughout this book, are needed to gain substantial increases in computing power and in speed of execution for very large programs.

There are many real (but largely unexplored) possibilities of building networks that perform with greater efficiency than do traditional serial computers. This may be possible because (as will be examined in subsequent chapters) a serial computer suffers from at least three kinds of inefficiencies that networks may be able to eliminate:

(a) When iterating through large sets of data, e.g., arrays, an appreciable overhead of instructions are needed to increment indexes and test for borders.

(b) The CPU is itself a collection of special-purpose processors, so that the fastest and most appropriate can be used for each operation. But all but one of these processors sit idle at any moment.

(c) Very large memories, special-purpose I-O processors, and multiprogramming operating systems are needed in order to keep the CPU as busy as possible. Today's standard measure of efficiency is percentage CPU utilization. But 99.99+% of the typical serial computer's logic gates and other resources sit idle, waiting to be used, so that the CPU will be delayed as little as possible. If we use a different measure of efficiency, e.g., the percentage of gates in the total system that are actively working at any given moment (see Uhr, 1981c), serial computers look surprisingly poor.

PART II

Arrays and Networks
Built or Designed

CHAPTER 2

First Attempts at Designing
and Organizing Multicomputers

Multicomputers tear apart the single-CPU serial computer. How to coordinate the multicomputer's many independent processors in ways both efficient and general is a key problem the many complexities of which we are just beginning to understand.

This chapter begins our examination of arrays, pipelines, and (other kinds of) networks. It explores some general issues of controlling, organizing, and programming a number of computers into an effectively working group structure. And it describes systems that were proposed or built in the primitive past before LSI technology (that is, more than ten years ago). [See Anderson and Jensen (1975) and Thurber (1976).]

It is important to survey the different ways a multicomputer can be used:

(a) Each computer can work on a separate problem (much like the ETHERNET personal computers).

(b) Each can multiprogram several problems, possibly including interactive terminal-based users and background programs.

(c) All can work on the same program, but essentially by first subdividing that program so that each can work separately.

(d) All can work on the same program, with increasingly closer coordination and intercommunication.

(e) Subsets can work on different programs (some may be single computers working in a personal computer mode, or multiprogramming).

(f) Those computers which are coordinated with others at some task might at the same time multiprogram, and in particular do (lower-demand) personal computing or/and background computing to make use of otherwise idle cycles.

This book concentrates on (c) through (f), with special emphasis on (d). Thus the key problem it examines is the efficient execution of a single program by a whole structured network of computers.

It is appropriate to mention here Flynn's (1972) widely used framework to distinguish between four major kinds of computer—essentially, all combinations of one versus many processors and one versus many sets of data:

(1) A "traditional 1-CPU serial computer" has only one processor working on one set of data, in a "single-instruction single-data-stream" (SISD) manner.

(2) In a "pipeline," each of a number of processors continually executes the same instruction as information flows through the pipe (much like an auto assembly line), each processor executing a different step in the longer sequence of instructions, giving "multiple-instruction single-data-stream" (MISD) processing.

(3) In an "array," all the many processors execute the same instruction, each on a different set of data, in "single-instruction multiple-data-stream" (SIMD) manner.

(4) In almost all "networks," each processor executes a different sequence of instructions on a different set or data, giving a "multiple-instruction multiple-data-stream" (MIMD) system.

As we shall see, these distinctions are made important by what underlies them: A single instruction means SIMD systems are synchronous, and virtually all contention among processors is eliminated. A single data stream means MISD systems need load instructions only once for very efficient pipelining.

The next several chapters examine the new SIMD arrays, MISD pipelines, and MIMD networks that have just been built, are under construction, or have been designed or proposed. We then examine the problems of developing parallel algorithms and programs, and good languages and operating systems for expressing and executing those programs on multicomputers. Finally, we explore a range of possible future designs that exploit the potentially enormous power VLSI technologies are today just beginning to put at our disposal.

Mapping Process—Information Graph
into Processor—Memory Network Flow

We code a computer program to express an algorithm, that is, a procedure for solving a problem. The program is designed to execute on some actual physical computer. The mapping of that program into a multicomputer, usually handled by a compiler and by the operating system, will determine how that computer actually executes the program.

There appear to be two major ways programs can be executed efficiently by networks:

(a) The network that executes the program can be decomposed into sub-networks, each executing a sub-network of the similarly decomposed program. This decomposition should attempt to minimize message passing, and to match the bandwidth of links between nodes and between sub-networks as closely as possible to the message-passing demands of the program.

(b) The program graph can be mapped onto a network that is as similar as possible. Then data should be flowed through it, as in a pipeline-like "pipe-network." This mapping should attempt to balance the load across all computer nodes and links through which information flows, in order to maximize the bandwidth of the total network.

A program consists of a set of processes that transform information input and then passed from process to process, as in data-flow graphs (McGraw, 1980) or Petri (1962, 1979) nets. This program must be mapped into a physical computer, which then actually carries out this moving of information through transforming processes.

It is instructive to examine this process with respect to the single-CPU serial computer.

All information about the program, and all information that the program will ever need, is fetched by the CPU from its randomly accessible main memory. This can be represented as a complete bipartite graph $G_{1,n}$ (i.e., with one processor linked to every one of n memories). Now the processor can obviously access whatever piece of information is needed, whether program or data (since it is directly linked to all memories). Clearly, to keep the processor busy we must make sure that whatever it needs will be in one of the n memories. Therefore we add hardware to make n as large as possible, and to effect parallel input of new information into that memory.

In sharp contrast, consider a special-purpose network of logic gates designed to execute a particular algorithm as efficiently as possible. Examples are Kung's (1980a,b) systolic arrays and also the fast adders and other carefully designed special-purpose hardware that make up a CPU. Here the structure of gates and wires is very close to a network flow graph (Ford and Fulkerson, 1962; Lawler, 1976). An optimal design would give each node just the capability and power to execute its given task, and each link just the bandwidth to transmit information so that no delays will occur.

The 1-CPU computer's $G_{1,n}$ graph can conveniently emulate the data-flow graph of any program that can be stored into its memory. But this process is very costly, since millions of links must be established to memories, and each instruction and each piece of data must be transferred across those links (and the instructions must be decoded). At the other extreme, output from one processor is immediately input to the next. Each processor need be linked only to those processors from which it receives or to which it sends information. Program instructions need not be fetched or decoded at all. Instead they are directly cast in silicon, the computer architecture actually embodying the isomorphic process—structure graph of the program.

It seems potentially extremely fruitful to describe a multicomputer as a network of processors that transform information flowing through them. We can then compare different architectures by asking how each executes particular benchmark problems, or sets or mixes of problems. We can further use these results to point to bottlenecks and underused channels and nodes to help us restructure our architectures in attempts to improve upon them.

For example, consider an SIMD array of the sort examined in Chapter 3, with one processor and its memory in each cell, and links between near-neighbor cells. When a program is coded to effect the same local operation at each cell (as is often the case in image processing and three-dimensional modeling) input data can be dispersed throughout the memories so that each processor is optimally linked to all the data it must fetch. Each processor is directly linked to all data stored in its own memory (just as in the serial computer, albeit typically with a much smaller memory). Each processor is also linked to its neighbors, so data can be shifted in from their memories, and more distant data can be got by successive shifts. If data are properly mapped into memories, as when a TV image is input without disturbing its topology, such a system will work efficiently to the extent that processes are local (that is, need data stored nearby).

In subsequent chapters we occasionally evaluate architectures from this information-flow point of view. The reader might attempt that as an exercise, especially when an intriguing architecture, e.g., an n-cube, star, tree, or ring, is described.

A Survey of Early (Pre-LSI) Multicomputer Arrays and Networks

Cellular Automata

John von Neumann defined "cellular automata" (in 1952, published in 1966) as arrays of computers in checkerboardlike cells, each cell connected to its "near neighbors." Each computer was viewed as a neuron-like module similar to the idealized neurons of McCulloch and Pitts (1943). Von Neumann, Thatcher, Codd, and others explored the theoretical characteristics of such highly parallel systems (see Burks, 1966; Codd, 1968). It is interesting to note that of the very large arrays and pipelines to be examined in the next chapter, Duff's (1976) CLIP and Sternberg's (1978) Cytocomputer were designed with these cellular arrays as a central background model, but it appears that most of the other architects were not aware of this theoretical development.

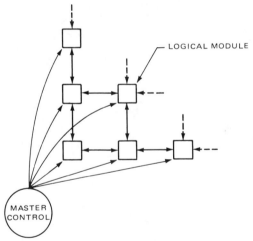

Fig. 1 The organization of Unger's Spatial Computer. The master controller decodes the program instructions and sends each instruction to all processors. [From S. H. Unger, A computer oriented toward spatial problems. *Proc. IRE* **46**, 1744–1750. Copyright © 1958 IRE.]

Unger's Cellular Array

Steven Unger (1958) described a two-dimensional array of simple 1-bit processors, each able to fetch information from its own small memory and to shift information from and to its four square neighbors.

All processors were to execute the same instruction (in SIMD mode) broadcast to them by the "master control unit" (Fig. 1). Unger's machine was a paper design, although it appears that one processor was actually built. Each processor is very simple, using about 30 gates. Figure 2 gives Unger's original logic design for each processor.

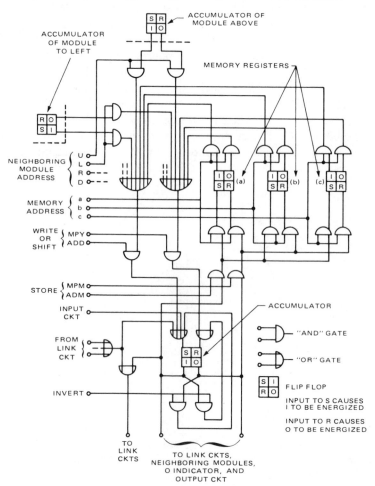

Fig. 2 Unger's logic diagram for one processor. [From S. H. Unger, A computer oriented toward spatial problems. *Proc. IRE* **46**, 1744–1750. Copyright © 1958 IRE.]

Unger defined a machine language and described and gave programs for a number of simple local picture processing operations (e.g., to detect corners and concavities) that this system would handle efficiently, in parallel. Unger (1959) also developed a number of interesting procedures for using this array for pattern recognition. As possible variations, he suggested using arrays of three or more dimensions, multibit processors, and polar-coordinate arrays.

The SOLOMON 1-Bit and 24-Bit Arrays

Daniel Slotnick (1967) and his co-workers designed several versions of SOLOMON, a "simultaneous-operation linked ordinal modular network" for number crunching with large matrices. SOLOMON I (Slotnick, *et al.*, 1962) was a large 32 × 32 two-dimensional array of 1-bit processors much like Unger's and the very large arrays that have recently been built. It was designed to be built in modules of 32 × 8 processors up to a maximum size of 64 × 32. Gregory and McReynolds (1963) describe the hardware in some detail. A prototype was built at Westinghouse and brought to the University of Illinois after Slotnick moved there. But there appear to be no reports describing results of programs run on it.

Figure 3 shows a portion of the array of processors in SOLOMON I.

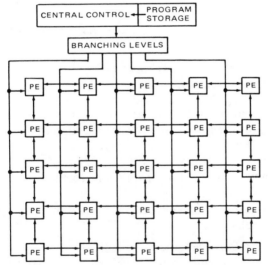

Fig. 3 The SOLOMON I controller and processors (PEs). [From Slotnick *et al.* (1962). Copyright © 1962 AFIPS Press.]

Figure 4 gives the block diagram for a processor.

SOLOMON II was to have much larger 24-bit processors for large number-crunching tasks. It was the precursor of the very different ILLIAC IV system, which was designed to have 256 processors but built with only 64 (because of the great expense of its very powerful 64-bit processors).

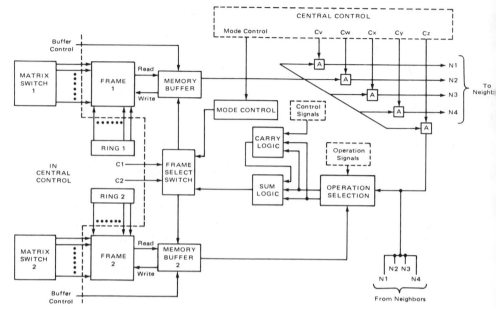

Fig. 4   Block diagram of a SOLOMON I processor. [From Slotnick *et al.* (1962). Copyright © 1962 AFIPS Press.]

ILLIAC III: Array Plus Graph Manipulator Plus Traditional Computer

Bruce McCormick (1963) designed and almost completed ILLIAC III, a very interesting system that combined three major subsystems of computers (see Fig. 5):

The "pattern articulation unit," a 32 × 32 array of 1-bit processors.

The "arithmetic unit," for stereoreconstruction, statistical analyses, etc.

The "taxicrinic unit," a processor that assembled, organized, and manipulated graphs.

Figure 6 gives the schematic of an ILLIAC III array processor.

ILLIAC III was designed with visual pattern recognition primarily in mind, and in particular the identification of significant events in high-energy-physics bubble-chamber photos. The array was intended to transform raw input images into line drawings. The taxicrinic unit would then use syntactic pattern recognition techniques (Narasimhan, 1966) to analyze the graphs into which the abstracted line drawings of images had been converted, and to recognize significant events, that is, patterns of lines.

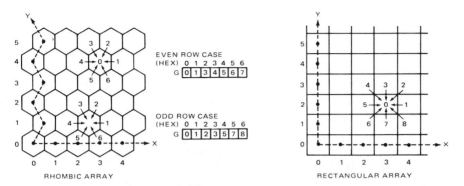

Fig. 5 One processor (called Stalactite), connected to handle either six-neighbor hexagonal or eight-neighbor rectangular arrays. [From B. H. McCormick, The Illinois pattern recognition computer ILLIAC III. *IEEE Trans. Comput.* **12**, 791–813. Copyright © 1963 IEEE.]

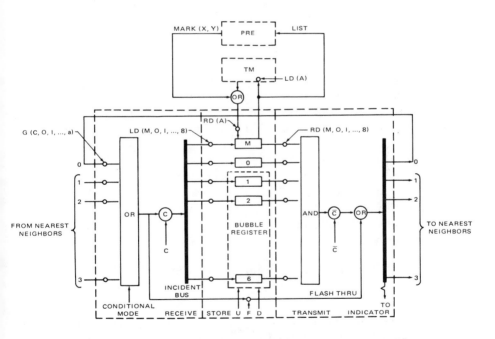

Fig. 6 One processor in the ILLIAC III array. Bubble registers and flash through are used for fast manipulation of data in the array (see McCormick, 1963, for a description). [From B. H. Mc Cormick, The Illinois pattern recognition computer ILLIAC III. *IEEE Trans. Comput.* **12**, 791–813. Copyright © 1963 IEEE.]

Because of the magnitude of this project using then-existing technology, it took many more years than had been expected. It was aborted when the almost-finished computer was destroyed by fire. It was an unusually interesting example both of a powerful array and a carefully designed total system of several different types of computers.

Additional Designs and Proposals for Array-Related Systems

John Holland (1959) described a basic module (a relatively conventional serial computer with its own controller) that would form one cell of a very large two-dimensional cellular array of computers designed to execute different instructions, in MIMD mode, but synchronously. Holland examined how large numbers of programs might be executed by such a system by moving data and intermediate results along paths of available computers.

Hawkins and Munsey (1963) sketched out a similar system where data (images) are viewed as streaming through the processors. They pointed out that a very large number of processors should be used, and suggested the system be built from optical components.

Rosin (1965) suggested using (cheap) drums with one processor for each of many tracks as an economical way to use standard technology. He estimated that such a system would need only two minutes (including loading and unloading the drum) for problems that would need ten hours on an IBM 7090. This is an interesting use of a one-dimensional array of processors to handle two-dimensional arrays of information. It was also anticipated by Zuse (1958), who in 1956 applied for a patent for an array of processors sharing a multihead drum.

A good deal of work has been done on using microcellular arrays as basic logic units, as well as memory stores. See Minnick (1967) and Mukhopadhyay and Stone (1971) for surveys that also examine some of the arrays of more complex processors.

Associative, Content-Addressable Memories (CAMs) and Logic-in-Memory

Fuller and Bird (1965) proposed building large associative memories whose contents could not only be accessed because they matched the string being looked for (which is what is meant by "associative memory") but could also be modified. They pointed out that this might be effected with simple extensions to the hardware needed to make the associative retrieval alone, and they showed how simple image processing algorithms could be executed by such a system. Shore

(1973) argues that this type of logic-in-memory system uses too many gates to be cost effective when compared with a traditional serial computer that packages all its gates into a single CPU. But his comparison was based on a particular system he designed using a particular technology, for a particular application where only small parts of the system would be active at any given moment.

I should mention that associative memories seem very attractive to many people, but they are rarely used. They are expensive, since they usually need extra gates for logic needed in the memory itself to test its stored contents for matches. But Fuller and Bird suggest the extensions they propose would cost relatively little more. And the price of such a component depends to a great extent on how much effort manufacturers make to develop and market systems that use them. To a great extent because traditional computers have been so successful, CAMs have been neglected.

Kautz (1971) suggested augmenting an associative memory with rather extensive arithmetic capabilities. His system used roughly 40 NOR (not−or) gates per bit, which Kautz estimated was about four times the number of gates used in the simplest of CAMs, and about two times the average.

Thurber and Myrna (1970), after conversations with Minnick (1966), designed a "cellular APL computer" that directly implemented the highly parallel APL programming language (Iverson, 1962). APL is a very attractive candidate language for a parallel computer, and such a system would need far less software development. This system was designed to have 16 32×32 arrays of memory words and a 32×32 logic-in-memory array that could access these arrays through hardware routing logic. Newton and Brenner (1971) designed a similar cellular APL computer, but with a one-dimensional vector-oriented array of 1024 logic-in-memory processors. (See You and Fung, 1977, for a survey.)

Super Multicomputers

The ILLIAC IV Supercomputer

Studies of the SOLOMON designs led to ILLIAC IV (Slotnick, 1967; Barnes *et al.*, 1968), a supercomputer that was to have 4 8×8 arrays of very powerful 64-bit processors, with a control unit for each of these quadrants. Each processor was reconfigurable, so it could also serve as 2 32-bit or 8 8-bit processors. This was in many ways among the most innovative and most ambitious of supercomputers.

Slotnick (1967) estimated that each processor would cost $10,000. But after design and construction costs mounted, the finally completed ILLIAC IV had only 64 processors, each with 9300 gates (Kuck, 1978). Falk (1976) commented: "a Government investment of more than $31 million. . . . Conceived as a machine to perform a billion operations per second, a speed it was never able to achieve, ILLIAC IV ultimately included more than a million logic gates—by far the largest assemblage of hardware ever in a single machine."

The PEPE ("Parallel Element Processing Ensemble") Supercomputer

PEPE (Cornell, 1972; Evensen and Troy, 1973; Martin, 1977; Vick and Cornell, 1978) was designed and built to handle radar data processing for ballistic missile defense. It is built from banks of very powerful "processing elements" (PEs). A full system has 288 PEs. Each PE is actually three processors, each specialized to handle its part of the total task for which PEPE was built:

(1) The "correlation unit" inputs data and associates these data with data already on file.

(2) The "arithmetic unit" has a conventional ALU repertoire, for fixed- and floating-point arithmetic and for logic.

(3) The "associative output unit" efficiently searches multidimensional files and orders and outputs data.

So up to 864 processes can be executed in parallel. PEPE has three controllers, one for each type of processor. PEs have their own local memories, and they share a large associative memory. They communicate only serially through the controller. This means PEs form an unstructured "ensemble" rather than an array. This independence gives flexibility in grouping and ordering processors. When processors fail, the system can keep working, and new processors can easily be added. For the applications considered interaction was not needed.

PEPE, like the ILLIAC IV, is an extremely expensive supercomputer. It has about 10,000 gates in each PE, and a relatively expensive associative memory. Each PE can execute a 32-bit floating-point multiply instruction in 800 nsec. It has been estimated that enough PEPE processing elements to execute the same number of instructions per second as a CDC 7600 (to which PEPEs are connected) would cost only about 10% as much as a 7600. Burroughs built an 11-PE PEPE system for the Army Ballistics Missile Defense Agency.

STARAN and ASPRO, Reconfigurable Arrays

Batcher's (1973, 1974, 1976) STARAN uses a large array (up to 32 banks of 256 one-dimensional arrays of processors linked to an associative memory through a partially reconfiguring "flip network").

The STARAN processor is among the smallest and simplest, using roughly 50 gates each. But STARAN is a very flexible system because the reconfiguring network gives the programmer a great deal of power to code a variety of manipulations of the data in the memory, as well as the relatively straightforward near-neighbor operations. On the other hand, this flexibility appears to make STARAN more difficult to program (or, very likely, appear to be more difficult to program until one has gotten over the first humps of learning how to think about reconfiguring data; see Rohrbacker and Potter, 1977; Potter, 1978). And STARAN is a rather expensive system. The use of fast off-the-shelf ECL hardware, an associative memory, and the flip network all served to drive costs up. [See Foster (1980) for a discussion of STARAN's potential promise.]

The largest configurations actually built contain 4 banks with 1000 processors. STARAN can be used very effectively to process two-dimensional images in a manner typical of two-dimensional arrays. The flip network not only handles the near-neighbor links of hard-wired arrays but also does a variety of useful shuffles of data, as for the fast Fourier transform.

ASPRO (anonymous, 1979; Surprise, 1981; see also Meilander, 1981; Batcher, 1982) is a new VLSI system with 1000 processors (32 plus a flip network on each chip) developed along the lines of STARAN. It will fit in a 6-in.-cube chassis, to be used in the nose of radar surveillance planes.

Organizing Computers: Clocks, Controllers, Operating Systems

The diverse and often complex functions of a fast, powerful electronic computer must be precisely and comprehensively coordinated at several hardware and software levels. Every step must be handled with completely unambiguous and detailed signals (see, e.g., Stone, 1980). This is a good point to examine how computers are synchronized and controlled, so we can, where appropriate, consider alternative possibilities.

Synchronizing Clocks and Asynchronous Gate Delays

Something must initiate actions: The CPU must be told to fetch and execute the next instruction. Input processors must be told to start executing a read instruction. This can be done "asynchronously," by having each gate signal that it has stabilized and its operation has completed. Simpler, and more common, is to have a clock "synchronize" (possibly with certain functions, e.g., memory fetches, asynchronous).

A very fast, very precise clock counts out cycles and pulses all the different parts of the computer to play at the same beat, much as a symphony orchestra would play to the beat of a precise metronome (rather than a human conductor). Every part of the computer is coordinated according to this clock. The instruction must be fetched (which means memory gates must be activated and their information transferred into the instruction register). Then that information must be decoded and used to activate the required special processor circuitry in the CPU, and to specify the addresses of the data to be fetched from memory (to serve as operands to be input to the activated processor). Each of these steps must be given enough time to finish, and then the next step initiated with minimum delay.

The basic clock cycle is typically much faster than the computer's instruction cycle, so that precise sequencing can be maintained. For example, a 10-nsec clock might be used in a computer that took 100 nsec to fetch a word from memory, several hundred nanoseconds to add two numbers and several microseconds to multiply two numbers. The exact times are chosen to fit the gate delays and electrical characteristics of the basic technology used. The clock must allow each circuit time enough to complete its actions and settle back into a quiet, stable state.

All instructions can be executed at the same speed, in "synchronous fixed" mode. But some instructions are much simpler and much faster than others, and this would mean that all but the slowest would sit around waiting for the next instruction cycle. So instructions are usually executed in "synchronous variable" mode, that is each one taking as long as it needs, followed immediately by the next.

The Controller Barks Out Commands

The processor transforms information fetched from and stored into the memory. The beat is set by the clocking circuitry, which is rather more complex than a metronome but far simpler than the conductor of an orchestra. There are a variety of organizing, coordinating, and controlling functions that controllers are built to handle. These vary greatly, depending upon the architecture and purposes of the computer.

First the controller must fetch and decode the program's next instruction and set up the execution of that instruction. For this it must have an instruction register into which that instruction is fetched, and circuitry that decodes that instruction and initiates fetches of the specified operands and activates the specified operation. A program counter is also needed, to point to the next instruction.

The controller grows much more complex when it must handle parallel I-O processors, special-purpose processors, hierarchies of several different memory devices, and the other kinds of hardware used to improve the performance of large computers.

An absolutely minimal bare-bones controller might need at least 1000 or 2000 devices—a relatively small amount with respect to 100,000-device chips, but large with respect to a simple 1-bit processor with only 50−500 devices. More typically, the controller will be as large as, or a good deal larger than, the rest of the CPU, on the order of 10,000−100,000 devices, or more.

In a multicomputer, one controller can handle several processors. The striking example of this today is the single controller that handles 4000−16,000 processors in the very large SIMD arrays. Now the cost of even a very large controller grows relatively small, when divided among and amortized over thousands of processors.

Today's computers have either a single controller for a single processor (the conventional SISD serial computer) or for all processors (the large SIMD arrays), or a controller for each processor (the MISD pipelines and the MIMD networks). But this need not be the case. A system might have several clusters of computers, or of components, possibly of different types, possibly configured in different ways, each with its own controller. In a multicomputer where several processors share a controller, the several parts of an instruction could be torn apart. For example, each processor might decode and execute its own operation but fetch data from locations broadcast to it by the common controller.

Since the controller must be linked by wires to everything it controls, one interesting possibility is to place the controller at the center of the physical region that houses the processors it controls (e.g., the IC board, or the VLSI chip, or a subregion on each). Each specific control function can be placed central to the devices it controls. The clock, which also must have wires linking it to all devices it coordinates, will coordinate the controller's processes also.

Getting many more or less specialized controllers that are dispersed throughout a large multicomputer to cooperate and work effectively together is one of the major problems we must confront.

The Operating System Runs the Show

The "operating system" is a (usually very large) program that coordinates the computer's input of programs and data, order of execution of programs, and output of results. It is designed to make the computer as efficient and as usable as possible, attempting to maximize throughput and also accessibility. In a very simple computer this might merely mean a program that inputs the next program, turns control over to that program (which then must include instructions to input needed data, output results, and terminate), and stops when all programs have been completed.

But that would be quite dangerous, since a program might never terminate, or it might have mistakes that crash the whole computer (e.g., it might make an error in calling for input and wait forever, or it might store information in the memory locations that contained the operating system). So typically the operating system is designed to handle input and output, monitor user programs and kill them when they run too long, and try to recover and restart the system when something less than fatal goes wrong.

Most computer systems have a number of "utility programs," like filers (for data and programs stored on disks), editors (for convenient keying in and changing of files), several compilers for different programming languages, and a variety of other programs and capabilities. These are usually handled by and considered a part of the operating system.

In a multiprocessing conventional computer the operating system must assign processes to processors, set and adjust priorities, and get each program done. The moment a system must handle several concurrent processors, a variety of possible contention, deadlock, and error-making situations arise.

Problems typically arise because several processes update the same variable or want to use the same resource (e.g., a disk or an input device). Each process needs some set of resources, including a processor to execute instructions; some memory (for program as well as data and intermediate results); at times, some input and output device(s); and, possibly, some disk or other bulk stores. The operating system typically decides which program to run by some priority system or by roundrobining all programs in turn (each until its present module, e.g., a procedure, is completed, or for a short time slice of a few milliseconds), or by a mixture of both.

Consider the following situation: A process is executing, and needs some resource (more memory, or an input or output device) to continue. But if that resource is not available, the process must wait. But since that process is itself using resources, some other process (which may even be part of the same program) may be similarly hung up. This is the classic deadlock situation. And, since Murphy's law is truest for computers, anything of this sort if it might go wrong surely will go wrong.

There are several ways to avoid such situations: e.g., a suspended process can be forced to relinquish control over all resources, or all resources must be requested and allocated at once. The operating system traditionally handles such problems, because they depend upon global interactions among processes.

If several programs use the same piece of data, it becomes possible that several might be changing it while others are fetching it, so that several are modifying one original value, and therefore unanticipated modifications are made. This means that processes must be forbidden access to data that are being worked on. A simple solution is to make the process call the operating system, which can then make sure no other process interferes. "We pay a price, however, for asking the operating system to do our work for us: on most machines, the overhead of switching to another task and back is orders of magnitude longer than the execution time of the two or three instructions actually needed (Stone, 1980, pp. 597−598)."

More economical solutions have been achieved. For example, a "compare and swap" (CS) (as introduced into the IBM 370 instruction set) can be used:

CS(OLD, NEW, ADDR)

(a) compares the contents of OLD and of ADDR,

(b) if they are equal, stores the contents of NEW into ADDR,

(c) if they are unequal, stores contents of ADDR into OLD,

(d) sets a condition bit to show what happened.

Now a program can set loc $:= a$; then execute: CS(a,b,loc); then test: IF $a =$ loc THEN {go forward} ELSE {try again}, going back to reexecute the CS instruction if some other process has changed things so that OLD and ADDR are not the same. Essentially, we must ensure that an operation effected by a whole sequence of instructions cannot go astray because it is interrupted in the middle. This might be done by turning the operation into a single instruction, or by giving it absolute priority (as is the case when it is turned over to the operating system), or by using a device like the CS instruction to keep enough information to detect and correct any modifications effected by an interrupt.

In a MIMD network the operating system must allocate programs to computers. Now a number of new, as yet largely unsolved, problems arise. A single program must be decomposed into sets of processes, and each set assigned to a different computer, in order to profit from the possibilities of parallelism offered by the network. The appropriate processes must be loaded into each computer, along with input data. Parameters, intermediate results, and coordinating messages must be passed between computers.

What subset of computers from the larger network should be used for each program? Today there is no consensus. Some people feel that any set will do, others that this should be considered a graph-matching problem, with the program graph mapped as well as possible into some suitable subgraph of the network. This decision interacts intimately with the kinds of network structures, programming languages, and operating systems different people consider appropriate.

In this very complex new situation it is turning out that the operating system's "message-passing protocol" usually takes thousands of instructions, and therefore makes the passing of data from one process to another take literally thousands of times longer than the passing of that data within the same process, or the execution of an instruction. If the flow of information among processes is not taken into account when assigning processes to computers, flurries of (thereby generated) unnecessary messages can completely overwhelm programs.

If the operating system does not map the program structure into as similar as possible a structure of hardware processors, the proper mapping must be achieved when the program actually executes, by potentially horrendous amounts of otherwise unnecessary message passing.

The Overall Coordination and Operation
of a Multicomputer Network

Multicomputers are so complex, and so new, that we are just beginning to get a feeling as to how to run and control them. Relatively few have been designed in any detail; even fewer have been built. Most of these have either been SIMD arrays that rigidly coordinate, or extensions to conventional computers to increase their power (e.g., large mainframes like the IBM 370/91 given several CPUs), or to broaden their uses (e.g., ETHERNETs to make possible the sharing of resources).

Virtually none have been tested, evaluated, or compared. Indeed we hardly know how to test these systems. To use meaningful benchmarks and program mixes we need to know what kinds of programs people will want to run on highly parallel multicomputers. But only as we develop these computers, and good languages for expressing and revealing parallelism, will researchers develop parallel ways of attacking problems, and parallel algorithms and programs that execute these algorithms effectively on parallel multicomputers.

How Might We Organize Assembly Lines of Robots?

Consider the following variation on this situation. Rather than networks of computers, assume we are developing networks of robots. This means to each computer we add motor effectors, so that the computer's streams of symbols are not only transduced to tapes, disks, or print on paper, but also initiate and control physical actions of arms, hands, fingers, legs, drills, dollies, or whatever output devices robots are built to use.

This differs very little from our present situation. A computer printer's type keys (or pins or jets of ink) are actually little armlike effectors that move physically to execute the symbol "strike."

How would we organize, control, and coordinate robot/computers? If we had one robot, we might use it like a slave, butler, amanuensis, sidekick, assistant, and/or co-worker. If we had several, we might assign them to several different sets of tasks—one to kitchen and food, one to cleaning, one to laundry and fixing leaks. Or we might post lists of jobs, and have each robot cross off and do the next one in turn. Simply consider the variety of ways we use and structure situations in which people or animals are put at our disposal.

Assume we set up a conglomerate and advertise for jobs, small, big, and enormous (among our businesses is a robot factory).

We might assign each job to a different robot, instructing it thoroughly (that is, giving it the needed program). But then somebody observes that much simpler and cheaper robots could do many of these jobs, and that carefully designed "specialists" or special-purpose robots could do at least some of these jobs far more efficiently. Somebody else invents the "robot assembly line" and begins to turn out "Model T Robots" at far lower cost than our "Customized Robot." Not only can simpler, dumber, defective robots man an assembly line, but also each robot can work more efficiently, since it repeats the same program, rather than learning to use new ones all the time. And the group can work more efficiently because every robot knows exactly what it is supposed to do.

We get jobs that must be done very fast, or immediately. So we assign large numbers of robots. Sometimes we tell each robot exactly what to do. But we usually do not know ourselves, so we ask the people with the job, and they do not know either and do not want to think about it; so we tell the robots to decide.

Many customers complain, or fail to return. Two approaches seem to work best:

(a) Assign but one robot to a job and tell customers to learn to wait.

(b) Build a completely coordinated, and rigid, assembly line.

When people insist that jobs be done fast, we tell them that it is impossible unless their jobs are divided up into separate smaller jobs for separate robots, or restructured to fit our assembly line. We have learned not to bother customers with such difficult and tedious chores (since they usually give up and rarely return), and we do not know how to do them ourselves, so we assign them to another group of robots.

This analogy is illuminating because it reminds us of the variety of different ways we organize (or disorganize) human workers, and also of the variety of different possibilities for organizing networks of computers. This situation is so familiar to all of us that there is no need for me to elaborate on the variety of issues involved. But it seems interesting to pursue the analogies, by on the one hand substituting "computer" or "program procedure" for robot, and on the other hand substituting "human being" and considering all the problems and the attempted solutions we have evolved for human organizations.

Networks of Computers, Assembly Lines of Well-Organized Workers

The analogy between a multicomputer network and an assembly line of workers seems rather obvious, yet it has rarely been made much less pursued. The assembly line is designed to efficiently execute some job or jobs much too big for any individual to complete in a reasonable amount of time. Raw materials flow into the line. Workers on the line successively transform and combine those materials. Finished products flow out. The whole purpose of the assembly line is to make as much as possible as cheaply as possible, to maximize the amount of useful work done in transforming raw matter into finished goods.

Computer programs work with information rather than material. But otherwise our goals, problems, and procedures appear to be very similar indeed. As we examine different computer networks, reconsider this analogy, and ask how suitable each network might be for factories whose products were material goods, whether books, bikes, stereos, computers, robots, or VLSI chips.

An assembly line organizes a steady progression of processes from raw material to finished product. All workers share the load so none is either rushed or forced to be idle. Each worker's output moves with minimum cost to the worker designed to handle the next steps.

Pipelines of computers are built this way. But they typically sequence all computer stages in a one-dimensional linear string. The flow of information through a data-flow network might start with several different inputs to several different workers. Then several outputs might be worked into larger ensembles at subsequent stages. Still other outputs might be scattered to a number of next stages. In general, a variety of graphs might be drawn to best express the flow of information as it is transformed by a variety of different algorithms. This would better be called an assembly network than an assembly line.

We can now set two goals:

(a) Networks that execute as efficiently as possible. This means a minimum amount of resources to output a given set of results.

(b) Networks that execute as fast as possible. This means increasing the speed and power of individual workers, or/and adding more workers, such that the whole network will output faster.

Summary Discussion

We have now examined the multiprocessor extensions to conventional single-CPU serial computers, the loosely distributed networks, and the variety of early multicomputer arrays and networks. They already pose many problems for the design and use of multicomputers. The coordination, control, and organization of such systems is handled at the several levels of hardware clocks, controllers and physical interconnections, and software operating systems and user programs. These were surveyed, and put in the context of the range of possible ways of organizing such networks.

A suggestive analogy to a network of computers executing a program that is a network of processes is an organization of robots (which are simply computers with physical effectors for output devices) designed to handle complex problems. This can be viewed as an assembly line (or, more exactly, an assembly network) of workers organized to produce finished goods. Such organizations can be optimized to produce as economically as possible, or/and as quickly as possible. In both cases it seems attractive to think of a network flow where bottlenecks are eliminated and the bandwidths of links between workers and the power of workers are adjusted for maximum overall throughput with minimum internal wasted time or effort.

This poses a variety of graph and network flow problems related to the following general issues:

How can an algorithm best be expressed in a data-flow or other graphical language that exposes that algorithm's structure and flow of processes?

How can a multicomputer network best be built to accept and execute such algorithms as efficiently, or as fast, as possible? This might be for a specific problem, a special type of problem, or a general mix of present and anticipated (or unanticipated) future algorithms.

How can algorithms be mapped into architectures? How can mismatches be eliminated, or handled with minimum inefficiency of wasted effort and needless message-passing overheads?

How can algorithms and architectures be constructed to be best suited one to another, and modified to make them better suited? This might involve a cycle of redesigns of either or both, reassignments and adjustments by the operating system, or/and reconfigurings of the hardware.

CHAPTER 3

Large and Powerful
Arrays and Pipelines

This chapter examines the very large cellular arrays that have been built and designed during the past few years, the very powerful pipelines that have been developed for image processing, and several smaller variants that combine features of arrays and pipelines. It concentrates on those systems that are the most modular and use the simplest, most flexible processors, since they appear to be the cheapest, most powerful, and most expandable with future VLSI technologies.

By "array" I usually (but not always) mean a "cellular array" with a large number of at least 1000 processors, each with its own memory, all connected to a small number of near neighbors (usually four, six, or eight), thus forming a two-dimensional grid. The processors are usually very simple (and flexible); they compute logical and arithmetic functions, but only 1 bit at a time. Each processor's memory is relatively small (from the traditional point of view), ranging from 32 to 4000 bits. But the total high-speed memory of the system is usually quite large, in several systems totaling 2,000,000 bytes. Thus the traditional single large processor has been torn apart into many small processors. And the total number of devices used for processors has been increased (because of the very large number of processors). Whereas most 1-CPU computers use 99% or more of their resources for memory, these arrays often use 50%, or more, for processors.

It is important to emphasize that virtually all of the arrays and pipelines are general purpose. Because they are designed to handle specific classes of problems efficiently (as are all computers that are actually

built to compete in the market places of the "real world"), they are
sometimes dismissed as "special purpose," in contrast to the single-
CPU computer and to the networks of processors we shall begin to examine in Chapter 4. They might better be called *"specialized,"* since
they are structured to handle special kinds of problems. But every
computer is more or less specialized; today our perspective is still so
limited that we can make no definitive judgments as to how much.

Make sure you are not misled by the descriptions of systems when
they detail what one single processor does. The power of these computers lies in the fact that *very large numbers of processors are all executing an operation at the same time, in parallel.*

The General Architecture of Parallel
Cellular Array Computers

A cellular array is a two-dimensional array of cells R rows by C
columns in size. Each cell contains one processor and attendant
memory and registers, that is, an entire computer (except for the controller). Each computer in the array has a direct hard-wired connection
to each of its N nearest neighbors (see Fig. 1).

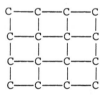

Fig. 1 A 2-Dimensional Cellular Array. [C = computer (processor + memory), |
and — = (sections of) links.] Note: This figure shows square neighbors. Some arrays
have direct links to diagonal neighbors also and some to six hexagonal neighbors.

A typical instruction specifies

(1) the subset of its neighbors from which each processor will get
information passed to it;

(2)   which subset of information in its own memory to use;

(3)   what operation it must perform on this set of information;

(4)   where in its own memory it should store the results.

Thus each computer has N neighbors and W words of memory in which it can store information. Each processor can effect any machine language process in the processor's repertoire (with the SIMD restriction that all processors effect the same process at each moment).

All arrays to date have either hard-wired links from each processor to its 4 square, 8 square-plus-diagonal, or 6 hexagonal adjacent neighbors, or (as in the case of STARAN) a switching network linking a one-dimensional array of processors to a large memory.

Most arrays connect the border cells with the cells at the opposite side, row by row or/and column by column (giving a cylinder or torus), or the next row (linking the array as a one-dimensional array also), or with reconfiguring switches that allow for a variety of flippings and shufflings of data.

Some arrays have hard-wired "windows" that allow for parallel fetches and operations on any subset of these neighbors. Others must effect such window operations serially, one neighbor at a time.

Several high-speed registers are always associated with each processor. Typically, two are used for the two operands fetched for an instruction, a third for temporary storage of the results, and, possibly, one or a few more to handle carries and other types of intermediate results. Therefore data are fetched from memory to registers, input to each processor, output to a register, and stored from that register to memory. When we say that each processor fetches data from its neighbor what typically happens is that these data are fetched from memory to the associated register, and then shifted to the register of the intended neighbor cell. Often this is done in one instruction cycle, so that processors "fetching from a near neighbor's memory" and "fetching from their own memory" take exactly the same amount of time.

Input to and output from the entire array are typically handled by shifting across all the rows in parallel: e.g., for input, data are shifted into the first column of each row of the array, and shifting is continued until the array of data has all been transferred in. Thus an $N \times M$ array typically needs M shift instructions to load a new $N \times M$ image into its registers, then one more instruction to store that entire array into the specified memory location. One cell of the image is now stored in each cell of the array. In some systems this shifting can be done in parallel with the execution of instructions by the processors (because extra registers and I-O controllers have been added for this purpose).

See Table I for the actual specifications of some of today's arrays.

Table I

Some Comparisons among Parallel Arrays

	Arrays of computers				
	CLIP4	CLIP3	Pilot DAP	DAP	MPP
Computers	96×96	12×16	32×32	64×64	128×128
Memory					
Words/computer	32	16	2048	4096	1024
Word length	1 bit	1 bit	1 bit	1 bit	1 bit
Total memory					
(1000 bits)	320	2	2,000	16,000	16,000
Registers					
(total bits)	5	4	5	5	40
Connected					
neighbors	8 (or 6)	8 (or 6)	4	4	4
Parallel fetch	8	8	1	1	1
Technology	CMOS	TTL	TTL	TTL	ECL
Speed (mhz)	2.5	10	10	10	40
Time/instruction	11 μsec	3 μsec	250 nsec	250 nsec	100 nsec
Rough Estimates					
Cost (in $1000)	100	—	—	1300	5000
Time/CLIP4-equivalent					
instruction	11 μsec	11 μsec	2 μsec	2 μsec	600 nsec
Year finished	1979	1974	1977	1980	1983
Devices on the custom chip					
Total/Computer	500	—	—	200	700
Processor	250	—	—	150	450
Parallel fetch	100	—	—	—	—
Memory	100	—	—	—	—
Registers	50	—	—	50	250

The Very Large LSI Parallel-Array Computers

Today's large parallel-array computers have been designed with specific kinds of tasks in mind, where large arrays of information must be dealt with. The two types of task on which most research has focused are (a) image processing and pattern recognition and (b) the solution of large sets of equations.

These different problems have sometimes led to somewhat different kinds of hardware—probably chiefly because numerical problems must work with large numbers, at least 32 bits long, for reasonable precision, whereas image processing and pattern recognition rarely work with more than 4- or 8-bit values (e.g., for intensity gradients or weights), and often work with only binary values (for black-on-white transformations of the raw image, or "present" versus "absent" judgments about features). And image processing must contend with very large arrays, on the order of 500×500 or 1000×1000 "pixels" (picture elements). But the similarities between a system designed for image processing and a system designed for number crunching are surprisingly great.

The Simple Processors and Structures of the Very Large Arrays

Almost all the very large arrays (with over 1000 processors) use very simple processors arranged in a two-dimensional array. [Exceptions are Batcher's (1973) STARAN, and ASPRO, its VLSI successor (Anonymous, 1979; see also Meilander, 1981; Batcher, 1982); the array of more powerful processors specialized for finite-element analysis proposed by Jordan *et al.* (1979), and the reconfigurable arrays being proposed by Siegel *et al.* (1979) and Briggs *et al.* (1979) discussed in Chapter 4.] Each processor has access to its own memory, which is part of a larger memory. Typically, the processor can fetch information from, and store information into, its own memory. And it can also fetch information from its (four, six, or eight) nearest neighbors. We can think of this as a little "window" through which the processor can look at and process a near-neighborhood of picture elements.

All processors effect the same instruction, synchronously, giving a single-instruction multiple-data-stream (SIMD) system.

In all cases actually built or designed, 1-bit processors have been used. A 1-bit processor seems rather strange from the perspective of traditional computers—weak, slow, and limited. But when thousands of processors are built the cost of each must be kept as low as possible. And in fact a 1-bit processor is probably the most general and flexible of all, since it can effect operations on numbers and symbols of any length without wasting bits. (What it does waste is time, since it must perform these operations serially, in "bit-serial" fashion. That is, the processors in the very large arrays handle numbers and symbols larger than 1 bit in length by working 1 bit at a time, storing intermediate results in special registers. This is inherently slow—as is any serial process. It also means that the process cannot be accelerated by using specialized hardware.)

Here is a curious situation: One (the traditional) type of parallelism—where a whole word of 32, 16, or 8 bits is usually stored, fetched, and processed in parallel—has been thrown away in order to make economically feasible another type of parallelism—where many many processors work together. One would hope that future economics, based on new technologies with different constraints and different costs, will make it feasible to build systems that make effective use of both types of parallelism.

An Examination of Today's Very Large Arrays

I here give the central features of today's major large arrays. A number of important details or features, ones that will sometimes appreciably improve processing power and speed, are omitted to keep these descriptions reasonably simple and clear.

CLIP4, A 10,000-Processor Array

Michael Duff's (1976, 1978) cellular logic image processor (CLIP4) is a 96×96 array of 1-bit processors. Each processor is linked, via three 1-bit registers for each processor, to either the six or the eight nearest neighbors. Designed and built at University College London, CLIP is a good example of a large array specialized for image processing. A specially designed VLSI chip was fabricated for CLIP4 with eight processors along with all memory (32 bits per processor) on each chip.

A basic CLIP instruction will

(a) fetch one word from the memories of *all* the processors in any program-specified subset of the eight nearest neighbors (or, if the program specifies a hexagonal array, the six nearest neighbors) plus the processor's own memory, plus (when desired) a second word from that processor's memory;

(b) effect a logical operation; and

(c) store the result in the specified word of the processor's memory.

This takes about 11 μsec. [Cheap, relatively slow CMOS chips are used. If faster TTL logic were used, as in DAP, an instruction would take only 1 or 2 μsec. Indeed, a much smaller 12 × 16 array (CLIP3) was built using TTL IC chips; it takes about 2 μsec for each CLIP instruction.]

Figure 2 gives the logic of the basic CLIP4 processor.

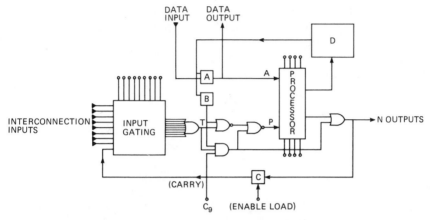

Fig. 2 The basic CLIP4 cell logic. Information is loaded from a processor's memory into the A and the B registers. The A register is linked to all eight neighbors, the input gating effecting a logical OR operation. The C register handles carries. The results of an instruction are output to the D register. [Reprinted from Duff (1977).]

Jelinek (1979) has shown that CLIP has a general and quite powerful basic instruction set. Arithmetic (or operations over whole strings of bits) can be done 1 bit at a time, using CLIP4's three registers. [A fourth "ripple carry" register helps speed up arithmetic. But the 32-bit memory severely limits CLIP4's ability to work with large numbers.]

Cordella *et al.* (1978) have shown that CLIP with N processors can achieve N-fold improvement or better (because of the window) over a conventional serial computer on a variety of simple but basic image processing algorithms including smoothing, thresholding, thinning, contour extraction, and perimeter finding.

DAP, A Very Fast 4000 Processor Array

Stewart Reddaway's (1978, 1980) distributed-array processor (DAP) is a 64 × 64 array of 1-bit processors each linked, via registers, to its four square neighbors. Each processor has 4000 bits of memory.

Designed at and marketed by ICL, England (the largest computer manufacturer in the world outside the U.S.), DAP is a good example of a very large array directed toward numerical problems.

Considering that DAP was designed (like SOLOMON) for number-crunching large arrays, while CLIP was designed (like Unger's machine and ILLIAC III) for image processing, and that neither Duff nor Reddaway (personal communication) was aware of the other's work, the similarities between CLIP and DAP are striking. Figure 3 shows the basic DAP processing element.

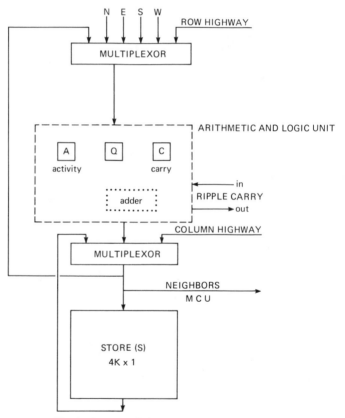

Fig. 3 Schematic of the DAP processor.

The DAP memory is shared with the mainframe, since it also is one of the high-speed main memory banks of the large ICL 2980 host computer (see Shooman, 1960, for a similar earlier design). Therefore either DAP can process 4000 (1-bit) words of information in parallel, or the mainframe CPU can fetch and process one 32-bit word from the same memory. Both can be working in parallel. But it is doubtful that the present mode of operating the total system makes that useful, since for efficient operation the very expensive main frame must multiprogram a large set of user programs.

The basic DAP instruction will fetch one word from any *one* of the four nearest neighbors or the processor's own memory in 200 nsec. The word can be fetched and any of a wide range of arithmetic and logical operations effected in 250 nsec.

The master control unit can fetch from or store into a whole row or a whole column of registers in parallel. In addition, data can be shifted between opposite edges in several different ways. These capabilities give the programmer a great deal of power and flexibility in remapping data and shipping data to distant processors. The master control unit can also be used to execute ordinary serial code interspersed with parallel code. Indeed DAP-FORTRAN, the higher-level language developed for DAP, includes FORTRAN as a subset.

A 32×32 "pilot" DAP, and the first production DAPs, were built with standard off-the-shelf TTL logic integrated circuits. But a new custom chip with four processors per chip has been designed and fabricated, and it will probably be used for future systems.

Marks (1980) has programmed the DAP to effectively perform a variety of basic image processing tasks.

MPP, A Very Fast 16,000 "Massively Parallel-Processor" Array

The massively parallel processor (MPP) designed at NASA—Goddard and Goodyear—Aerospace—see Fung (1977) and Batcher (1980, 1982)—is a very fast 128×128 array of 1-bit processors. Each processor is linked, via registers, to its four square neighbors. MPP is a good example of a system designed with both image processing and number crunching in mind. Each processor has a fast 32-bit register, which greatly speeds up arithmetic, along with several 1-bit registers.

An LSI chip (delivery in 1982) has eight processors, and all registers, on each chip. There are also external RAMs, to give each processor 1024 bits of memory. The MPP was delivered in April 1983. A prototype 16×16 system was designed at NASA—Goddard using off-the-shelf IC components, and simulations are being run at the University of Illinois to check and test the design.

Built from very fast logic, the MPP runs with a 40-mHz clock. It executes its basic instructions (much like the DAP instructions, fetching and operating on one word from *one* of the four nearest neighbors'

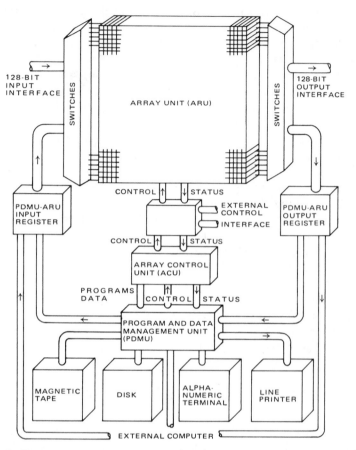

Fig. 4 The massively parallel processor (MPP). The switches and PDMU serve to reconfigure data while effecting fast input−output to all rows in parallel. [From K. E. Batcher, Design of a massively parallel processor. *IEEE Trans. Comput.* **29**, 836−840. Copyright © 1980 IEEE.]

or its own memory) in about 100 nsec. The large 32-bit register will also appreciably speed up processing for multibit arithmetic and string matching operations.

The MPP can input and output images in parallel to processors' operations. It can reconfigure data by outputting it to buffer memories, then inputting these data with a different mapping.

It is important to note that no network to date has actually been built with more than 8, 16, 30 or 50 processors—except for this kind of SIMD array, containing 1000 (STARAN, ASPRO), 4000 (DAP), 10,000 (CLIP4), and 16,000 (MPP).

Pipelines of Processors

When processors are connected into a one-dimensional "pipeline" array that transforms information being flowed through it, very fast execution speeds can be attained. For example, a pipeline of ten processors will have processor ten effecting the tenth step on the first element of an array in parallel with processor nine, eight, ..., one, effecting the ninth, eighth, ..., first instruction on the second, third, ..., tenth element of the array.

Figure 5 shows how the information stored in the cells of an array can be pumped through a pipeline.

Data array (numbers indicate their coordinate/locations)

```
1,1  1,2  1,3  1,4
2,1  2,2  2,3  2,4
3,1  3,2  3,3  3,4
4,1  4,2  4,3  4,4
```

Pipeline (the first five processors are shown)

Time 1:   [1,1]→[ ]→[ ]→[ ]→[ ]...
              (Only the first processor is processing data
              at Time 1, the data stored at,
              and fetched from, location 1,1.)

Time 2:   [1,2]→[1,1]→[ ]→[ ]→[ ]...

Time 5:   [2,1]→[1,4]→[1,3]→[1,2]→[1,1]...

Time 15: [4,3]→[4,2]→[4,1]→[3,4]→[3,3]...

Time 18: [ ]→[ ]→[4,4]→[4,3]→[4,2]...

Fig. 5   Pipelines.

Each processor in the pipeline continues to execute the same instruction, each time on the next piece of data fetched from the array and piped through. This can be a very efficient procedure:

The output from one processor is the input to the next processor, so transfer of this information can be handled with great efficiency, using a minimum of intermediate register/way stations.

The program instruction must be shipped to the processor, read, and decoded only once for the entire stream of data; no time is needed to read and decode the next instruction.

In programs where the stream of data is obtained by some serial input device (e.g., a television camera or a disk) the whole pipeline can process at the rate of that input stream. Therefore processing and transforming information does not take any extra time at all over the time needed simply to input and transfer that information! An array must serially input the array of information digitized from a TV camera or transferred from disk, then have its processors begin to execute instructions on that array. In contrast, the data simply flow through the series of processors in the pipeline, and each processor repeatedly executes its assigned instruction or instructions.

The pipeline can be made arbitrarily long. But the problem appears to be whether effective algorithms can be constructed that use similarly long sequences of code that can make do with the limited memory of the processor-stages in the pipeline.

Two unusually powerful systems of this sort have been built that can execute a very wide range of image processing and pattern recognition operations: Bjorn Kruse's PICAP (1973, 1976, 1978) at the University of Linköping (Sweden) scans the array using several pipelines and other specialized processors. Stanley Sternberg's Cytocomputer (1978) at the Environmental Research Institute of Michigan (Ann Arbor) uses an extremely long pipeline.

PICAP, a Well-Integrated System of Pipelines
and Special Processors

PICAP I uses a pipeline of processors specialized to compute useful picture processing operations over a local 3×3 window. Two general types of powerful operations are built very efficiently into hardware: numerical operations (e.g., weighted sums, differences, or products of subsets of the 3×3 window) and logical operations (these can get nonlinear patterns). PICAP I effects these 3×3 window operations in less than 1 μsec each, while scanning over a 64×64 array (in 30 msec).

Much of PICAP's power comes from a capability to zoom, pan, and in other ways move about in choosing the 64×64 subarray of the larger TV image to process at each point in a program. For example, the programmer can have the program make a high-level survey of the entire picture, using very coarse resolution (with local neighborhoods averaged), examine the highlights and clues found, and then decide which subregion(s) to zoom in on and look at in more detail.

This is possible because at each point in the scan of the array where something of interest is found it is entered into a table that can then be processed by routines using the host computer. Thus a program can move up and down in a pyramidlike structure, and also make good use of both the parallel capabilities of the pipeline and the serial capabilities of the traditional host.

PICAP II (Kruse *et al.*, 1980) uses several pipelines and a number of other new processors specialized for picture processing. These are integrated into a beautifully architectured larger system, one that encourages the programmer to use the array processing pipeline and the traditional host computer together in a way that powerfully exploits the special capabilities of each. All the pipelines and specialized processors are on a common bus, and can be used in parallel by the system (see Fig. 6).

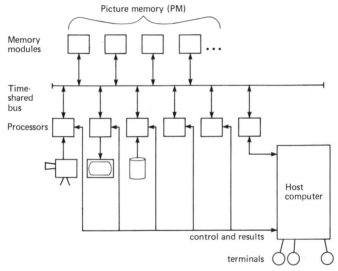

Fig. 6 The system architecture of PICAP II. Several programs can be run simultaneously, each using several processors in parallel.

Both PICAP I and PICAP II are built from off-the-shelf bit−slice hardware. Kruse (personal communication) estimates that PICAP II is roughly 25 times as powerful as PICAP I, because of the carefully chosen and engineered new special-purpose processors and the system's ability to use several of them in parallel, as well as somewhat faster, newer hardware.

The Cytocomputer, a Very Long and Powerful Pipeline

Stanley Sternberg's Cytocomputer is a very powerful system that scans over a 1024 × 1024 array with an unusually long pipeline of 113 powerful processors. Such a pipeline can be extremely efficient (see Lougheed and McCubbrey, 1980, for an interesting examination of this issue). The Cytocomputer's basic instructions perform logical and

arithmetic operations (much like those of CLIP and PICAP) over either
1-bit symbols or 3-bit numbers. Look-up tables are used to effect these
operations.

Figure 7 shows the way the pipeline extracts a window over three
successive rows of the larger array.

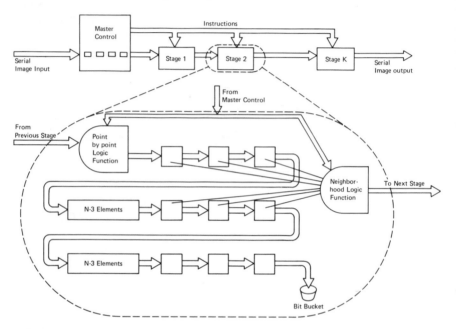

Fig. 7   The Cytocomputer window processing pipeline. [Reprinted from Sternberg
(1978).]

A programmed operation (using 3-bit as well as 1-bit numbers or
symbols) over a 3×3 window can be executed in about 700 nsec. The
large number of processors in the pipeline means that 113 instructions
can be executed at each point in the array. Therefore if the pipeline
can be filled with a sequence of instructions, each operation takes only
about 6 nsec per pixel-window.

A new VLSI chip has been designed that puts an entire processor
stage on a single roughly 16,000-device chip. This should make possible a variety of new architectures, since a system can be extended simply by plugging more chips into the boards. Therefore Sternberg (personal communication) has contemplated building customized Cytocom-
puters, including ones with several hundred or even several thousand
processor stages in the pipeline. (It is estimated this can be done at
roughly the cost of the processor chip—which it is hoped will be about
$70 per stage.)

Sternberg is also investigating the possibility of parallel pipelines, and is considering building systems with two, five, ten, or more pipelines, all working in parallel. He is also exploring the possibility of building circular pipelines, so that data can cycle several times through.

Systems That Combine Array, Pipeline, and Specialized Hardware

An attractive structure is one that adds specialized or special-purpose hardware to a larger serial system (see Estrin, 1960).

The "systolic arrays" that Kung (1980a) and his associates are developing, for example, the Kung—Song (1982) convolution chip, are elegantly designed array-embedded complex pipelines which one would interface to a high-speed bus of a serial computer. Figure 8 shows a systolic array for matrix multiplication (Kung, 1980a). The matrix product $C = (c_{ij})$ of $A = (a_{ij})$ and $B = (b_{ij})$ is computed by the recurrences

$$c_{ij}^{(1)} = 0,$$

$$c_{ij}^{(k+1)} = c_{ij}^{(k)} + a_{ik} b_{kj}, \qquad k = 1, 2, \dots, n,$$

$$c_{ij} = c_{ij}^{(n+1)}.$$

Floating Point Systems, Hewlett—Packard, and a number of other companies have built and are marketing special-purpose pipelines for fast floating-point arithmetic, including multiplication, over large arrays of data. For example, the Floating Point System pipeline takes 167 nsec for each 32-bit floating-point multiplication. These pipelines are typically put on the high-speed bus of a conventional serial computer, which uses them when appropriate to process arrays of numerical data.

Several special-purpose systems have been built for particular image processing functions (only a few are mentioned here). They tend to have the general structure of one powerful processor built to effect the desired class of operations through which a stream of data (usually an array) is piped. Sometimes several processors are linked as stages in the pipe (though far fewer stages are used than in the Cytocomputer's pipe). Sometimes processors are arranged in a small array (though much smaller than CLIP's) or other parallel structure, as on a common bus (like PICAP II).

Golay's (1969) GLOPR effects hexagonal-window operations, and is used commercially to identify and count different types of white blood cells. Mori *et al.* (1978) have built a microprogrammable parallel picture processor (PPP) with special hardware for two-dimensional convolution, gradient and feature detection, histogram generation, and other

(a) Input matrices. The elements in bands A, B, and C (each c_{ij} initialized to zero, or, for the problem $C = AB + D$, to d_{ij}) pump through the array of Fig. 8b.

$$
\begin{bmatrix}
a_{11} & a_{12} & & \\
a_{21} & a_{22} & a_{23} & \\
a_{31} & a_{32} & a_{33} & a_{34} \\
& a_{42} & \cdot & \\
& & & \\
& & & \cdot
\end{bmatrix}
\begin{bmatrix}
b_{11} & b_{12} & b_{13} & \\
b_{21} & b_{22} & b_{23} & b_{24} \\
& b_{32} & b_{33} & b_{34} \\
& & b_{43} & b_{35} \\
& & \cdot & \\
& & & \cdot
\end{bmatrix}
=
\begin{bmatrix}
c_{11} & c_{12} & c_{13} & c_{14} \\
c_{21} & c_{22} & c_{23} & c_{24} \\
c_{31} & c_{32} & c_{33} & c_{34} \\
c_{41} & c_{42} & \cdot & \\
& & & \\
& & & \cdot
\end{bmatrix}
$$

A   B   C

(b) Hexagonal array of inner-product processors. Arrows indicate the directions of data flow.

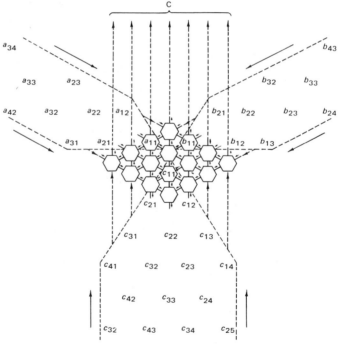

Fig. 8 A systolic array for matrix multiplication.

basic image processing operations. Gerritsen and Monhemius (1981) have also built a programmable system with several special-purpose pipelines for image processing functions. Ledley *et al.* (1977) have built a special-purpose system designed to search for and analyze textures at TV scan rates. Granlund (1981) has built an interesting system that uses a general and powerful programmable operator to effect image processing functions.

Comtal, De Anza, Grinnell, Vicom, and a number of other graphics terminals have been built with specialized hardware, including short pipelines, for image processing operations. These commercial systems are beginning to incorporate the specialized and special-purpose features of the experimental systems described above. Typically, one buys a set of special cards, each performing a different operation, e.g., local averaging, "zoom" and "scroll" (that is, the linear transformations of magnification and translation), histogramming, or computing a convolution (that is, a weighted linear function of a local window).

Some of these special functions are programmable. For example, the program can specify the window size and the weights for a convolution operation. Operations are effected at the rate of the television scan, that is, 30 msec to effect each instruction at every point in the 512×512 array. (For convolution this usually means only one step per scan, but by increasing the length and the power of the pipeline this can be speeded up, with several operations per scan.)

Reeves and Rindfuss (1979) have designed a small 8×8 parallel array much like CLIP that serially iterates over the larger image. The array is combined with a host serial computer in such a way as to give relatively fast processing, considering the array's small size. Reeves's expectation (Reeves, personal communication) is that with VLSI technology larger arrays can be fabricated, probably on a single chip, and the speed of such a system increased accordingly. This system is an interesting combination of array and specialized hardware for fast iteration, and an interesting contrast with the other types of parallelism obtainable with a pipeline system, one which puts processors one after the other, rather than side by side.

Several very expensive number-cruncher systems incorporate arrays. The Burroughs BSP (Kozdrowicki and Theis, 1980; Kuck and Stokes, 1982) uses 16 48-bit arithmetic processors linked through reconfiguring networks to a large memory. This system must be connected to a large B7800 front-end computer. It can achieve 50,000,000 48-bit floating-point multiplications per second. DataWest is marketing an array to be used with a Univac 1100/84 (Sugarman, 1980). An $8,000,000 system with two such arrays can execute up to 250,000,000 floating-point operations per second.

Summary Discussion: Arrays, Pipelines, Parallelism, VLSI

Arrays, Pipelines, and Specialized Hardware

An array that is parallel at the hardware level (like CLIP, DAP, or MPP) takes no more time to process the entire image or other type of array than to process one cell.

In contrast, if the processor must scan the array, it needs its basic instruction time multiplied by the number of cells it must scan, and by overhead time to index through cells. We can minimize overhead and instruction decoding by building a pipeline. In theory we can make this pipeline arbitrarily long. But it seems unlikely that we can use pipelines with thousands of processors as efficiently as we can use large arrays.

By using a parallel set of pipelines we can further speed up processing. Indeed, the two-dimensional arrays are often used to process one-dimensional sets of data, e.g., flowing a row of data through successive columns. (This is a very effective way to add or multiply numbers on CLIP or DAP, since much larger numbers can be handled in parallel.) This turns the two-dimensional array into something much like a one-dimensional array of pipelines (except that more transfers of data must be made to registers and memory than in a well-designed pipeline).

By specially designing a processor for pattern recognition purposes (which seems, essentially, to mean having it fetch small amounts of information in parallel from nearest neighbors, and execute appropriate logical operations) we can cut time by two orders of magnitude.

Using words longer than 1 bit and fetching, storing, and processing the entire word in parallel (as is routinely done by the traditional 32-bit, 16-bit and 8-bit 1-CPU computers) is another very effective way of speeding up processing by introducing parallel input−output and logic.

Most of today's "supercomputers" get much of their power by adding highly specialized processors that themselves usually have hard-wired parallel (including pipeline) hardware that speeds up processing.

But by using increasing thousands of processors we can cut time by increasing thousands. Here a parallel array of simple computers, like CLIP, DAP, or MPP, becomes increasingly attractive and appropriate, since the larger the number of processors we use, the cheaper and easier it will be to mass produce each single processor.

When all its processors can be used effectively (and this is very frequently the case for large number-crunching and image enhancement tasks, and also for the "lower" and at least some of the "higher" levels of scene analysis), an array like CLIP4 or MPP can increase speeds by four orders of magnitude, because of its 10,000 or so processors, when compared to a single processor built from the same technology. Its crucial advantage is that it will continue to process increasingly larger arrays of information in exactly the *same* amount of time. In contrast, a scanner like PICAP or the Cytocomputer needs 100 times more time for 1,000,000 as opposed to 10,000 cells in the picture.

It is quite straightforward to increase the size of the arrays. As more and more devices can be packed so that hundreds, or even thousands, of simple processors plus their attendant memories can be packed onto a single chip, it should be possible to build larger and larger arrays, up to whatever size appears useful. Arrays that are 1000×1000, 4000×4000, or even larger may soon be technologically feasible. One attractive possibility is a family of arrays of different sizes and shapes, with different sizes of memories assigned to each processor—all easily linked to a host (and, possibly, a larger network), and all programmable with the same programming language (simply by declaring the array size being used).

The Importance of LSI and VLSI Chips for Large Arrays

CLIP4's large 96×96 array of (roughly) 10,000 hardware processors was feasible only because a special chip was designed to contain 8 processors on each single chip. Similarly, the very large 16,000-processor requirement for the MPP led to the design and fabrication of an 8-processor chip.

The use of LSI chips has several important consequences:

(a) Once such a chip is checked out and produced in large quantities it will be relatively straightforward to build arrays still larger than 96×96 or 128×128.

(b) Chips are cheap (the CLIP4 chip, which is rather large and is produced in relatively small quantities of a few thousand, costs about $50; if it were produced in quantities of hundreds of thousands or millions its cost would drop to only $5 or $2).

(c) Therefore this type of computer is, potentially, cheap. CLIP4 could be marketed (albeit with little monetary profit) for roughly $100,000 even if only two or three were built.

(d) A 1000×1000 array could be built for $5,000,000, and soon for much less.

(e) If built and sold in large quantities (which, potentially, seems quite plausible, since there are a wide variety of important image processing tasks, e.g., to enhance satellite and medical images, for vision-controlled robots, and in surveillance systems, among others) an array with 10,000 processors could be built and marketed for $50,000, or, in a few years, substantially less.

CHAPTER 4

More or Less
Tightly Coupled Networks

A wide variety of different architectures have been proposed for large networks of computers. These include systems using buses, cross-point switches, rings, lattices, stars, snowflakes and similar clusters, clusters of clusters, binary trees, n-ary trees, pyramids, and binary n-cubes—among many other interconnection patterns. This chapter describes most of these architectures, and begins to put their different structures into a common framework and to explore the striking similarities between many of these on-the-surface-different configurations.

The survey that follows examines today's networks, which (with very few exceptions) contain only a few dozen computers. But networks are being explored with an eye to the relatively near future when VLSI technology will open up the possibility of far larger numbers, in the hundreds, thousands, or millions.

It is important to emphasize that most of these systems exist only as paper designs and are still being thought through. In contrast to the arrays only a handful of small MIMD networks have been built or even designed in full detail.

Simple Structures: Buses, Rings,
and Cross-Point Switches

The simplest way to build a network is to link all computers directly together. This is impractical for more than 4 or 8. But it can be approached by putting them on a common "bus," on a circular "ring," or by linking them through a general "cross-point switch" (see Fig. 1).

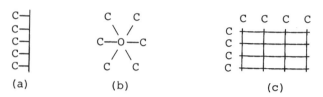

Fig. 1 Examples of a bus, a ring, and a cross-point switch [C = computer (processor + memory); | = (sections of) bus; + = switch; O = ring; −, /, = (sections of) connector links].

A bus is simply a wire or cable onto which all the computers are interfaced. Any computer can send information to any other computer (or input, output, or memory storage device) connected to the bus. But a bus carries only one message at a time. Therefore too many computers passing too much information can result in long delays.

A ring is a circular bus that allows a message to continue to cycle until it reaches its sender, or a degree-2 polygon graph (that is, with two links from each node, forming one cycle), where each node must receive and send.

A cross-point switch is an array of switches arranged as follows: Put each computer (and each other device on the cross-point switch) starting one row, and also one column, of a square array. Now run connectors along each row and along each column. Put a switch at every point where a row connector and a column connector cross. Thus a cross-point switch connecting n computers and other devices needs an $n \times n$ array of switches.

Buses, rings, and cross-point switches are feasible only when fewer than 32 or so computer nodes are involved (or very little message passing is needed, or allowed). For as more nodes are added, the buses give excessive delays, and the cross-point switch needs too many switches (n^2) to be economically feasible.

Buses, rings, and cross-point switches have been the most common techniques for building networks to date. See systems like the CDC 6600 and the IBM 360/67 for buses. Farber and Larson (1972) designed and built a ring of processors, terminals, and peripherals for distributed computing. Hanaki and Temma (1982) have proposed a ring of processors to pipeline images, and Sato *et al.* (1982) are building a ring with 32 computers. Rieger *et al.* (1980) are building ZMOB, which will have 256 Z80 microcomputers, each interfaced to a shift register fast enough (because Z80s are relatively slow) to ship one instruction or word of data to each of the 256 computers in the time the Z80s need to execute one instruction, and therefore causing no delays (within the strict limit of one word of data). Wulf and Bell (1972;

Wulf *et al.*, 1981) built the cross-point switch-based C.mmp network. Widdoes (1980) and his collaborators are developing the very powerful S-1 computer system, with plans to use VLSI technology that will allow them to build 16, or 64, computers, each with roughly the power of the Cray-1, and connect them with a cross-point switch.

Several systems have been developed where all computers use a common memory. Thus C.mmp has all computers connected by the cross-point switch to a single blackboardlike memory (each also has its own local memory). Some of the systems being proposed to handle "data-flow" programs (see Chap. 8), by Dennis *et al.* (1980) and by others, also link many processors to a common memory.

Lattices, *N*-Cubes, Trees, and Stars

Lattices

As defined formally, by von Neumann, Codd, Thatcher, and others (see Burks, 1966), a cellular array might be *n* dimensional. But the cellular arrays that have actually been built have all been two dimensional. So I shall use the term "lattice" for three-dimensional structures over processors or computers that have the simple regular characteristics that we think of as cellular. (We can also generalize lattices to *n*-lattice structures that, when two dimensional, will include, among other structures, the cellular arrays.)

A lattice is a structure of computers arranged in *n* dimensions, with each computer linked to its near neighbors.

A good example is WISPAC, developed by Cyre *et al.* (1977). Each computer links to its six square neighbors, two in each of the lattice's three dimensions. (In contrast to today's large two-dimensional arrays, which are SIMD systems, WISPAC is planned as a three-dimensional lattice of MIMD computers.)

An interesting variant is Wittie's (1978; Wittie, *et al.*, 1982) MICRONET design—an *n*-dimensional lattice with each computer connected only to two buses, with the interconnection pattern appropriately staggered (see Fig. 2).

Binary N-Cubes

Consider a three-dimensional cube that has eight vertices, each with three edges leading into it. Put a computer at each vertex, and link it to its three near-neighbor computers (vertices). This gives a 3-cube of

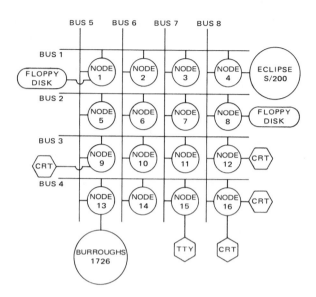

Fig. 2 An example of a possible MICRONET configuration. Each computer node always shares two buses. Here, each bus is shared by four nodes. [From Wittie (1978).]

eight computers, each computer with three links. Now generalize this to an n-dimensional cube. A binary n-cube (see also Fig. 3) has 2^n nodes, each with n links. Its diameter (shortest path between most distant nodes) is n. For example:

A 20-cube has 1,035,476 nodes, each with twenty links, and diameter 20.

A 10-cube has 1024 ten-link nodes; diameter 10.

A 4-cube has 16 four-link nodes; diameter 4.

Fig. 3 A binary N-cube in two dimensions (a 2-cube).

Squire and Palais (1963), Sullivan *et al.* (1977), and others have worked out preliminary designs for such an organization, envisioning millions, or billions, of processors. A curious example is the n-cube network of 16, 81, or 256 microprocessors that IMSAI, a manufacturer of personal/hobby computers, advertised around 1978. The 81-computer system was connected in the form of a 3^4 cube (hence it was more like what I call a lattice), with a selling price of $400,000. These networks were marketed without an adequate operating system, software, or suggestions how to use such a network, and apparently nobody ever bought one.

Trees

Connect a set of computers into a tree, e.g., a "binary tree" (see Fig. 4). That is, link the "root" computer to two (or more) "offspring"; link each offspring to two (or more) offspring; continue. Goodwin (1973) has proposed tree architectures for large networks. Rumbaugh (1977) has investigated how to use trees to implement data-flow computers. Mago (1980) is exploring trees for efficient execution of parallel functional languages.

Fig. 4   A binary tree [C = computer (processor + memory store); |, −, /, = links].

Trees pack a relatively large number of nodes within a given diameter; for example, a tree with degree 10 (one link toward the root and nine to offspring) and diameter 10 has 59,049 nodes.

Stars

One of the largest running networks, *Cm** (Swan *et al.*, 1977) connects local clusters in the form of stars (see Fig. 5). Finkel and Solomon (1977) defined stars as having p computers interconnected via a bus, with each computer linked to two buses. Such a system would be built by putting p computers on one bus, then linking each computer to another bus with $p - 1$ new computers (see Fig. 6). If repeated n times this procedure gives a "level-n star."

Note that these stars can be represented by n-ary trees whose nodes are buses linked by computers. An equivalent representation would replace each bus by $n(n - 1)$ links, to join all pairs of computers on that bus. Thus a bus effects virtual links between the set of nodes on that bus. But increasing the number of nodes linked to a bus, and the amount of information sent over that bus, will increase contention problems and delays. In general, buses are convenient devices for connecting small clusters, but in order to avoid contention and delays as

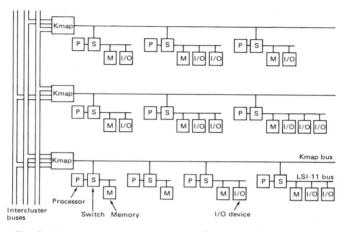

Fig. 5   Three clusters of a *Cm** system. [From Swan *et al.* (1977).]

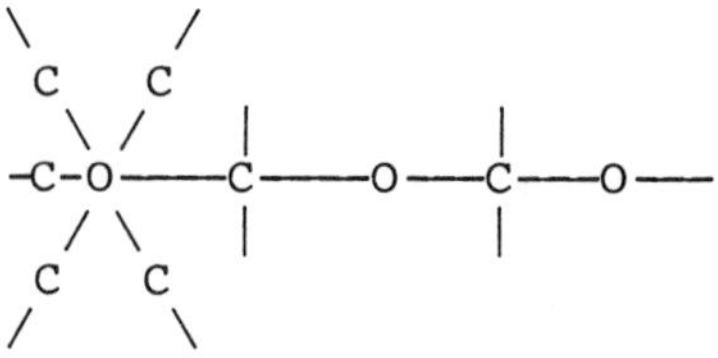

Fig. 6   A star (only 1 complete ring is shown) [C = computer; |, −, /, = links; O = ring].

the clusters grow large a single bus must be replaced by several smaller buses, or by direct links.

Augmenting Trees for Density, Connectivity, and Structure

Augmenting a Tree to Make an X-Tree

Since a tree has only one path between pairs of nodes (that is, it is a tree), if any single computer or link fails the network graph is broken and the whole system fails. Therefore it seems very desirable to add connections, to give at least two paths between any pair of computers. This Despain and Patterson (1978) propose to do by adding "lateral" connections, that is, links between offspring at the same depth, in terms of distance from the root computer. They call these augmented trees "X-trees" (these configurations now have cycles and are no longer really trees).

An elegant augmentation that uses only three links per node is the "hex-tree" (suggested by Carlo Sequin) in which downstream nodes are linked to near neighbors to form hexagonal rings (see Fig. 7).

```
            C
          / \   /
     C—C     C—C—
         |   |
     C—C     C—C—
          \ /      \
            C
            |
```

Fig. 7 A simple augmented X-tree (a "hex-tree") [C = computer (computer + memory store); |, — = links]. [From A. M. Despain and D. A. Patterson, X-tree: A tree structured multi-processor computer architecture. *Proc. 5th Ann. Symp. Comput. Archit.*, pp. 144–151. Copyright © 1978 IEEE.]

With four links per node a variety of interconnection patterns exist, including "threaded trees," "ringed" and "half-ringed" trees, with and without "shuffle" (and also, as we shall see, the simple cellular array). (See Figs. 8 and 9.)

 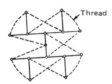

Fig. 8 Threaded, double-threaded, and half-ringed trees. [From A. M. Despain and D. A. Patterson, X-tree: A tree structured multi-processor computer architecture. *Proc. 5th Ann. Symp. Comput. Archit.*, pp. 144–151. Copyright © 1978 IEEE.]

  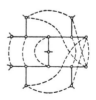

Fig. 9 Half-ringed + shuffle, ringed, and ringed + shuffle trees. [From A.M. Despain and D.A. Patterson, X-tree: A tree structured multi-processor computer architecture. *Proc. 5th Ann. Symp. Comput. Archit.*, pp. 144–151. Copyright © 1978 IEEE.]

An especially attractive augmented tree is the Goodman−Sequin (1981) "hyper-tree," which has regular links between nodes at each ply of the tree (Fig. 10).

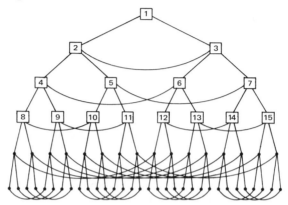

Fig. 10 Goodman and Sequin's "hyper-tree" (1981).

A simpler and more specialized system, for database applications, has been proposed by Bentley and Kung (1979). It is interesting to compare the former, with each computer using a projected 200,000-gate CPU plus I-O processor and 1,000,000-gate memory on a single VLSI chip (Patterson *et al.*, 1979) with the latter, where it is planned each computer will have a 512-bit memory, and a processor that uses equivalent area on the chip (roughly 500 to 2000 gates).

Pyramids and Cones

A "pyramid" (or "cone") is a layered structure that starts with a "base" array of computers R rows by C columns in size. A next layer contains a smaller array, of R/SHRINK by C/SHRINK computers ("SHRINK" is a small integer, usually 2 or 3), each serving as a "parent" computer connected to a subarray of SHRINK by SHRINK "offspring" computers. Each node is connected to near neighbors in its own and adjacent layers.

A number of variations are possible:

Two-way links are usually used, but one-way links might be used.

Connections between nodes in the same layer can be eliminated (they are still connected via parents if two-way links are used).

Connections can be of the form P−P; P−M−P; P−R−P; P−S−X[P,P] (that is, processors can be connected directly, or via memory, or via registers or intervening switches).

The base can be hexagonal, circular, or some other shape, and each successive layer suitably SHRINK-ed, giving different kinds of pyramids and cones.

Handler (1975, 1977; Handler *et al.*, 1979) has suggested pyramids of up to seven layers, with 1, 4, 16, 64, 256, 1024, and 4096 MIMD processors in each layer. Vorgrimler and Gemmar (1977) have proposed similar, much smaller "cascaded" networks of processors. Giloi and Berg (1978) have suggested trees of arrays. Uhr *et al.* (1980) have designed a two-layered "array/net" system consisting of a 16×16 SIMD array (of 8-bit processors) and a 4×4 MIMD array of 64-bit processors (each connected to its near neighbors, and all on a bus), each looking at the same memory as its 4×4 subarray of 16 "slaves." Dyer (1981, 1982), Tanimoto (1982), and Uhr (1981b) have also proposed multilayered converging SIMD arrays, stacked into a pyramid. Boeing Aerospace is building a 64×64 pyramid of 1-bit processors designed by Steven Tanimoto and Warren Snapp (personal commmunication). Brinch-Hansen (1978) has suggested a hierarchical tree or pyramid shape for a set of processors and their hierarchy of memories that would be appropriate for assigning different concurrent processes to different processors.

Stars with "Sutures"

Finkel and Solomon (1977, 1980b) extended their "stars" by adding links (which they call "suture buses") to form "cross-stars" connected at their lowest-level buds), "novas" (built from *p* identical stars "sutured" together at the edges), and "lenses." The suturings are reminiscent of the schemes for threading X-trees, except that new connections are always added at the ends of paths, rather than dispersed throughout the tree.

Miscellaneous Interconnection Patterns

A variety of other interconnection patterns have been proposed and explored. These might best be thought of as variants on binary *n*-cubes, lattices, stars, and trees, where some of the edges have been eliminated for economy, and/or some new edges have been added to give alternate paths and/or shorter paths. For example, Sullivan *et al.*

(1977) place a cluster at each node of an n-cube; Preparata and Vuillemin (1979) replace each node in an n-cube with a pipeline ring of n
nodes embossed into that node (that is, each joined to one of the original links). Halstead and Ward (1980) are trying to develop systems,
based on Hewitt's (1973) concepts of actors passing messages, that
"diffuse" a program's procedures out to nearby processors in any
"modular" network.

A number of more or less special-purpose networks have been built
for military or other specific applications. Each has an interconnection
structure that has been built to conform more or less with the structure
of the problem being executed. Among the largest and most successful
are the roughly 80 PDP-11s that are included in the enormous SHIVA
laser-based nuclear fusion system at Lawrence Livermore Labs, and the
EMMA networks of roughly 60 microprocessors designed and built for
the Italian and French Post Offices to effect pattern recognition of portions of envelope addresses (Manara and Stringa, 1981).

Possibly the most powerful of all will be the system Gilbert *et al.*
(1981; Gilbert and Harris, 1980) are building to handle three-
dimensional reconstructions of x-ray CAT scans in real time. This system will in 11 msec stack 60 to 240 thin adjacent cross sections of a human being, and repeat this 60 times a second. It will use about 30 specialized processors operating at subnanosecond speeds, and will execute
two to three billion instructions per second. Another example of interest is the 30 or so special-purpose dedicated microprocessors that will
shortly be found very loosely coupled through the engine and body of a
typical mid-1980s automobile.

Reconfiguring Network Topologies and Component Parts

Switches can be introduced between components so that they can be
reconfigured under program control. This can, potentially, be done at
any level. For example, the ILLIAC 4 and Cray-1 can reconfigure each
64-bit to two 32-bit processors. CLIP reconfigures to either a six hexa-
gonal- or eight square-neighbor window. The array/net reconfigures to
either 16 32-bit (or 24, 16 or 8-bit) MIMD processors, or 256 8-bit
SIMD processors. Koren (1981) proposes a reconfigurable array for a
fault-tolerant VLSI implementation of binary trees and other topologies.
Agrawal (1981) proposes a reconfigurable MIMD network for real-time

motion analysis. Kartashev and Kartashev (1979; Vick *et al.*, 1980) are exploring reconfigurable pipelines and dynamic architectures in general. Kartashev and Kartashev (1981) examine reconfiguring networks that can generate tree, star, and ring configurations.

Rather than use a fixed set of interconnections, a (more or less general) interconnection network of switches can be interposed between nodes (processors—processors, memories—processors, etc.). Such a system can then be "reconfigured" under program control. This can be used to physically change the network's topology. For example, links in an array might be broken and reconnected to form a tree. (Note that a processor or memory with three or more links can serve this purpose, but reconfiguring interconnection networks usually consist of additional switches added for just this purpose.)

Most researchers have concentrated on systems where an entire set of data can be shuffled around among processors, as for fast Fourier transforms, as the result of throwing whole sets of switches. Note that this is a specialized (but highly useful) type of simultaneous message passing, where M pieces of data are shuffled among n processors. At the other extreme, "message passing" refers to the movement of one piece of data to one other processor (or, in the case of broadcasting, to n processors—I know of no research into issues of moving 2, 3, ..., pieces of data simultaneously).

Clos (1953) and Benes (1965) did the original work on reconfiguring, for telephone networks. A Benes switch can be implemented with $O(n \log n)$ switches ["$O(...)$" means "of order ..."]. It needs $O(\log n)$ switch delays to transmit data, and $O(n \log n)$ time to set up the switches (this is very slow, so it can be done only rarely, as at the beginning of the program, to configure the appropriate network).

Other somewhat more economical (and also more limited) networks have been designed. For example, Lawrie's (1975) Omega network uses $O(n \log n)$ switches, and can be controlled and can transmit data in $O(\log n)$ switch delays. Pease (1968) and Stone (1971) describe the "perfect"shuffle that it implements (see Fig. 11).

Pease (1977) describes a network that implements an n-cube, and shows how it can be used (albeit with some delays) to handle near-neighbor and distant-neighbor shifts in square and hexagonal arrays—along with several other types of data remappings. See Kuck (1977), Marcus (1977), and Thurber (1978) for surveys. Wu and Feng (1978), Siegel (1979), and Agrawal (1980) have shown that many of these networks are equivalent.

The "flip network" that connects processors to memory in the STARAN and ASPRO systems (Batcher, 1976; Anonymous, 1979) is a

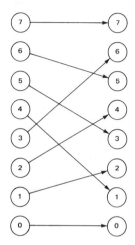

Fig. 11   Perfect shuffle. Each number refers to the same processor on both left and right.

good example of a reasonably economical network for partial reconfiguring. Briggs *et al.* (1979) compared a delta network (Patel, 1978) with a full cross-point switch, and found that for 256 processors the cross-point switch costs roughly twenty times as much but has only roughly twice the bandwidth (162 vs 77 messages per cycle).

Several researchers have begun to apply graph theory to these interconnection networks. Masson (1977) and Hwang and Lin (1979) have shown how a network with a set of desired properties can be constructed. Agrawal (1980) has shown that all networks using binary switches that link all pairs of nodes in log N steps are topologically equivalent.

Siegel *et al.* (1979), Briggs *et al.* (1979), Lundstrom and Barnes (1980), Lipovski (1977), Lipovski and Tripathi (1977), Bogdanowicz (1977), and others have proposed building reconfigurable arrays of powerful MIMD processors (see, for example, Fig. 12).

Gemmar (1982) has built a 16-processor system that uses a data-exchange processor to reconfigure. It is specialized for image processing, and in particular for use as a parallel array, or as a cascading (tree-like) set of processors for computing correlations. Lipovski is building a 16-processor system. Briggs *et al.* and Siegel *et al.* appear to have 1024-processor systems as goals, but probably with less than complete reconfiguring capabilities, because of the costs of the switches.

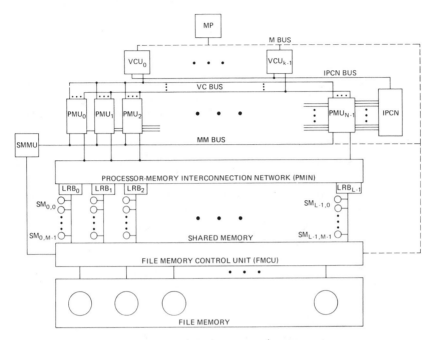

Fig. 12 The architecture of the reconfigurable PM[4]. (MP: monitor processor, VCU: vector control unit, PMU: processor-memory unit, IPCN: interprocessor communication network, SMMU: shared-memory management unit, LRB: line-request buffer, SM: shared-memory module.)

Network-Structured Programs and Algorithms

Programs for Arrays

Arrays are often useful for number-crunching large matrices in many domains of mathematics and simulation. They are also widely used for image enhancement and other simple image processing problems. I merely mention these programs here, since they are the obvious motivating forces behind such arrays as CLIP, DAP, and MPP.

Lattices

Very important meteorological and geological problems involve the modeling of large three-dimensional masses of moving matter, e.g., the earth's atmosphere or an ocean or volcano. WISPAC is proposed as a three-dimensional lattice, with each of its processors assigned a local portion of such a mass. With a suitable future technology, such a system could be made successively larger, thus giving successively greater resolution of successively larger masses.

Pyramids

A number of "pyramid" structures have been simulated on serial computers by Levine (1978), Tanimoto (1976, 1978), and others (see Tanimoto and Klinger, 1980, for a good collection of this work). Usually the pyramid computes simple averaginglike functions over a 2×2 or 3×3 region, and maps nicely onto a network with processors that link to the four or nine processors or memories at the next lower level.

They therefore suggest structures quite similar to internally augmented stars or trees with four or nine downward links, but each node also given several upward links, giving overlap to processes. The pyramid shape results from the fact that cells uplink only toward the center.

Klinger and Dyer (1976), Dyer (1979), Rosenfeld (1979, 1980), and others have investigated such pyramid structures, and have shown how a variety of simple algorithms can be effected.

Cones

Uhr (1981a) has suggested building layered converging arrays of processors. Each processor layer "looks at" (that is, operates on) data in its input buffer memory, and merges its outputs into its output buffer memory. This procedure continues until all information has been merged into the output buffer memory of the single processor that forms the apex of the cone.

This system is proposed for pattern recognition and scene description. The raw sensed scene (e.g., a television picture) is input to the input buffer memory (call it the "retina") of the lowest-level layer array of processors. Each layer of processors might be a (successively smaller) cellular array of CLIP4-DAP-MPP-like structures, or it might be a portion of one plane of the WISPAC lattice or a set of processors serially scanning over a larger array. It might well be that the processors would differ at the different layers to make the most appropriate processes available where needed.

These structures have been simulated on the serial computer (e.g., Uhr, 1972, 1976, 1978; Uhr and Douglass, 1978; Douglass, 1978), and therefore a variety of different interconnection patterns at a distance could be handled, as well as nearest-neighbor connections only. When actual cones are built it will be necessary to use nearest-neighbor connections (e.g., to the four, six, or eight nearest-neighbor memories that contain the outputs of the offspring at the next lower level). Then interactions at greater distances would be obtained by waiting until a sequence of converging transformations brought sets of information distant from one another into near-neighbor proximity.

Summary Discussion and Preliminary Comparisons

It will not be clear which of the great variety of network architectures that have been proposed is "best" until large mixes of programs have been run on each. This will not happen for a long time. Nor can we ever expect a "definitive" answer, since new classes of problems and new algorithms will continue to be uncovered as new computer architectures move us closer to the possibility of handling them. It seems most likely that network designs will evolve in the marketplace, as new networks are found useful for new applications that in turn suggest improved networks. Thus a mixture of casual observations, uncontrolled experiments, marketing decisions, recessions and booms, and luck are likely to determine our actual future from the enormous range of possible futures.

Given the expense of building large networks, and of coding programs with appropriately structured parallel algorithms that make good use of their potential power, it may be more reasonable to build and use experimental networks with rather general and regular structures, with enough reconfiguring capabilities so that data can be collected about the particular structures that are found useful enough to actually build commercially.

The general problem appears to be one of allocating resources among processors, reconfiguring switches, links, and memory stores so that they best fit the (mix of) problem(s) that the network is designed to handle. But there is some reason to conjecture that specializing the connection pattern at best improves performance only moderately (except when the specialization and the programs the system is tested on are very carefully chosen to fit one another well).

It may well be that any judicious choice of architecture for a general-purpose network—and that would seem to mean a sensibly designed pyramid, lattice, or augmented tree—will be just about as efficient as any other judicious choice. If this is true, then the particular configuration should be chosen primarily for economy, modularity, and ease of building into a large network, rather than to squeeze out a few extra drops of efficiency. But there appear to be some general design principles of interest:

An "everything-linked-to-everything" complete graph of processors minimizes distance between processors (it is always 1), but at the expense of impossibly large numbers of ports to each processor, and interconnections. Cross-point switches, buses, and rings can be used to link computers in a (virtual) complete graph. But as the number of computers, and the amount of message passing, increases, delays grow excessive. These may continue to be viable alternatives for the next few years, when only a few dozen computers are linked together or (as in the case of ZMOB) a special set of restrictions is imposed. But they do not seem feasible even for a few hundred, much less thousands of computers.

At the other extreme is a one-dimensional array.

We can increase to two, three, ..., *n* dimensions, folding and moving computers closer and closer together. We can put as many computers as we wish along each dimension.

The local interconnection pattern can be varied in a number of ways (e.g., link each computer to two buses; directly to four, six, eight, sixteen, or even twenty-five near neighbors; or/and to computers at interleaved distances).

People designing interconnection strategies for multicomputers have tended to be intrigued by, and to focus on, *n*-cubes, stars and clusters of stars, and a variety of procedures for combining trees and augmenting trees. But the structures that have been achieved by people taking a more graph-theoretic approach (see Chapter 14 and Appendix A) already appear to be substantial improvements over these. They are much denser, packing substantially more nodes within the same diameter graph, and they often are more highly connected, more symmetric, and with appropriately structured local clusters. The structures of relatively general classes of algorithms also appear to suggest network structures that combine well with these more formal graph-theoretic properties. For example, problems where information interacts as an inverse function of distance map nicely onto arrays and pyramids.

As described in subsequent chapters, we can build pyramids by layering successively smaller arrays. We can further embed such three-dimensional structures in more regular lattices, spheres, and toroidal lattices and spheres. And we can lace such structures with further interconnecting structures that reduce their diameters, as deemed desirable.

In general, a system should be rich enough in links between nearby computers. And it should have an overall shape that makes it easy to pack programs into compact subsections whose interconnection structures and overall shape match these programs well. This seems to suggest three-dimensional toroidal spheres, lattices, and pyramids—or possibly networks of *n* dimensions when it is important to link computers more closely.

CHAPTER 5

Parallel Parallel Computers

Many different architectures are being explored for computers with more than one processor, all working in parallel. Each particular architecture usually reflects the (sub)class of problems for which it was designed. It is instructive to survey the range of possibilities in the immediate present, and to explore how we might combine several different desirable architectures into a single system, much as several different specialized hardware processors are often combined into a single CPU. This chapter begins to consider the present and near-future possibilities. Subsequent chapters examine additional possibilities, moving into the more distant future.

The Great Variety of Possible Array, Pipeline, and Network Structures

Arrays and networks can vary in many different ways, some of which will be enumerated here. This book describes specific architectures chiefly in terms of the overall interconnection pattern, but they can vary in many other ways. Especially important are the size, power, and structure of the individual processor, the number and type of link between computers (e.g., processor−processor, processor−memory−processor, etc.), the type of coordinating control, and the structure of programs being executed.

Parallel architectures appear to fall into several general types:

(a) large arrays (usually 1-bit processors with nearest-neighbor connections);

(b)   closely coupled networks (typically 8-bit, 16-bit, or 32-bit processors, with a variety of interconnection patterns);

(c)   pipelines (typically one pipe using relatively powerful but specialized processors);

(d)   systolic arrays where each executes a particular function in a specially designed two- or three-dimensional pipeline;

(e)   distributed networks (each processor works on a different program, but processors can communicate and pass data).

Several aspects of architecture are important, including:

(1)   the individual processor: its power, generality, specializations, bit size, etc.;

(2)   the interconnection pattern tying processors together;

(3)   communication between processors (whether via registers, one or several shared memories, communication-controlling processors, or direct links);

(4)   tightness of coupling between connected processors;

(5)   reconfiguring capabilities (to change interconnection patterns, under program control);

(6)   control (from SIMD, with each processor executing the same single instruction on a multiple data stream, to extreme MIMD, with each processor executing its own instruction);

(7)   the number of processors (in existing and announced computers, this ranges from 1 to 16,000; but computers with millions of processors will soon be technologically feasible);

(8)   the nature of problems to be executed (their size; their structures; whether there are real-time compatibility constraints, etc.);

(9)   the nature of the technology (especially IC versus LSI versus VLSI economics and design strategies).

This chapter explores a set-of-resources computer system that might make effective use of each type of architecture when it is called for, multiprogramming in a "parallel parallel" mode.

Such a structure may seem inelegant, since it puts together a variety of different types of computers in a single system. But it is actually, on a much larger scale, the structure we see in the CPU of traditional serial computers, with their several specialized processors. It is reminiscent of ILLIAC III (McCormick, 1963) and PEPE (Evensen and Troy, 1973), each a supercomputer with several major subcomputers, each a parallel network or array. This is also the structure of commercial image processing computers which typically are given a different specialized microprocessor board for a number of different functions, and of Kruse's (1980) PICAP II, with its set of (each very powerful) different kinds of processors.

The Value of a Parallel Set
of Parallel Resources

Each type of computer appears to be especially suited to different classes of problems. Indeed, it would be very surprising if one class turned out to be best for everything. The single-CPU serial computer was merely the simplest to start building, programming, and using.

But we have only a vague idea of how efficient the different kinds of parallel (serial) computers really are.

First, many people have been kept from using computers, either because their problems were not amenable to standard serial programming or because their algorithms did not map efficiently onto such single-CPU serial computers.

Second, how do we compare computers? For example, how do we measure the efficiency of a computer that has a large number of special-purpose processors sitting idle while one of the processors is working, and an enormous amount of memory that is needed to store intermediate results that accumulate because only one processor is working, but usually sits idle?

Instead of choosing one particular computer, it seems attractive to build a system to which a growing number of different, more or less specialized, computers can be added, so that they can be used together, when called for, by a single program. All computers might be put on a common bus, or linked using a connection topology of the sort described in Chapter 4.

For example, when logical operations that combine features in large image arrays are needed, a suitably sized array of 1-bit processors would be called for.

When 32-bit floating-point multiplication is needed over an array, a floating-point pipeline like the Floating Point Systems processor might be used.

When convolutions are desired the convolution systolic array of Kung and Song (1982) might be used.

When different sequences of processes are to be executed on different subsets of data (e.g., to look for different kinds of objects in different regions of an image) several more powerful processors in a network could each be assigned a different region and a different set of processes to execute.

Dynamic pipelines (Kartashev and Kartashev, 1979) and other types of adaptive reconfigurable components (Vick *et al.*, 1980) might also be included.

A network that can be reconfigured under program control (e.g., Batcher's STARAN, 1976, and ASPRO, 1982; Lipovski's varistructured array, 1977; PASM, Siegel *et al.*, 1979) is probably a better choice for the initial network in such a system than any specific design. The extra cost of the switches needed for reconfiguring should justify themselves to the extent information is gathered by the system as to the kinds of configurations programmers actually use. Then, once it becomes clear that some specific network architecture would efficiently handle some sufficiently large set of problems (from the economic point of view) that particular network could be added.

Several specific network designs appear attractive, and (when and if they are produced commercially at a reasonable price) might be added to such a system if there appeared to be sufficient local user interest. (But note that only after a great deal of time and effort—in changing people's thinking to a much more parallel mode, in reformulating problems and discovering parallel algorithms for them, and in developing better parallel programming languages and programs coded in those languages—will we finally know how useful each network architecture really is.)

Such a network could today have only a few hundred independent MIMD processors at most. Among the most attractive network architectures for possible inclusion might be the Cyre *et al.* (1977) WISPAC three-dimensional and the Wittie (1979) *n*-dimensional lattice; the array/network pyramid of Uhr *et al.* (1980); the Despain–Patterson (1978) X-trees, some of the compounds to be examined in Chapters 7 and 14 and in Appendix A (Leland *et al.*, 1980, Bermond *et al.*, 1982), and pyramids (Dyer, 1981, 1982; Tanimoto, 1982; Uhr, 1981b).

These might be used separately, or combined, e.g., as follows: The lattice might nicely be used in conjunction with an array, by using the array as a first surface layer on one side of the lattice. A tree or a pyramid might be combined with an array at its buds. The bottom buds might be built of 1-bit processors and form a traditional array. Note that this further suggests that processors might become successively more powerful as we move upward in the tree. We might start with the 1-bit array processors, then move to 4-bit, 8-bit, 16-bit, and finally 32-bit processors. An array of more powerful MIMD processors might surround and interconnect the various layers of a pyramid of SIMD arrays.

The pyramidal array/network is a two-layered system with this general design. The first layer is a 16 × 16 array of 8-bit processors and the second layer is a 4 × 4 network of 64-bit processors. The first layer can be used either as one SIMD array, or as two, three, ..., sixteen different

SIMD arrays. The second layer is really the "master" level, each processor having access to the memories of all its 4×4 subarray of sixteen "slaves."

There are many issues and problems in the ways such different processor groups might be connected (e.g., through common memory or parallel communication links). These are major problems, since the fact that a structure is parallel means that each of its many processors needs links to input and output data. The following is a relatively simple first step:

Suggested Requirements for Parallel Parallel Systems

The total system should have the following capabilities:

(1) One should be able to code programs in a good general-purpose programming language. The system would then compile and execute each program, the system choosing and using the most appropriate particular resource for each part of the program.

(2) Coding, debugging, and testing of a program should be made as easy, as quick, and as painless as possible. This means that programmers should be able to do this interactively, working at a terminal that gives reasonably fast response, and using special-purpose devices (e.g., graphics terminals, television cameras) when needed.

(3) The system should be able to run several programs at the same time, in a "multiprogramming" mode, in order to keep all its resources as busy as possible. (It is likely that computers will become so cheap that every user can have a personal computer. But this type of highly parallel computer is likely to be so powerful and so expensive that it should be used by, and its costs amortized over, more than one person.)

(4) The system should contain as many of the different kinds of parallel structures as users find desirable (and can be afforded).

(5) Users should be taught and helped to understand the virtues, and the vices, of each type of system, and should be encouraged to explore and evaluate the usefulness of each type of system.

(6) The system should keep good records of how much each type of structure is used, and, to the extent possible, when, in combination with what other structures, and why. These records should be broadcast and interpreted to users, in a way to help them improve their own program choices and evaluations.

This set of considerations suggests the following specific system:

(a) A traditional single-CPU host to handle the traditional program development, editing, compiling, etc., functions.

(b) A very large array—similar to CLIP, DAP, or MPP, except with the kind of limited reconfiguring capability described below.

(c) A network of networks of more powerful processors (in 1983 on the order of PDP-11/44s or M68000s; by 1985 on the order of VAXs or Cray-1s). This network should be expandable in number, and might therefore best be started small, e.g., with four, or sixteen processors. It should be given enough reconfiguring capability (discussed below) to keep whatever specific interconnection pattern is chosen from being too constraining. Possibly specific architectures, like the hyper-tree, WISPAC, and the array/network pyramid should be used for subnetworks of this network. If it seems indicated, a multilayered system that combines (parts of) the array with (parts of) the network, might be used. This might be along the lines of an array as the buds of an augmented tree or one face of a lattice, or the base of a multilayered array/network pyramid.

(d) Ideally, the operating system should be designed to allow future additions of processors of this sort to include different (newer, more powerful, cheaper) processors, and different interconnection patterns.

(e) A pipeline optimized for image processing; e.g., PICAP (Kruse, 1976); or the Cytocomputer (Sternberg, 1978).

(f) A pipeline optimized for 32-bit floating-point multiplication; e.g., the Floating Point System.

(g) A variety of special-purpose systolic arrays, and/or processors of the sort used in PICAP II.

These different systems should be as nicely interconnected and reconfigurable as possible. As a first step, the following is suggested:

(h) The very large array should be made of subarray "facets" (we assume that each facet resides physically on one board or chip). Assume a 128×128 array built from 16×16 facets. This would give only 64 facet nodes in the reconfiguring network—a large number considering the large number of processor links needed, but reasonably small compared to complete reconfigurability.

(i) Shipping data between processors is a major problem, given the high data throughput of parallel computers. Several alternatives seem interesting, and should be explored:

(1) The array might be made one of the host's memory banks (just as the DAP is a memory bank for the ICL 2980 main frame). This gives the host, as well as the array's processors, direct access to

the array's memory stores. The pipelines might also be given this direct memory access.

(2) The array and pipelines might be interfaced to the host's highest-speed bus (e.g., the VAX SBI bus).

(3) All structures might be interfaced to the host's standard bus for peripherals (e.g., the VAX Unibus).

Summary and Conclusions

Which of the large variety of possibilities is the "right" or "best"? There have been very few comparisons made, either theoretical or empirical. But a rough picture is beginning to emerge:

For one particular kind of problem a carefully designed specific network will be best. But over a mix of problems these differences may well grow small, or wash out entirely. Interconnection patterns that appear quite different on the surface often turn out to be (almost) identical. The adding of nonregular connections at a distance, or of reconfiguring capabilities, appears to give only 5, 10, or 20% increases in throughput over tested mixes of problems (Finkel and Solomon, 1980b; Reddaway, 1980). This may not be worth the additional cost in hardware and operating system, especially when more computers can be added to increase throughput.

This suggests that particular classes of problems, those that have a specific structure, are best handled by a system with an "anatomy" which reflects this structure. But a general-purpose computing center will not be able to handle a wide variety of different types of problems efficiently unless it uses the most appropriate particularly structured processors for each. They might all be put on a common bus, or a tree, or some other simple but slightly more specialized structure.

Those structures that are today known to be useful for specific sets of problems—and these include large arrays and pipelines—should be included. A reconfigurable network is of special value in its potential for helping us to learn how to build, program, and use parallel (serial) systems.

Several specific networks appear to be attractive, for reasonably large classes of programs. These include lattices, augmented trees, and pyramids. They might be combined in a variety of ways, e.g., an array as the buds of an augmented tree, or one side of a lattice, or the base of a pyramid.

The overall system might best be built as an experimental vehicle for program development, testing, evaluation, and comparison. The system should collect and organize data about the types of systems built, the physical resources used, and how well each performed. Thus such a system could be an effective experimental test bed for learning about and developing better new architectures. It seems likely that experience with parallel computers will suggest new structures, and (potentially more important in the long run) new problems that are amenable to attack and that suggest still other new structures for sets of processors.

CHAPTER 6

Converging Pyramids of Arrays

This chapter introduces and explores architectures that could be achieved today for very large arrays of processors that stack successively smaller arrays into pyramids or cones. (Chapters 16−18 examine such architectures in more detail and extrapolate into the future.) Such a structure gives, in effect, a two-dimensional array of pipelines. [Tanimoto (1982), Dyer (1981, 1982), Dyer and Rosenfeld (1981), and Uhr (1981b) have examined and proposed pyramid-structured systems.]

If data flows "inward" (from the largest to the smallest array), the pipelines flow together and converge. If data flows "outward," the pipelines branch and diverge. Inward flow appears useful for many image processing, pattern recognition, and matrix manipulation applications, where very large arrays of data are gradually reduced to a relatively smaller number of pieces of information. Outward flow is useful when exploring expanding trees of possibilities, as in top-down parsing or heuristic search.

Traditional two-dimensional arrays (e.g., CLIP, DAP, MPP) link each processor to its four (or six or eight) nearest neighbors in the plane. "Stacked" arrays also link processors to the two nearest neighbors in the third dimension. "Converging" stacked arrays link processors to one "parent" processor in the inward direction and to a subarray of "offspring" processors in the outward direction.

Each layer executes the same single instruction on the different data fetched by each processor, in an SIMD mode. Each layer can execute a different instruction, and therefore has its own instruction register and controller. As layers become smaller, processors can become more powerful, and also can be given their own controllers, so that they can operate asynchronously, in a MIMD mode.

Such an architecture has a number of advantages (when used for problems for which it is appropriate):

The array processes large amounts of data simultaneously, with enormous potential speedups.

The pipeline further speeds up processing, at relatively small additional cost.

The converging structure, when fully and properly used, means that processors in a single large array in which successively smaller arrays of data are processed do not sit idle, and shifting to the smaller array is handled efficiently.

Input and output can be done where appropriate, with efficiency. For example, output of a much reduced enhanced or recognized image can be effected at a higher layer or at the apex of the pyramid, rather than by shifting out from the traditional large array.

A Pipeline of Converging Stacked Arrays

As a first step, consider a traditional large array like CLIP, DAP, or MPP. For simplicity, assume that each processor is connected to its four square neighbor processors (rather than, as in the case of CLIP, to its six hexagonal or eight square and diagonal neighbors). Thus each processor has access to information stored in its own memory, and in the memory of its four nearest neighbors.

Now give each processor a link to its parent processor in the next/adjacent array. In this way a three-dimensional array is constructed, with each processor (except for those on the surface) linked to its six nearest neighbors, two in each of the three dimensions. This stacks a set of two-dimensional arrays into a three-dimensional array.

Finally, make each layer smaller than its adjacent layer: Divide row (r) and column (c) size by some small integer like 2,3,...,7, and replace each $r \times c$ rectangle of nodes by a single node (keeping all links). Thus each parent has rc offspring linked to it; each offspring has one parent linked to it.

Each array/layer has its own controller. Thus within a layer every processor executes the same instruction sequence, but the different layers can execute different instructions.

The first (largest) layer has 1-bit processors, each with 1k, 4k, 16k, or 64k bits of memory. As layers grow smaller, processors will increase in size to 4, 8, 16, and 32 bits.

The first layer must have some relatively fast input. The traditional method of shifting data in through one edge-column of the array seems appropriate. The last layer must have input—output lines (but here the amount of data to be transferred has been much reduced). It is simplest to have all layers shift in and out through one edge. But when necessary on large pyramids, input−output channels could be put at every Nth column. When large numbers of computers are put on a single chip, input−output can go through one column, or two, or all four edges of each chip. For example, a 16×16 array of 256 1-bit computers on each chip could use 16, 32, or 64 pins for parallel input or output.

The first layer of such a pyramid should be at least 32×32, and it might be (technology and costs permitting) as large as 1024×1024. Table I gives a range of plausible configurations.

Table I

Some Reasonable Example Configurations[a]

	Array Specifications											
layer	rows	bits	rows	bits	rows	bits	rows	bits	rows	bits	rows	bits
1	32	1	64	1	128	1	256	1	512	1	1024	1
2	16	2	32	2	64	1	128	1	256	1	512	1
3	8	4	16	4	32	4	64	4	128	4	256	1
4	4	8	8	8	16	4	32	4	64	4	128	1
5	2	16	4	16	8	4	16	4	32	4	64	4
6	1	16	2	16	4	8	8	8	16	4	32	4
7			1	32	2	16	4	16	8	8	16	8
8					1	32	2	32	4	16	8	16
9							1	32	1	32	2	32
10											1	32

[a] The above are all square arrays, with convergence to one-half row and column length at each layer. Rectangular arrays, with varying convergence, might also be used. Probably hexagonal arrays, with 7-to-1 convergence, are the best of all.

Consider the 64×64 seven-layer array, since this is large enough to have appreciable power yet small enough to be feasible. It has a pipeline seven layers long although it uses only one-third more processors for seven layers than it would need for one 64×64 layer. If all its processors were 1-bit processors it should cost only marginally more than the one-layer two-dimensional array. For programs that used its arrays and converging data paths fully it would not only speed up processing sevenfold but would also save what might be large numbers of shiftings.

A variety of "overlapping" pyramids where nodes have more than one parent are also possible and will be examined in later chapters.

An Example of a Potentially Powerful
yet Economical Pyramid

We now examine a feasible system, one that can be built with existing technology at a reasonable cost for image processing and pattern recognition as well as numerical problems and takes advantage of such a layered converging structure:

Such a system should cost (excluding the host) from $50,000 (or less) to $500,000 (at the very most). It should have at least a 32 × 32 array of processors in its largest layer, and preferably a 64 × 64 or even 128 × 128 array. It should have four to eight layers *in toto*. It should be nicely integrated with a host computer (which might itself form its inmost apex/layer). A user should be able to code, edit, compile, test, and run programs that use both the host and the converging array. This means that the operating system must be able to handle multiple users at alphanumeric terminals, along with graphics terminals and television input.

The host might well be a small minicomputer like a PDP-11/44, LSI-11/23, or 68000-based SUN (which also includes a graphics terminal), costing $5,000−$50,000. But it should be a computer with a good operating system that allows for continuing user development of software packages and programs. This strongly suggests a computer like the 32-bit PDP-11/VAX, for $50,000−$200,000.

It appears that the VAX is becoming the preferred 32-bit computer, both within university Computer Science departments and for a wide variety of uses. And the VAX should decrease in price as LSI technology matures. So assume that the VAX, with its two standard buses— the UNIBUS (which it shares with the family of 16-bit PDP-11 computers, and which has a data rate of 1,300,000 bytes per second), and the SBI bus (13,700,000 bytes per second)—is the host of choice (except for the cheapest systems).

We can consider adding a large array to a host VAX by using a direct memory access (DMA) interface to the SBI bus. This makes the array's memory look exactly like any other part of the VAX's memory. (Note that this is an appreciable amount of memory. Assume that each of the array's processors has one 16k chip as its memory. Then a 32 × 32 array will have 2,000,000 bytes of memory; a 64 × 64 array will have 8,000,000 bytes. If 64k memory chips are used, multiply these figures to 8,000,000 and 32,000,000 bytes. In 1984 VAX memory cost about $2000 for 1,000,000 bytes.)

All of the very large arrays (that is, arrays with at least 1000 processors) that have been built or even announced to date use 1-bit processors. Most of them use, or plan to use, LSI chips, with several processors on each chip. Thus CLIP4 has 8 processors (plus 32 bits of memory per processor) on each chip; DAP has 4, MPP has 8, ASPRO has 32 per chip. It therefore seems most reasonable to use such systems of 1-bit processors (whether starting at the level of chips, boards, or whole arrays will depend upon a number of technical and economic considerations) for the largest array/layer, and, probably, for several adjacent arrays.

Each processor chip might cost from $2 to $200 (depending upon the technology, but even more important upon the size of the order — these prices would go down appreciably if there were a market for such chips to be made and sold by the millions). Each memory chip might cost from $2 to $10. Thus each processor-plus-memory computer might cost from $3 to $35. This is far below the cost in a completed system, with support chips, wires, boards, chassis, etc. But it gives a rough idea, and it suggests that $10 – $200 per processor is a reasonable range for such arrays.

In contrast, in an MIMD network, even an 8-bit microcomputer with reasonable power would cost in the range of $500 – $2000, and a 16-bit computer built around a 68000 chip or an LSI-11/23 might cost from $1000 to $10,000.

Therefore even an 8×8 layer of 64 processors would be quite expensive (given our budget constraints) if built from traditional microcomputers. This means that 1-bit processor arrays should probably be used (purely on economic grounds) for the 64×64, 32×32, 16×16, and 8×8 layers. The final 4×4 and 2×2 layers might reasonably be built as a network of more powerful processors, each working independently in MIMD mode.

A Combined Array/Network Architecture

Another economically feasible alternative is the following: Build special arrays of processors more powerful than 1-bit arrays, thus using fewer of the 1-bit layers before data are moved to the more powerful and more traditional processors. Here a good example is an array/net (Uhr, *et al.*, 1980) designed and explored at Wisconsin.

Our goal was a system with enough processors to substantially speed up performance on problems involving large parallel arrays, including television camera images for pattern recognition and scene description, and also numerical problems. We were also constrained in funding to a level around $50,000 – $150,000. And we were not in a position to use any but off-the-shelf components (nor were any of the 1-bit processor chips potentially available).

We therefore decided to use bit−slice technology, which allows one to design and build quite powerful microcodable processors and controllers of any desired size divisible by 4 (since four bits worth of processor resides on each chip). This was to be our entire system, and we wanted some MIMD network capability, as well as a large enough (therefore necessarily SIMD) array to handle interestingly large problems. So we designed a rather unusual mixture of SIMD and MIMD computers:

As a first step, take a 4×4 array of sixteen "slave" processors, all working under a single "master" controller. Each slave is an 8-bit processor with its own 16,000-byte memory and connections to each of its four nearest neighbors.

Next, put sixteen of these master−slave groups into a 4×4 array.

This gives a 16×16 array of 256 8-bit processors, each with 16,000 bytes of memory. It also gives a 4×4 array of 16 rather unusual computers that can be used as 8-, 24-, 32-, or 64-bit processors, each one with 256,000 bytes (the same memory as its sixteen slaves). The sixteen master controllers can each work on a separate program, and each can fetch data from and store data into any of the memory locations of any of its sixteen slaves.

Thus this system converts itself back and forth into either a 256-processor SIMD array or a 16-processor MIMD network. It can also be used in a third mode, where each group of sixteen SIMD slaves can be executing a different program. This is possible simply because each master−slaves module is quite separate (largely for economic reasons, since 4×4 arrays of slaves-plus-controller are packed onto a single board, and the program must reside in the memory accessible by each master controller).

Our relatively firm estimate was that we could finish the design and build this system for $170,000 (including design and manpower as well as hardware costs). Since about half our cost would have been for design and debugging it seems reasonable to estimate that such a system could be built for $100,000 or so in quantities of five or ten.

Our estimate is that this system, when used over the kind of mix of logical operations (for combining feature planes in pattern recognition) and 4-bit and 8-bit arithmetic as well as occasional 32-bit floating-point arithmetic would perform a good bit better than a comparably sized array of 1-bit processors. But our 16×16 array would probably not be as powerful as a 32×32 DAP or MPP, especially if the differences in cycle time were taken into account: DAP 250 nsec, MPP 100 nsec, our system 400 nsec (because of cheap, hence slow, memory). This, however, is very hard to estimate, since it depends upon how well the programmer can restructure algorithms to map nicely onto a 1-bit, or an 8-bit, architecture.

Possible Mixtures of N-Bit Processors

This, then, suggests that we might consider using 8-bit processors for the 8×8 and even the 16×16 layers of a larger multilevel pyramid. Here I suggest several alternatives:

(a) Simply use a 1-bit array (either 32×32 or 64×64) plus this array/network, as just described. Processors in the 1-bit array would now be linked to the array/network with eight processors inputting 1 bit each to form a 1-byte word in the 8-bit processor's memory.

(b) Use two layers of 1-bit arrays (64×64 then 32×32), feeding the second into the array/network.

(c) Use either one or two layers of 1-bit processors, and then a new array/network modified as follows: Link each master to one additional 16,000-byte memory, which is (part of) the memory of a new 16-bit processor. This now gives a new layer of sixteen MIMD processors. These in turn would link into a 2×2 layer, giving two network layers plus the array layer of 8-bit processors. Possibly the twenty MIMD processors should be connected in some pattern other than nearest neighbor, or with some capability for being reconfigured. But note that although this seems interesting it is much easier said than done. The design problems are major, and the costs, considering that each processor will probably need a separate board, might well add another $100,000 or so for the twenty new processors.

Summary Discussion

A converging multilayered pyramid of arrays and networks offers, potentially, a well-structured two-dimensional pipeline for processing problems that involve near-neighbor computations and a regular reduction and convergence of data. A number of programs for pattern recognition, scene analysis and description, and other visual perception tasks, as well as many programs for database management, artificial intelligence, and numerical problems, appear to have such a structure.

The largest base layer probably should use the simplest and weakest processors, purely for economic reasons (since from 1000 to 1,000,000 processors are needed). The successively smaller layers, moving inward, might be given more powerful processors, richer interconnection and reconfiguring capabilities, and more independence among processors—as economics permit.

It is not at all clear that the processors in the large base array should be simpler and weaker; on the contrary, high-resolution inputs contain large many-bit numbers. (We can loosely equate the base with the first layer of the retina of the visual systems of human beings and other higher animals. But nature appears to use the same kinds of neurons at all layers.)

It seems more certain that the near-neighbor connections are indicated, especially at the large base and lower layers. If processors were connected at a distance in a variety of ways, connection costs would rise and the problems encountered by the operating system and programs in handling different kinds of connections would increase. Connection patterns should reflect the flow of information through processors, and the structure of the whole network should reflect the structure of the problem(s) it is built and programmed to solve. The near-neighbor links and array/pyramid structure meet these desiderata, at least for a large range of problems and algorithms.

The base array of processors can, at least when images are being handled, be considered as the interface where information about the external environment is input, or sensed. The subsequent layers, moving inward, transform that information.

To anticipate some of the structures we shall explore in the last few chapters of this book:

We can generalize such a system by enclosing it in its base. Think of the pyramid as an umbrella (with many flexible struts to the base). Now let the umbrella spread out (as though collapsing in the wind) until struts go in all directions, stretching the base until it entirely encloses the system. The base is now the skin of a sphere whose center is the apex.

An array of 1-bit processors might be plastered on every surface of a lattice. Or a sphere might be built with 1-bit processors at the input-sensing surface. When considered from this perspective, a converging pyramid is one section of such an array/network lattice or sphere.

CHAPTER 7

Comparisons among Different
Interconnection Strategies

This chapter begins to explore some of the similarities and differences among the many different interconnection strategies used in networks and arrays. It also surveys the relatively few evaluations and comparisons that have been made. A variety of criteria have been suggested and used—empirical, structural, and formal (see Appendix A, on graph theory) to construct and evaluate multicomputers. But we are just beginning to understand which are the most important.

Comparisons between Arrays and Networks

The "more general" MIMD networks have not yet been built to contain more than 8 or 16 or [in one or two cases only, e.g., the Swan *et al.* (1977) Cm*; the Manara—Stringa (1981) EMMA] 50 processors. Plans for even a few hundred are still quite vague, with the exception of ZMOB, with its 256 Z80 microcomputers on a fast shift-register ring (Rieger *et al.*, 1980; Kushner *et al.*, 1980). And they are expensive (on the order of $500—$20,000 or more per processor as packaged into the complete system, as opposed to $10—$200 for the arrays).

Even worse, the problems of developing operating systems, programming languages, and efficient programs for such networks appear to be enormous. It seems likely that, say, 1000 traditional processors connected in a "general" network pattern may well not increase throughput more than 10 or 50 times, unless they can be built or reconfigured into a structure that closely mirrors the parallel structure of the processes being effected.

The arrays are far simpler than the networks (see Appendix C). All processors execute the same instruction, synchronously. Each (1-bit) processor has from 50 to 500 gates, rather than the 2000−20,000 or more in each network processor. Each processor has only from 32 to 4000 bits of memory, rather than the 64,000−64,000,000 or more bits in a network. Thus the arrays' processors are much smaller and simpler (and more flexible), and the arrays allocate a far higher percentage of their devices to processors, as opposed to memory. As the DAP architecture makes apparent, the arrays scatter (tiny) processors uniformly throughout the memory.

Only one controller handles all processors, rather than a controller for each processor. This not only cuts down substantially on the hardware requirements, but also means that an instruction must be stored only once, and is fetched only once. The resulting SIMD lock-step behavior makes things much simpler—and also restrictive and often difficult to program for.

The regular near-neighbor connections enormously simplify (and also limit) message passing. But the arrays are large and relatively poor at moving information among distant processors. Connections that wrap around at the edges cut the diameter in half. Reconfiguring networks, like the flip network on STARAN, can improve this situation drastically—but at a relatively high cost in additional hardware.

The single controller, fixed array size, and near-neighbor connections can severely limit the range of processes that can be handled efficiently. The programmer must condition, bend, and sometimes break, his problem to fit such a system well. But often this can be done with little or no loss. Such systems can be made less restrictive, with independent processors and reconfigurable interconnections, and more powerful, with multibit processors and far larger memories. But nobody has done this yet, and the costs might be very high.

Comparisons among Networks Using Formal Criteria

Networks are sometimes compared by using formal characteristics—e.g., the longest shortest path between nodes (diameter), average distance between nodes, number of nodes, number of links, number of links per node. But a well-written program will be just that program that puts processes that must communicate frequently as close together as possible. The operating system should, as Despain and Patterson point out, assign processors to work on large banks of data that are linked as closely as possible to the mass store containing that data. More generally, physical processors should be assigned to a program's processes so that information is passed as efficiently as possible from one to the next.

So there may be little or no need ever to send messages except locally, and worst-case possibilities may well be quite irrelevant. If we could assume some general rearrangement of the program to move communicating processes closer together, and to build a program structure that maps as nicely as possible into a subgraph of the total network structure, then some measure of average message-passing distance among nearby processors might be the most pertinent.

Interesting attempts to compare network architectures have been made by Despain and Patterson (1978), Finkel and Solomon (1977, 1980b), Siegel *et al.* (1979), Sullivan *et al.* (1977), and Wittie (1978). They have uncovered a number of instances in which two networks that look very different (often because of the ways they have been defined, constructed, and pictured in graph figures) are actually identical or very similar; for example: trees and stars; binary n-cubes and certain kinds of augmented stars.

Even simple graphs can be drawn in several ways that make the same graph look very different. The very small 10-node Petersen graph (see Appendix A) is a good illustration of this—it can be drawn as a highly symmetric star-within-pentagon, a tree with rather strange rings, and a variety of very regular and irregular shapes.

Finkel and Solomon have examined how well (randomly generated) messages are handled by different interconnection patterns. They found that suturing stars and snowflakes gives relatively small (10−20%) improvements in message handling. Their results (and also Wittie's) appeared to suggest that n-cubes, lenses, and stars (which, remember, are trees) were the best, and rather comparable. But (as Table I shows) hyper-trees, de Bruijn networks and other augmentations of graphs are substantially better still.

These groups (along with others) found the following problems:

Trees and stars have uneven link traffic that grows increasingly heavy (an undesirable characteristic) toward the apex/root.

n-cubes have the undesirable characteristic that the number of links to each processor increases as additional processors are added, and only certain fixed sizes (the powers of 2) are possible.

Unless given additional links, trees, stars, and snowflakes become disconnected if a single node or link fails.

Despain and Patterson found that hex-trees (among the simplest of augmented trees) have a greater average path distance than trees in networks with more than 64 nodes. This is apparent from the following exercise (see Fig. 1): starting with the root of a (binary) tree, take two offspring-of-offspring and combine them into one node (which now has two links to parents and only one to offspring); treat this as a root; and iterate this process. We have eliminated (by merging) a number of nodes for the same linkages, to form the hex-tree.

Fig. 1   Losing nodes changing tree to hex-tree [C = computer (processor + memory store); |, — = links].

They also augmented trees, giving them four ports, to get "threaded trees," "double-threaded trees," "half-ringed trees," and "half-ringed trees with shuffle." ("Rings" connect nodes the same distance from the root, that is, laterally. "Threads" link distant nodes across levels. The "shuffle" was Stone's (1971) "perfect shuffle" but only at the buds.) They ran simulations to compare these with each another, and also with a square array, and showed there were only very small differences among them in message traffic or average path distance.

With five ports, a "full-ringed tree with shuffle" appears (from their computer simulations) to almost halve the average number of messages per node and the average path distance.

The Petersen graph (see Appendix A) may well be preferable to the small clusters of processors (usually eight to sixteen on a bus, or in a star) commonly proposed for networks. A local cluster with the Petersen graph structure would have ten nodes grouped in a highly symmetric manner, with a diameter of two, only three links per node, and high connectivity and cohesion. This gives configurations that are locally "dense" (that is, with a large number of nodes for a given degree and diameter), and with good message passing and addressing capabilities and a high degree of fault tolerance.

A compound of 11 Petersen graph clusters can be built (Uhr, 1980) by linking each of each graph's 10 nodes to a node in a different graph, giving a network with 110 nodes, degree 4 and diameter 5. [Note that this is substantially better than the best previously reported in the literature by Storwick (1970) — see Appendix A, Table I — which was 62, but substantially worse than the Moore upper bound, which is 485.] Several other compounding techniques have also recently been devised to give a variety of dense new graphs (Leland, *et al.*, 1980; Bermond *et al.*, 1982). Imase and Itoh (1981) proved that de Bruijn networks (de Bruijn, 1946), which are commonly known as shift registers, and can be drawn as trees with additional links, give the densest graphs found so far as graphs grow larger.

It is instructive to compare the networks designed by network architects with the graphs discovered with the maximum number of nodes, since there has been virtually no overlap between these two lines of research [Arden and Lee (1978) appear to be the only researchers designing networks who have seriously explored any of their graph-theoretical properties. And density appears to be the only property explored]. Table I shows the packing densities (with respect to diameter) of several of the major proposed network structures. In many cases much denser graphs have been identified. (Chapter 14 examines these issues more fully, and Table I of Appendix A gives many additional dense graphs.)

The augmented trees can have a variety of different structures. And, since many of the best graphs found so far have been built by completing trees, augmented trees can be made quite dense.

Wu and Feng (1981) have shown that any permutation can be achieved by some sequence of Stone's (1971) shuffles. All of the multistage reconfiguring networks appear to be equivalent to the "indirect n-cube," which is a realization of an n-cube using two-input-to-two-output switches (Agrawal, 1981). Therefore the n-cube figures given above would appear to apply to them, except that they establish distances for passing N messages simultaneously, rather than one message at a time.

It would seem of great value to investigate some of the best structures that have already been found, and to make a greater effort to find still denser structures. Three general techniques appear to be emerging, from the recent work in Japan and at Wisconsin:

(a) Complete trees, using a heuristic search program. This gives the densest graphs up to about 525 nodes.

(b) Compound good, dense graphs. Several compounding operations have been discovered by Leland, Li, and Uhr which in general involve replacing the nodes of complete graphs, and of complete bipartite graphs, with local cluster graphs. This gives densest graphs between roughly 500 and 15,000 nodes. Possibly more important, these graphs are highly symmetric and have good local properties.

(c)   De Bruijn networks (shift registers), which can also be viewed as trees whose degree is raised by one, are densest starting around 15,000 nodes.

Table I

Packing Densities for Selected Network Structures[a]

A. Graphs of degree 3 and 4; diameter 3,4,...,10 (diameter)

Graph	3	4	5	6	7	8	9	10
degree 3								
n-cube	8							
tree		10		22		46		92
Best	20	28	36	60	66	90	138	216
Bound	22	46	94	190	382	766	1534	3070
degree 4								
n-cube		16						
tree		17		53		161		485
Lens	8			64			384	
Petersen compounded			110					
Best	35	40	62	114	188	320	566	996
Bound	53	161	485	1,457	4,373	13,121	39,365	118,097

B. Additional graphs with interesting degrees and diameters (degree, diameter)

Structure	2,2	5,5	6,6	7,7	8,8	9,9	10,10
Petersen	10						
n-cube	4	32	64	128	256	512	1,024
Compound		240	1,152	4,512	25,200	382,500	2,000,000
de Bruijn net					65,536		9,765,625
Best		262	1,152	4,512	65,536	382,500	9,765,625
Upper bound		1,365	19,531	335,923	6,725,601	153,391,689	3,922,632,451

[a]See Appendix A, Table I for all Best (reported in the literature) and Bound (the Moore upper bound) to degree 10, diameter 10.

Tests, Simulations, Estimates, Models

Empirical examinations of networks have been few. This is not quite as surprising as it may seem, since scarcely any networks have actually been built, and it is too time-consuming to simulate a nontrivial network on a one-processor serial computer.

Networks Sharing Bus, Common Memory, or Ring

Lehman (1966) simulated the behavior of networks sharing a common memory. Compared to one, four processors needed 27.1%, eight processors 15.8%, and sixteen processors 11.7% as much time. Connell (1976) estimated that, to handle telescopes scanning the night sky, ten minicomputers on a bus, costing less than $200,000, gave the throughput of one Cyber 76 or Univac 1100, costing over $2,000,000.

Bisiani (1980) found that a four-processor C.mmp executed the Harpy speech recognition system 2.5 times as fast as a single computer. When a special-purpose "memory computer" (built from AMD 2900 bit−slice hardware, but emulating active logic-in-memory) was added, speed increased to 3.5.

Fuller (1976) compared the C.mmp network of minicomputers with a larger one-processor computer, and concluded that the network was 3 to 4 times better, in terms of the economics of hardware at the time of the comparison. This seems rather encouraging, especially considering the simple and expensive interconnection strategy used. But the comparison depended heavily upon DEC pricing policies that may have little to do with actual hardware costs.

Jafari *et al.* (1980) found in simulations that ring-structured networks increased relatively little in performance as the number of processors was increased.

Networks with Star, Tree, or Other More Complex Topologies

Jones and Schore (1980) found that an eight-processor Cm*'s performance, using a synchronous parallel method on partial differential equations, was almost tripled by making all code and variables local rather than global. They comment (p. 140): "The rather obvious conclusion is that performance degradation is substantial if program reference patterns are not considered in making resource allocation decisions."

Fishburn (1979, 1981) showed that a sufficiently parallel algorithm could be devised for "alpha beta lookahead" (an efficient and widely used technique to explore large trees of possibilities) to give linear increases in throughput on a network interconnected as a tree.

Goodman and Despain (1980) presented procedures for efficient use of an augmented X-tree for merging large databases. Goodman and Sequin (1981) compared hyper-trees with de Bruijn network shift registers, using average distance, and found hyper-trees to be about 10% denser.

Supercomputers with Pipes, Arrays, and Specialized Hardware

A large amount of work has been done to evaluate ILLIAC IV. Kuck (1977), Stone (1973), Thurber (1976), and others all find speed increases far greater than "Minsky's conjecture" (1971) that only log n improvement can be achieved with n processors. Thurber (1976) points out that there are certainly algorithms that ILLIAC IV, along with the other parallel computers, can handle far faster than Minsky conjectures possible. But ILLIAC IV is too controversial a computer to be easily judged. It cost so much (although with VLSI technology the economics would be completely different); it has been judged hard to program; it is very hard to keep busy. Probably most important, one must think hard to restructure a problem that does not obviously fit its structure.

The very large arrays and, to a lesser extent, the pipelines, already offer increases of several orders of magnitude in speed and power. DAP and the MPP can be more powerful than even the largest of supercomputers, at least on the array-processing algorithms where their parallel power can be used effectively.

DAP has been compared with a variety of number-crunching super-computers, with what appear to be good results. For example, DAP was 5 times faster on simple relaxation than ILLIAC IV, 10 times faster on many-body simulations and data reorganization than the CDC 7600, and 3 times faster than the Cray-1 on a table look-up problem (Flanders *et al.*, 1977a,b). But these comparisons were made by the DAP group. They may well have chosen programs suited to DAP, and others might choose programs that would give different results. Stanley and Redstone (1979) found DAP 10 to 100 times faster than a Floating Point Systems processor on convolutions and a variety of other image processing tasks.

In terms of pure number crunching (in "megaflops"; that is, millions of floating-point multiplications per second), the Burroughs BSP (Kuck and Stokes, 1982) can compute 50, the ILLIAC IV 80, the Cray-1 160, and the Cyber 205 200 (Kozdrowicki and Thies, 1980; Sugarman, 1980). The three-dimensional real-time CAT scanner of Gilbert *et al.* (1981) is expected to compute 2000 − 3000 megaflops.

Lundstrom and Barnes (1980) simulated and analyzed Burroughs's FMP (Flow Model Processor), a system with 512 processors linked via reconfiguring switches to shared and local memories. They found it could achieve 1000 megaflops for an aerodynamic-flow model problem.

It is not clear whether the 1-bit arrays like DAP and the MPP are competitive with the number crunchers at 32-bit floating-point arithmetic. (DAP is slower, but also much cheaper. The MPP will execute 373 megaflops.) But they achieve successively increasing superiority at 8-bit arithmetic and 1-bit logic. The MPP will execute over 6,500,000,000 8-bit additions, and over 1,800,000,000 8-bit multiplications per second (Batcher, 1980). When executing 1-bit logical operations, which are extremely useful for edge detection, feature extraction, and feature compounding, the 1-bit arrays are even faster. For example, DAP can compute the logical-AND of three features and store the result (which indicates the presence or absence of a compound feature) all in 1 μsec. The 64 × 64 DAP does this 4000 locations at once, and therefore executes 4,000,000,000 such extremely useful instructions per second. A larger DAP would be commensurately faster. The MPP should be 10 times faster still.

Lee (1980) found in simulations using Kuck's (1977, 1978) parallelization procedures that MIMD systems were less than 2.5 times faster than SIMD systems. If this indicates a general relation, it suggests the power and economy of the much cheaper SIMD approach.

Some Problems in Allocating Resources and Specializing Networks

Such figures should not be taken too seriously, since they depend upon all the processors being busy. That cannot be expected over a whole program, and it is not clear what kind of mixes of programs, and of instructions in individual programs, these arrays will be given, and should be given.

Shore (1973) points out that most parallel systems increase the percentage of gates used in processors (because they increase the number of processors), but therefore make it more difficult to write programs that keep all processors busy at useful work. There are extremely interesting issues, largely unexplored, as to how logic gates should be allocated, whether to memory or to processors, and whether to one processor or to many (see Uhr, 1981c).

The best evaluation of these systems will, in the long run, be in terms of how effectively they execute large programs that have been written for them. Unfortunately, very few papers describing complex programs discuss timings fully enough to make clear the enormous increases in speed that a large array can achieve. For examples of programs run on arrays and pipelines, see Kruse (1978, 1980) for fingerprint, circuit layout, and other applications on PICAP; Marks (1980) for image processing on DAP; and Duff (1982) for general programming principles using CLIP. It will be especially interesting to compare these programs for arrays with programs executed by the proposed pyramids (Dyer, 1981; Tanimoto, 1982; Uhr, 1981b), and by systems structured to run other types of perception programs (Briggs *et al.*, 1979; Siegel *et al.*, 1979; Rieger *et al.*, 1980; Agrawal, 1981).

Potentially Useful Tools: Network Flow and Queueing Theory

Graphs, Petri nets, and data-flow models exhibit the general structure of processes and information flow in a program, and also of processors and their interconnection topology in a multicomputer. Additional specifications are needed, of the channel capacity of each link versus the load of information flowing through it, and the power of each processor versus the load of processing it must handle. A power can be assigned to each node by putting new nodes at all joins between links and nodes (thus turning nodes into links) and assigning a nonnegative integer to each new link formed in this way.

Network flow and queueing theory models can be used to describe, examine, and analyze networks where loads and capabilities are specified. Conventional computers, communication networks, and loosely distributed networks have been studied with great success from the standpoint of queueing theory by Kleinrock (1975, 1976), Baskett *et al.* (1975), Samari and Schneider (1980), and many others (see Denning and Buzen, 1978).

The assumptions of random arrival time and distributions of messages that make analysis possible appear to be far less plausible for more tightly coupled networks with complex topologies. Here network flow models (Ford and Fulkerson, 1962; Stone, 1977) may be more germane. But the general problem quickly becomes intractable. Chu *et al.* (1980) point out that the min-cut solution Stone (1977) used for two processors can handle the allocation of processes in a program graph to at most four processors. They suggest exploring integer $0-1$ programming (which has handled problems with fifteen processors and twenty-five modules) and heuristic methods like that of Gylys and Edwards (1976).

Network flow and queueing models should be extremely useful simply as descriptive models, even if large multicomputer topologies are too much for them to handle. Subsets of the general problem might be tractable, including many structures of practical interest, since powerful topologies are likely to be symmetric and regular. And several different multicomputers could then be compared and, possibly, evaluated, analyzed, and improved.

Architects of skyscrapers and bridges build only structures they can analyze or test. Network topologies can similarly be chosen not from all possible graphs but from those that can be understood.

Some Structural Similarities among Different Networks

A tree is very similar to a star (except that the star has a bus at the root; a tree has a processor). A snowflake is simply a star with some of its branches pared away. A pyramid is a tree with interconnections between offspring (that is, lateral augmentations) that impose a two-dimensional array structure over the base array and, possibly, other array layers of offspring.

Consider a square two-dimensional cellular array: pick it up by a central cell and you have an augmented X-tree, with many lateral interconnections. A three-dimensional (or n-dimensional) lattice can similarly be picked up centrally, forming an augmented tree with many more links between nodes.

Take a binary n-cube by one vertex, and slice off its lower half. Now you have a mangled snowflake or X-tree with many inner (some lateral) connections and many branches removed. Judiciously regularizing these lateral connections will give a variety of (some old, some new) networks.

It is interesting to look at a tree that spans all the nodes in an n-cube (see Fig. 2 for the 3-cube).

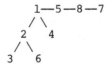

Fig. 2 A tree that spans all nodes in a 3-cube (to get the cube add the links $3-4$, $3-7$, $6-5$, $6-7$, $4-8$).

Compared with a balanced tree, a number of branches have been eliminated. The links not shown (but listed at the bottom of Fig. 2) can be thought of as augmenting, much like "rings" and "threads" but, if anything, with a less regular or potentially useful pattern.

X-Trees, Hex-Trees, N-Trees, Arrays, and Lattices

Let us look more precisely at the relations between trees (usually binary trees), Despain—Patterson augmented X-trees (like hex-trees), cellular arrays, and lattices. First, I redraw a binary tree (Fig. 3).

```
C———C———C
|       |
|       C
|
C———C
|
|
C
```

Fig. 3 A binary tree, redrawn to suggest an array [C = computer (processor + memory store); |, — = links].

Figure 4 shows a similarly redrawn hex-tree, plus another *n*-tree.

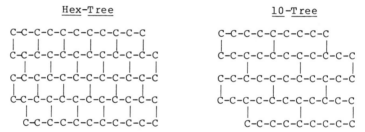

Fig. 4 A hex-tree and a 10-tree drawn like square arrays [C = computer (processor + memory store); |, — = links].

Bulge each hex-tree rectangle into a hexagon (by moving the nodes connected by a vertical bar closer together), giving a typical picture of an hexagonal array—which is a hex-tree! Note that links could be added between every other pair of nodes to give a complete square cellular array. Thus a hex-tree is obtainable from a square array by eliminating half the links in one direction, or one-fourth the total number of links. (A program that processed nearest neighbors would now have to use three instructions to shift information where links were missing, plus more instructions to turn on only those processors missing those links.)

Any of the nodes can be chosen as the "root."

More links can be removed, to give higher-order trees, like the 10-tree shown. This means that some nodes will have only two links to other computer nodes—which, potentially, frees ports for links to other resources, like input and output devices and mass memory stores. All nodes on the edges of the hex-tree have fewer than three links, and could also be so linked.

Take a tree with n^2 offspring and link each group of offspring to form a two-dimensional $n \times n$ array—thus achieving a pyramid.

Moving from an *N*-Cube to an *N*-Lattice

Take a binary n-cube and put new processors on each edge, one to an edge. Now increase the number of processors per edge, as indicated in Fig. 5.

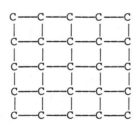

Fig. 5 One face of a binary *N*-cube with processors added to edges [C = computer (processor + memory store); |, — = links].

Now generalize from the edges to the surfaces, as shown in Fig. 6, thus giving arrays of processors on each surface.

```
C——C——C——C——C
|   |   |   |   |
C——C——C——C——C
|   |   |   |   |
C——C——C——C——C
|   |   |   |   |
C——C——C——C——C
|   |   |   |   |
C——C——C——C——C
```

Fig. 6 One face of a binary *N*-cube generalized to an array [C = computer (processor + memory store); |, — = links].

Finally, generalize the surface arrays into the interior, giving a three-, four-, ..., or n-dimensional lattice (a two-dimensional lattice is the same as a two-dimensional array). As the dimensionality of an n-lattice is increased, the added links pull nodes together, diameter decreases, and degree increases.

Now consider other n-figures, e.g., an n-hexagon, n-pyramid, n-sphere, or some other appropriately shaped n-volume. We can add and disperse processors at its vertices, along its edges, on its surfaces, and/or within its body. (Note that the distinction between vertices and edges becomes increasingly indiscernible as we move toward a sphere.)

Processors near the surface would appear to be especially useful for collecting data (as in image processing, perception, and database management), while inner processors serve to process data and to pass messages.

Thus we might consider whether different types of processors might be appropriate for these different regions and purposes. One attractive proposal is that simple 1-bit SIMD processors of the sort used in the large arrays of today might serve best at and near the surface: many are needed, so expense is important; the low-level sensing operations are simple, regular, and local.

The Need to Build and Evaluate Networks, and to Handle Messages

This chapter has compared networks on a variety of formal, empirical, and structural grounds. Relatively few comparisons of this sort have been made, and much more thorough examinations are needed. Networks must be built: The actual performance of networks cannot be compared until meaningful problems and mixes of programs are run through them in empirical tests. They are too complex to be analyzed, and too large to be simulated.

A number of new classes of graphs have been found recently that have good structural properties, especially density. But the problems of choosing relevant properties, and developing techniques for finding good graphs, are just beginning to be attacked.

Arrays differ from other kinds of networks chiefly in, and because of, their much larger size. Their near-neighbor connectivity is strongly suggested by the kinds of picture processing and numerical problems for which they were designed. That is, they are very strongly user oriented, and therefore specialized. Their near-neighbor connectivity is not very different from trees and other kinds of networks. Indeed, they can be viewed as (highly) augmented trees. The use of a single controller, in SIMD mode, eliminates almost all the problems of message-passing overheads—if the programmer is willing to structure algorithms as sets of local functions.

There are a number of striking structural similarities among networks that on the surface appear quite different, which this chapter has begun to examine. These similarities are explored throughout this book, especially in Chapter 14, and used, especially in Chapters 16−18, as the basis for the development of several potentially powerful network architectures.

Certain network topologies are better than others (although the criteria for judgment are not yet well understood). Crucially important are the problems of information flow: message passing quickly dominates. Here several widely held assumptions have enormous effects:

(1) Having the operating system map program into network, allowing and using little or no help from the programmer, invites mismatches that can be overcome only with great flurries of message passing.

(2) Assuming random message passing unrealistically invites the operating system to expect no structure from the programmer.

(3) Networking many traditional serial computers together can only degrade performance, since to all their old inefficiencies are added new overheads for coordination and message passing. Each additional computer can potentially compound the I-O bottleneck of every other computer.

Integrated designs of multicomputers, the size and power of each processor and each memory fitted to their roles in the larger network, can, potentially, eliminate a large amount of no longer necessary hardware. For each processor is now far less important; the whole computer does not depend upon keeping that processor busy. Good mappings of programs into architectures, so that processors work on local information, and information flows with little or no extra shuffling about through the program's processes, can eliminate large amounts of no longer necessary message passing.

PART III

Developing Parallel Algorithms, Languages, Data Flow

CHAPTER 8

Formulating, Programming,
and Executing Efficient Parallel Algorithms

H. S. Stone (1975, pp. 372,373): "The major issue is that of partitioning a problem into many processes that can be executed in parallel on an MIMD computer. For a small number of processors, say two to four processors, this problem is not a significant one because the parallelism available is not significant. For several processors, say sixteen or thirty-two, the problem is extremely difficult. Programs without a specific iterative structure are seldom so complex that they have sixteen to thirty-two distinct sub-processes. Programs with an iterative structure are likely to be better suited to SIMD computers and execute with somewhat lower efficiency on MIMD computers because of resource allocation and synchronization overhead."

This chapter examines techniques for restructuring serial code into more parallel form, and also ways in which we might begin to rethink our algorithms, our programming languages, and even our approach to problems, to make fuller use of network architectures. The next four chapters examine some of the problems of developing whole programs that are complex structures of algorithms, and of designing and using programming languages for parallel programs that are executed on parallel arrays and networks.

The MIMD networks pose extremely difficult problems with respect to developing efficient programs. For they allow asynchronous processes, which means that there is no way to avoid the possibility of one procedure sitting idle for a potentially infinite amount of time before other procedures finish generating the operands it needs. The SIMD arrays are synchronous, since all processors execute the same instruction. But only when all processors can be used fruitfully do they really have near-perfect load balancing and avoid the delays that plague the asynchronous systems. In all cases, the key appears to lie in the programmer's formulation of algorithms that lead to programs that balance the load across all processors as well as possible.

As Stone suggests, today we find that the parallelism is relatively apparent and regular (e.g., an iteration through a FOR...DO... loop), in which case it is best suited to an SIMD array; or it is not of a very high order; or it is quite hidden in places where we do not yet know how to dig.

Flynn and Hennessy (1980) point out that there are three levels at which parallelism can be detected and a program can therefore be restructured to take advantage of it:

(1) the algorithmic specification of the tasks;
(2) the higher-level source language representation;
(3) the machine language specification of state transitions in the host computer.

Programming Languages and Operating Systems
for Parallel Programs

Almost certainly (except for obvious parallel structures, like large arrays), programs must be reorganized at the level of the algorithm to be made significantly more parallel (that is, sped up by more than a factor of $2-8$). And almost certainly the programmer must do this, rather than the compiler or the operating system. ". . . it is mandatory (for efficiency reasons) to take into account the target architectures" (Baer, 1980a). Support tools and techniques are needed to help:

(a) Languages should invite, suggest, encourage, or even (gently) pressure, the programmer to be parallel.

(b) Interactive program-development systems should help the programmer explore and pin down the possibilities of parallelism.

(c) Today we try to train our students and ourselves to be good top-down structured programmers and serial problem solvers. We will need radically new kinds of training and experience in parallel structuring, thinking, and conceptualizing.

(d) The compiler and the operating system can do a significant amount of the work at the machine language and source language levels. But, I will argue, at the crucial level of the algorithm they can at best give the programmer some help. This suggests that the crucial problem is in ourselves: we must learn how to construct parallel algorithms.

We may well (at least for the near future) need to give the compiler and operating system information beyond that required by today's serial programs about the parallel flow of processes. The operating and support system should encourage the programmer to specify the structure of the desired architecture and flow of information, and the operating system and programmer, by working together, can most efficiently handle reasonable amounts of interaction among nearby processors. Information from the program and the programmer can replace the otherwise necessary brute (and expensive) force, of faster I-O processors and larger buffer memories.

The Danger of Excessive Message-Passing Overheads

M. Stonebraker (1981, pp. 415, 417):*

Although we have never been offered a good explanation of why messages are so expensive, the fact remains that . . . most operating systems the cost . . . is several thousand instructions. . . . operating system services in many existing systems are either too slow or inappropriate. Current DBMSs [database management systems] usually provide their own and make little or no use of those offered by the operating system. It is important that future operating system designers become more sensitive to DBMS needs.

Message-passing bottlenecks can quickly become major problems for any kind of multiprocess program. Today's operating systems for MIMD networks take milliseconds to send messages only a few bytes long. For example, using LSI-11/23s (machines whose typical instruction takes 10 μsec), Arachne takes 16 msec to send its standard 40-byte message [see Finkel and Solomon (1980a, pp. 40–56), for a description of the message-passing protocol]. A 1-byte message takes 12 msec. The MICRO operating system for MICRONET (Wittie and van Tilborg, 1980) takes 35 msec for an 80-bit message, using LSI-11/03s. Wittie estimates that a speeded up revision will reduce this figure to 2–10 msec. Spector (1980) appears to have achieved 500 μsec, or less, per message.

*From M. Stonebraker, Operating system support for database management. *Comm. ACM* **24**, 412–418. Copyright 1981, Association for Comuting Machinery, Inc. reprinted by permission.

But if a system takes 5000 or even 100 times as long to send a message as to execute an instruction, interaction among computers must be kept very low, that is, virtually nil; otherwise there will be excessive degradations in performance. The operating system's overhead can too easily throw away all the time gained by parallel execution.

These high overheads are needed because thousands of instructions must be executed to fend off possible overloads and to handle possible contention, synchronization, and worst-case problems. This kind of operating system also assumes that a message might pass from any processor to any other, no matter how great the distance. It makes no use of knowledge about flow of information among procedures. It does not help, or even let, the programmer fit the program to the network. It assigns a program to processors with little or no regard for the program's structure. It takes the same amount of time to send a message between any pair of processes, whether they are executed by distant or by neighboring processors, or even by the same processor.

In striking contrast, the SIMD arrays send messages between nearest neighbors in nanoseconds or a few microseconds at most; that is, as fast as they execute a simple fetch from memory. For example, DAP takes 250 nsec to fetch from each processor's own memory and execute one instruction; it takes no longer to fetch from one of the four near neighbors and execute—which means pass a message (the data) and gobble it up.

Message passing is the way these multicomputers effect I-O. It is at least as important that this be handled efficiently as that the I-O and fetch and store operations be handled efficiently in conventional single-processor computers. There I-O processors, cache memories, and the operating system all serve to anticipate data needs before delays can occur. Such traditional techniques, and also new techniques, are needed for networks.

Attempting to Map Program Information Flow
to Architecture Information Flow

Should the operating system be asked to offer such a wide range of services or, to speed up program execution, should it give the programmer a range of options for help, including the bare minimum? The proper structuring of the program's flow of information (messages) and the proper mapping of that structuring onto an appropriate physical sub-structure of the computer network should make it unnecessary for the operating system to handle unanticipated messages across great distances.

The crucially important operating system service appears to be the best possible mapping of the program structure into the network structure. Otherwise the operating system must make the needed links between processes and data by handling messages while the program runs, in effect building the proper structure virtually, at the very heavy cost of the message-passing protocol.

Bokhari (1981) points out that "it is important to map problem modules onto processors such that modules that communicate with each other lie, as far as possible, on adjacent processors," and shows this "mapping problem" to be equivalent to the very difficult problem of determining whether two graphs are isomorphic. No exact algorithms are known, or likely to be found (except for restricted classes of graphs) that determine isomorphism efficiently, in polynomial time. But actual programs and actual multicomputer networks form a relatively small subclass of graphs. Programs can be coded, and networks can be designed, with one another in mind, and with this mapping problem in mind.

Bokhari explored heuristic techniques for finding better mappings. Kung and Stevenson (1977) have also explored techniques for mapping algorithms into networks. Heuristics, techniques for restricted classes of graphs, and the construction of architectures into which classes of algorithms map well, are all avenues worth pursuing.

An interesting example of an appropriate mapping is Stone's (1971) shuffle-exchange network used for reconfiguring. It is a direct embodiment of the computations used for the fast Fourier transform (Pease, 1968).

Without such a reconfiguring capability, arrays like CLIP become enormously inefficient at such tasks.

These structural issues become clear the moment we think of our programs as structures of processes that transform information flowing through them, as in network flow (Ford and Fulkerson, 1962; Lawler, 1976) or data-flow (Petri, 1962; Dennis, *et al.*, 1980) formulations. Brantley *et al.* (1977) from this perspective found that C.mmp must be given programs with low-interaction data-flow graphs. Just as materials should flow smoothly through a factory, so, ideally, information should flow through a multicomputer. Its structure should invite good mappings; the programmer and the operating system should achieve, use, and respect good mappings.

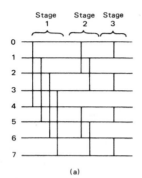 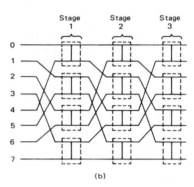

Fig. 1 (a) FFT communication structure; (b) shuffle realization. An eight-point fast Fourier transform (FFT) has three stages, i, $i = 1,2,3$, that combine data from two nodes differing by 2^{3-i}, as shown in (a). (b) shows the shuffle-exchange network superimposed over a topologically equivalent redrawing of (a).

Converting Our Programs and Ways of Thinking from Serial to Parallel

To handle really large problems, we shall soon be forced to convert our ways of thinking to a more parallel mode. We must also be willing to change our algorithms. That does not mean we must be willing to compromise and settle for less. Often we shall arrive at a more powerful algorithm for a larger set of problems.

In almost all cases we have a problem that we want to attack. Occasionally we can pose that problem precisely and definitively, and develop a solution procedure that we can formulate and code for a computer. In such a case the situation is rather clear-cut, since it becomes a matter of developing more parallel, but entirely equivalent, algorithms.

But much more often we have a fuzzy problem, really a whole set of problems or a whole realm of obscure and unknown issues about which we would like to know more. We want to better understand and predict the weather. We want perceptual systems for robots to recognize, grasp, and manipulate objects. We develop programs to hack away at these problems, to give partial solutions, or simply new information and greater insight. In such cases we should go back to the problem set, to our realm of inquiry, to start afresh in developing more parallel approaches to the problem. Sometimes we shall succeed only at the price of losing some of the information that our original algorithm provided. More often we shall develop a parallel algorithm that gives us more, although sometimes more of *A* but less of *B*.

Let us look at some obvious and some not-so-obvious examples of situations where a traditional serial program is probably executing what is really a parallel process, or could be converted into a parallel process with little or no loss of information, and possibly with gains. These include

(a) restructuring a loop, even when it contains many conditional decisions and is ended by a conditional decision, into a parallel procedure;

(b) restructuring long sequences of relatively independent code that could have been sequenced in a variety of different ways into more parallel code;

(c) reformulating a procedure as a tree of processes, and restructuring this tree so that it has a serial depth plus a well-balanced parallel breadth at every ply that best fits the constraints of time and the architectural structure of the available hardware.

Today this is a very difficult enterprise. And it might well be that the underlying amount of parallelness that can fruitfully be exploited with more parallel programs for more parallel computer networks is not nearly so great as people like me would like to believe. (I think there is a very strong argument from human perception and cognition, which—conclusively in the case of perception—takes amazingly little time, using at most from thirty to a few hundred serial steps in what is, essentially, a highly parallel process. But this is an empirical argument, and we do not know enough about the human mind/brain and how it works to be able to weigh its import.)

There is certainly at least a substantial amount of parallelness in a surprising variety of problems. The proper total architecture for a computer network is almost certainly one that can assign processors to work on a program in parallel when that is indicated, and in serial fashion when that is indicated. Both are needed; only the future will tell how much each is needed for what.

Converting Serial Iterations into Parallel Procedures

When a program has a FOR...DO... loop, so that it iterates through all the instances within the range of that loop, and no conditional tests can be used to break out of that loop (this is the case with a well-structured language like Pascal where GOTOs and EXITs are discouraged), this FOR...DO... loop can quite straightforwardly be converted into a parallel procedure, with a different hardware processor assigned to each loop in the iteration. (This is not always true, since the code within the loop can change things for subsequent iterations; but that is the dangerous kind of clever trick best forbidden.) If there are more loops than processors, the processors can work in parallel, but iterate as many times as needed. [For example, if the serial program would need 300,000 loops and there are only 4000 processors, 71 iterations of this set of 4000 would be needed (the last iteration using only 2000 of them)].

A FOR...DO... loop is simply a convenient structure for expressing and handling a parallel process on a serial computer, so this is an obvious observation. But it is an important one, as FOR...DO... loops are very common. More important, we can often recode some other structure (e.g., a REPEAT...UNTIL... or a WHILE...DO... loop as a FOR...DO... structure with little if any loss, and often with a simpler and clearer resulting program.

This in itself is of sufficient value to justify the development of parallel computers, since many programs make extensive use of FOR...DO... loops that iterate many many times. This of course includes programs that work with large arrays, whether of picture images or numbers. But it also includes many other kinds of programs.

Loops That Can Often Be Reformulated into Parallel Processes

Let us look now at loop structures that cannot so easily be converted to parallel processes, but often are promising candidates.

Often there are two major approaches to a problem that involves the exploration of a large body of information (e.g., a picture image, or a large database, or a traveling-salesman network), or of a large set of potential deductions (e.g., a lookahead tree for chess or Go, or a problem in logic):

(a) The program might make a serial search through the body of information, or through the tree of possibilities or deductions. At each step in this search it might examine and evaluate what has just turned up to see whether this is the goal it is looking for. The program might further reevaluate its search procedure in light of this new piece of information, redeciding what to do next.

(b) The program might look at the whole set of alternatives at once (as when an array examines a whole image), examining all in parallel, checking whether each is what is being looked for.

I have described two extremes of completely serial versus completely parallel processing. Usually the programmer must strike some compromise that moves these two extremes toward one another:

(a') The cost of a full reevaluation of the search strategy is too great to do this after each new piece of information is examined. Rather, only a quick-and-dirty reevaluation may be made, or this reevaluation may be made only after every Nth new piece of information has been gained.

(b') The physical array may not be big enough, so subarrays are processed in series.

This suggests a reconciliation of the two approaches:

(c) As many of the pieces of information as can be looked at appropriately by the parallel set of processors at the programmer's disposal should be looked at. Then the program should check (again using parallel hardware) whether what is being looked for has been found. Then the program should evaluate (again using parallel hardware working on all the information that has been gathered in this iteration) what has been found, and redecide what to do next.

The size of the chunks the program bites out of the problem should be determined by the size of the parallel hardware at its disposal for each step, by the amount of information the program needs to assess to make a valid decision, and also by any real-time constraints within which the program must finish executing. The programmer should look at all the steps together to see if they are compatible (in size, in mapping of data and program to processors, in flow of data from one step to the next).

Such an exercise is not radically different from traditional practice. We often, especially when we try to optimize a program and make it as efficient as possible, decide to iterate through an entire set of data and only then test, rather than make some relatively expensive test at every iteration. With parallel hardware the iteration is made in parallel, in one step. Then the test can be made everywhere (sometimes this is more appropriate than a serial sequence of tests, sometimes wasteful).

Such a procedure may seem wasteful to the traditional serial programmer. Often many pieces of information will be obtained when more serial testing for the needed piece would have ended processing and made that unnecessary. But repeated reevaluations of the situation, conditional tests, and intermediate decisions are often a very expensive (in terms of time) addition to a program. Nor is it even clear how useful have been such careful repeated attempts to tailor processing to the information as it emerges. I would argue to the contrary that procedures that try to do serial edge following, region growing, or other perceptual tasks spend a great deal of time trying to decide what to do next, are often wrong, and even today do not do as well as more parallel programs.

The serial program takes a lot of time deciding and redeciding. The more parallel program spends time (when executed on a serial computer) looking at subregions that the serial program (sometimes correctly, sometimes wrongly) decides not to bother looking at. Sometimes the serial program will reevaluate or redo something, wasting time recomputing something that the parallel program would have stored and kept around (using, sometimes wasting, space). Thus there are trade-offs, and there is probably no "right" way; rather, the whole set of resources at one's disposal and their associated costs should be taken into account, and an algorithm developed that uses them as efficiently, as economically, and as wisely as possible.

Exploiting the Parallel Aspects of Functions

Expressions in a programming language in the form of (nested) functions can often be converted to more parallel forms. Basically, we can convert each function into prefix form, and convert the whole program into a single function in prefix form. For example, the simple function $(A + B) / (C * (D + 1))$ can be converted to "polish prefix" form: $/+AB*+D1C$ [i.e., divide(add A B)(multiply(add $D1$)C)].

This can also be represented as a tree that specifies the flow of data through processes (see Fig. 2).

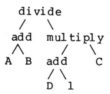

Fig. 2   A tree representation of a set of processes.

This function can be executed by starting with the data at the bottom buds of the tree, applying to them the procedure specified by their parent nodes, applying to the result the procedure specified by their parent nodes, and continuing in this fashion until the procedure at the root of the the tree outputs the final result.

We can map such a tree directly onto a parallel—serial tree of processors, and then flow the data through them in the manner just specified (see Mago, 1980).

This is a simple and attractive way to convert from serial to parallel. But, as we can also see from the example I have just given, it will not (usually) give any very high degree of parallelness. We can assign a different processor to work, in parallel, to each of the processes at the same ply (that is, the same horizontal level/row of the tree when drawn with a vertical axis from the root to the center of the buds).

A number of procedures have been arrived at for converting arithmetic expressions to equivalent forms that reduce the height of the tree needed to execute them, increasing the parallelness at each step (see Kuck, 1977, 1978; Sameh, 1977). The laws of associativity, commutativity, and distributivity are sufficient to transform expressions to close to the lower bound. For example: $x = a(b + c(d + e(f + gh)))$ can be transformed (by distributing its atoms) to $x = ab + acd + acef + acegh$. Four processors can compute this expression in five time steps. But the expression can be cut and distributed to give $x = a(b + cd) + ace(f + gh)$, which needs only three processors and four time steps.

An arithmetic expression with n atoms (constants or variables) can be evaluated in at fastest $O(\log n)$ time steps using $O(n)$ processors. But n is rarely more than 2 or 4 in most arithmetical expressions. And, to quote Stone (1975, p. 373), "these techniques do not consider the high overhead of process synchronization and processor reassignment. . . ." But as programmers are trained and helped to code arithmetical expressions more amenable to automatic parallelization and the synchronization problems are overcome, these procedures may give significant increases in speed. Lee (1980) has empirically determined that with unlimited numbers of processors the increase can be $O(p / \log p)$.

Specific Kinds of Algorithms That Have Been Formulated to Date

A wide range of parallel algorithms have been explored. See Kuck (1978), Kuck *et al.* (1977), and Kung (1980a) for numerical algorithms; Tanimoto and Klinger (1980), Preston and Uhr (1982), and Hanson and Riseman (1978), for perception programs; and Boral (1981) and Fishburn (1981) for database and artificial-intelligence problems.

Kung (1980a) examined a variety of algorithms for MIMD, SIMD, and systolic systems, when structured as linear arrays, two-dimensional square or hexagonal arrays, trees, or shuffle-exchange networks. He comments:

Systolic algorithms are most structured and MIMD algorithms are least structured. For a systolic algorithm, task modules are simple and interactions among them are frequent. The situation is reversed for MIMD algorithms. Systolic algorithms are designed for direct hardware implementations, while MIMD algorithms are designed for executions on general purpose multiprocessors. SIMD algorithms may be seen as lying between. . . . Using the central control, SIMD algorithms can broadcast parameters and handle exceptions rather easily.

Kung points to the success of pipeline algorithms for arithmetic in the Cray-1 and other supercomputers, and describes several interesting examples of systolic algorithms for filtering, matrix multiplication, and other operations using two-dimensional square and hexagonal arrays. He finds trees most suitable for searching and shuffle-exchange for sorting. MIMD systems work best when problems can be decomposed, each part a reasonably large and independent module.

Gannon (1981) explores the use of an array plus an omega network to converge to smaller arrays (much as in a pyramid) for adaptive algorithms for partial differential equations.

Functional Applicative, Production, and Data-Flow Languages

*Functional Languages as a Possible
Basis for Parallel Languages*

It may be of value to use an "applicative" language (that is, a language composed of functions), as a framework within which to develop parallel languages. Backus (1978), in his FFP language, Friedman and Wise (1978), and Flynn and Hennessy (1980) have made proposals in this direction (see Appendix E), suggesting that such a language might be based on LISP (McCarthy *et al.* 1962) or APL (Iverson, 1962). [LISP is based on recursive functions (Kleene, 1936) and the lambda-calculus (Church, 1936); APL's basic data structures are two-dimensional arrays.] A functional language can have a very clean structure, without GOTO or assignment statements (which Backus calls the "von Neumann bottleneck," since they limit the flow capacity and bandwidth of serial computers and languages). Rather than iterate (as in a FOR loop) it handles structures recursively (that is, with functions that call functions including themselves).

Both recursion and iteration are serial processes. Both must make a conditional test after each step, and go through a good bit of business to increment addresses and (in the case of recursion) again call and set up the function. In sharp contrast, parallel code executed on a parallel machine effects in a single instruction the entire iteration or recursion over an array or other set of data.

It seems most reasonable to handle such parallel processes with a new kind of construct, one that instructs all processors to execute. Iteration and recursion are often simply techniques for simulating parallel constructs when there is no parallel computer. Our parallel constructs should reflect and reveal their parallel structure, not the vestiges of the serial constructs used to simulate them. It is not clear what will turn out to be the best expression of parallelism. But a FOR loop seems, to my intuition, to have a more parallel look than does a recursive procedure. Today's first-generation parallel languages use a FOR — ALL loop, and/or assume that operations are executed over all members of a set of data that has been declared as a parallel type. It is not clear whether these constructs can be improved.

LISP may not be the best basis for a functional language, since it so strongly encourages serial searches and recursive programming (or at least it appears to do so, from the kinds of programs almost every LISP programmer writes). A good language should encourage and tempt one to be parallel, offering parallel data structures and suggestions as to when and how to use them. As Arvind *et al.* (1978) point out, "Pure LISP . . . and FFP languages . . . even though their semantic bases are elegant and functional, neither caters to asynchronous and history sensitive operations."

Reeves (1979) embedded a language for parallel arrays in APL, which is not as "pure" a functional language as is pure LISP (but both languages have been extended and made "less pure" to make them more easily usable). Since APL was designed explicitly for parallel processing (albeit concentrating on numerical processes) on arrays of numbers it gives the programmer primitive data structures that are parallel, and primitive parallel operations for transforming data so structured. Since it was designed to handle arrays (albeit using serial computers) APL's parallel constructs can quite nicely be mapped onto hardware arrays.

This may well be a key—to give the programmer procedures for effecting parallel processes on whole structures of data (that can be processed in parallel). Then a physical network with these structures (or reconfigurable to these structures) can quite straightforwardly execute these procedures in true parallel fashion.

Extending IF−THEN Production Languages for Parallel
Systems

Almost nothing has been done to extend production languages to execute parallel code, and I am aware of no work at all implementing production languages on arrays or networks. But this type of language may well be appropriate. Production languages are based (more or less loosely) on Post productions (Post, 1943), which give an alternate but equivalent formulation of Turing machine (1936) universal computability that underlies the general-purpose computer. SNOBOL (Griswold *et al.*, 1968) is probably the most powerful example of a production language. Waterman (1975) and others have been developing simpler, more modularized production languages without the control structures (including the many GOTOs) of SNOBOL.

Essentially, a production system looks at a field of information for some specified pattern. IF the pattern is found, THEN some set of actions (including changes to that pattern) are effected. Almost all production systems look for ordered strings, apply only one production at a time, and look only for one location where that production succeeds. But in a powerful language like SNOBOL this is a sufficient basis for conveniently looping through all occurrences of one production, within an outer loop through a whole set of productions.

Forgy (1977) explored how to speed up a production system (but one that still has the above serial restrictions) using parallel hardware. Uhr (1979c) has developed a prototype system that applies a whole set of productions in parallel.

The next step (which it appears nobody has yet taken) would apply each production in parallel to all subsets or regions of some set of data (e.g., an image array or a modularized database). The programmer could be given the capability to declare (more or less arbitrary) graph structures (including arrays and records) as data types, and to designate subgraphs as "window" data types. Then a parallel construct could look through that window positioned at a specific location for a parallel set of variable values, or/and look through that window positioned everywhere as indicated by a step size. (See Chapter 11 for descriptions of languages that are beginning to have some of these capabilities.) Possibly the most serious problem with such a system is helping and forcing the programmer and the program to load balance, so that asynchronous processes will not idle too high a percentage of the processors.

*Languages That Express the Parallel Flow
of Data through Structure*

Data-flow computer systems are being explored by Dennis *et al.*
(1980; Dennis and Weng, 1979), Arvind *et al.* (1978), and others, and
languages are being developed (see Appendix E) within which one can
express algorithms in terms of the flow of data through processes (see
Ackerman and Dennis, 1979; McGraw, 1980) much like Petri (1962,
1979) nets. This is a very attractive approach, since it attempts to
structure programs explicitly in terms of the flow of information
between procedures or, to express the concept more suggestively,
through structures expressly designed to best handle that flow. The
structure of procedures minimizes data flow. Ideally this should be
mapped onto an appropriate multicomputer architecture where flow of
data and other messages between processors similarly remains minimal.

This suggests a physical network of computers whose interconnection pattern mirrors the flow of data. But that would be a very special
topology, since virtually every program would have a different flow. So
the actual data-flow computers that have been built or proposed instead
store each instruction of the program, append its operands as they are
input or computed, and (when all operands have been got and this instruction has therefore been activated) assign this instruction to a processor, which outputs the results (which are in turn distributed as
operands to further instructions). Thus at the "anatomical" level of
the physical machine the topology of the "physiological" information
flow is destroyed. That is, the physical machine is structured as a complete bipartite graph, in typical multiprocessor serial computer fashion.
And it appears to suffer from the additional overheads of storing each
instruction's results in memory (as operands for subsequent instructions) rather than keeping them when possible in the CPU's high-speed
registers and fetching sequences of instructions that transform them.

Unless a data-flow system has the appropriate structure at the physical level of hardware, operands and procedures will have to be shuffled
around as though through a random topology, with excessive message-
passing delays. There are several possibilities worth investigating: A
relatively general topology, like an array or tree, might be explored to
see how well typical mixes of data-flow graphs can be embedded, and
data-flow languages might be structured to make programs written in
them more easily mapped onto this topology. Limited reconfiguring
can also be used to improve this mapping. Kung's (1980a) systolic arrays, in their elegant pumping of information through algorithm-
structured hardware, and also the proposed pyramid/cone and lattice architectures, may well be closer to the spirit of data flow.

Language Development Systems to Specify
Data Flow through Structures

Let us assume a physical network that has a topology relatively close to the topology of a typical program we might ask it to execute. It might be a partially reconfigurable network, with the programmer and/or the operating system specifying the desired anatomical structure. Or it might be one of several specialized networks, with the program written expressly with those structures in mind. What kind of language, and language-development environment, might we want?

Best might be an interactive graphics scope-oriented language that would encourage the programmer to draw the tree of processes directly on the scope. For example, the system could begin by showing the programmer the features of the architecture—the number of processors available, their structure, links, speeds, etc. If reconfiguring were possible, the system could describe the range of possibilities, and give examples, as requested. The system might then prompt and help in the construction of an appropriate structure.

We therefore can write the actual code in terms of these processors, e.g.:

```
PROCESSOR A1;
CONST
   .
   .
   .
END;
PROCESSOR A2;
   .
   .
   .
```

Among the programmer's rules of thumb will be

(1) Find out how many physical processors there are, their speed and power, their interconnection architecture, and the extent to which they can be reconfigured (and the costs of this reconfiguring).

(2) Within the limits imposed by the hardware, choose a serial depth of layers and a parallel breadth within each layer compatible with the program (that is, that executes the program fast enough but with input, output, and each processor spending as nearly as possible the same amount of time, so that all processors work as much as possible).

(3) This structure, and the assignment of sets of processes to the processors, must be orderly (in the sense that the output from one processor can input to the appropriate processor at the next level in an orderly fashion, with minimal message passing over minimal distances).

(4) Try to develop a program whose processes can be executed directly, with minimum information flow, by a processor structure that can be realized on this multicomputer. That is, the programmer should try to generate an algorithm that gives isomorphic algorithm−graph and architecture−graph.

(5), (6),... A number of more specific rules of thumb about how to convert FOR...DO... and WHILE...DO... loops and other common structures for serial programs into appropriate parallel formats. These should be printed for programmers in a manual and also, possibly, displayed on the terminal at the start of an interactive programming session (unless the programmer chooses to turn down the level of verbosity). Each should also be displayed when the compiler decides that the programmer needs advice of that sort.

On Fitting Problem, Algorithm, Program, and Network to One Another

The programmer should structure the algorithm so that the program's procedures map nicely onto the hardware structure of processors. The operating system should take advantage of this correspondence. It should allocate the program properly, for efficiency, and eliminate the overhead of more general (and slower) methods that are needed when program and hardware structure do not fit well. It should warn the programmer when there may be problems, e.g., in excessive message passing between distant nodes, or in contentions or poor load balancing and idled processors.

Since a single loop on one processor can stop all the other processors in an entire MIMD network, the operating system must have options that terminate overly long or potentially endless loops. Programmers should be told how this is handled, and how to handle it themselves, e.g., by successively approximating and saving key intermediate results so their programs can recover reasonably gracefully. The programmer should thus be given the opportunity to improve things, and information that might be useful in making possible improvements.

At least for the present, a background stream of low-priority jobs might be the simplest way to sop up idle processor time. Each of these might best be executed by only one processor, in traditional serial mode. Sets of processors could then be assigned for parallel execution of one foreground program that is given top priority and virtually complete control.

Restructuring Our Ways of Thought

Hardware is becoming very very cheap. To the extent that we move toward harder and harder problems (which need bigger programs with a larger total number of instructions), time becomes more valuable. To the extent that we develop a tool powerful enough to be used to handle more and more problems with stringent real-time demands, time becomes more valuable still, often even crucial. So the curves of the future would seem to point strongly in the direction of the parallel solutions.

I would even argue that, because human time becomes increasingly valuable and costly, while computers become cheaper, parallel computers are again indicated—because they are far easier to program. But this is not the case today; it will only become true after we have developed better languages and interactive systems for programming and debugging parallel programs, and have changed our ways of thinking to a more parallel mode.

I strongly suspect, however, that parallel algorithms and programs will turn out to be more "natural" (in the sense that they much better fit the parallel structure of the underlying set of processes and the flow of data through that structure) and much "simpler" (in that they will lead to much shorter, and often more straightforward, programs). People like Michael Duff and Stanley Sternberg, and others who have worked in their labs with CLIP or the Cytocomputer, or with other parallel computers, argue that once one has got over the humps of using such systems, parallel programming becomes much easier than traditional serial programming. From my personal experience I tend to agree.

As we learn to formulate parallel algorithms and write parallel programs, shall we also learn to think, at the conscious as well as the unconscious level, in more and more parallel modes?

CHAPTER 9

Development of Algorithm−Program−Architecture−Data Flow

This chapter explores the kind of well-structured, slavelike, drone-like organization that, at least in our present state of ignorance, seems the best hope for getting networks to do as much work as possible. [Appendix F surveys a variety of more interesting but for the present less appropriate ways in which groups (of neurons, animals, human beings, and computers) have been, and might be, organized.] It then examines how the pattern-perception problem can be formulated for and conditioned to a converging cone- or pyramid-shaped network, as an example of how we might design program and architecture together. The next chapter looks at other examples of processes that might fruitfully be reformulated for parallel networks.

Developing Program, Mapping onto Network, and Program Flow

I shall posit a dogma (one that I am inclined to believe at the moment, but one that I hope will turn out to be false), and sketch out an approach to programming multicomputer networks that follows from this position.

The dogma: Since getting many processors to execute a single program efficiently is extremely difficult, we must ask the researcher/ programmer to do as much of the work as possible, and to give the system as much help as possible. This further means that the programmer should be fully aware of the network and how it works, as well as the programming language and operating system and how programs are mapped into and executed by the network.

The problems in developing parallel algorithms that map efficiently onto suitably structured networks are so great that the programmer, who knows more about the problem and its structure than anybody else, should be drafted to help. But note that I am talking about a research programmer trying to develop good new parallel algorithms and techniques. All programmers should be given as much help as possible, and applications programmers should be burdened as little as necessary.

Good Interactive Programming Tools
  to Ease the Increased Burden

It is not clear whether this is asking too much of programmers, in an era when the dominant trend (for good reasons) is to try to make things easier. But, at least until we have made some progress in understanding networks and how to program them efficiently, this may just be necessary. Good interactive facilities that give very powerful and satisfying immediate feedback can be used to lighten the programmer's burden.

To give an example, the programmer should try to think through whether enough memory store is available for the intermediate results associated processors must generate. Otherwise the compiler should be able to output comments of the sort "Warning; It looks like you are now assigning too many variables to this array of stores. If so, the delay in swapping will be approximately x msec." The delays will usually be no worse than those we typically endure when traditional computers swap memory in and out. But arrays and networks can (potentially) execute so many thousands of times faster than traditional computers that these delays become incommensurate, and quite intolerable.

The Crucial Need to Develop Our Powers of Parallel Thinking

The greatest problem of all lies in changing our ways of thinking, to a more parallel mode. This is the key reason that we should involve programmers with the network and its structure, asking and forcing them to consider together problem, program, network structure, mapping of program onto that structure, and flow of data and other kinds of information through that structure. I strongly suspect that thinking about the architecture's structure can itself become a powerful tool for developing more powerful algorithms and programs.

Just as many of us "think in English" we should someday reach the point where we shall "formulate in silicon." That is, we shall move from problem to an anatomy—physiology formulation that embodies our solution procedure into an architecture through which an algorithm flows information.

Parallel versus Serial versus Appropriately Structured Parallel—Serial

Under many circumstances, n processors working in parallel can execute n instructions, each needing k moments of time, on n different sets of data in k time.

For example, each processor can examine a different cell or window in an array containing an image or a three-dimensional model. An image enhancement program can execute a sequence of i instructions on an $x \times y$ image in ki moments of time using a multicomputer with $n = xy$ processors. A serial computer would need $ckixy$ moments (c is a small constant for the iteration overhead). For a parallel computer with fewer processors than cells or regions of data to be processed, where $n = axy$, $ckia$ time is needed.

Each processor might be executing a different instruction; that is, this might be either an MIMD or SIMD multicomputer. Both serial and parallel computers need the same total memory and the same input and output capabilities. The serial computer, since its processor is linked to every word in memory, needs a bigger address register and controller. Thus the parallel computer needs one moment of time and axy area (a is the area of one node). A serial computer needs $ckin$ time and axy area.

So long as there is no interaction among data at distances greater than the window can handle, such increases in speed are possible. But most problems appear to need interactions at greater distances (at least as formulated today, with serial computers in mind). For example, local window operations might detect simple features which then combine into complex features or objects. Or the program's job might be to output not the entire enhanced image but only the coordinates of cells that meet some criterion (e.g., have been identified as containing red, or smoke, or angles, or buildings).

To choose one or more items of interest (e.g., the n most strongly implied items in a scene, or all implieds above some threshold), a serial processor needs $lcxy$ time (l is the time needed to choose one item). A parallel computer needs $\log_d xy$ time (d is the degree, that is, the number of links fanning in through a window).

In many programs processors search through large sets of data for items of interest, sort or rearrange large sets of data, or compare, combine, or cull large sets of data. In all these cases there is a need for several pieces of data to interact and often (at least with today's typical serial algorithms) these data may be relatively distant from one another. When these data are ordered (or can be assumed ordered) the maxima can be found much faster.

Karp and Miranker (1968) showed that, for typical single-item search problems which serial techniques like binary search handle in log n time, parallel techniques can at best give log n speedup. "Minsky's conjecture" (Minsky and Papert, 1971; see Thurber, 1976) that n processors can achieve only log n speedups is based on assumptions that one item is to be output as a function of a whole set of data, which must therefore interact with one another. But these are basically serial optimum-solution-finding assumptions.

The problem appears to arise when data must be reduced to a single output, and processors can work on only this one program. Although all processors may be able to work at enhancing or otherwise transforming the different pieces of data in a large set, the moment data or results must be compared, chosen among, or in other ways examined for interactions, not all processors are needed. If the data are to be reduced (typically by many-to-one "fan-in" up a tree) to one, or even a few, final results, successively more processors become idled.

There are several ways of avoiding this situation:

(1) Processors can multiprogram. Possibly simplest is to give each processor a stream of background jobs to be handled serially.

(2) Algorithms can be designed to take advantage of the largesse offered by parallel systems. Often we look for a single maximum or solution because we cannot afford the luxury of exploring other regions of our data that might be of interest. With parallel networks we can write programs that find many local maxima and other kinds of results of possible interest.

(3) Image enhancement programs output the whole enhanced image, because that is the most useful representation for subsequent human processing. Similarly, the results of a scene description or modeling program can be output in array formats that may well present more information more informatively. The array might be output with

objects and their features named right where they were perceived, instead of today's typical output of a list of objects and their coordinate locations. The behavior of the working model might be exhibited directly, by displaying its successively produced array states.

(4) The multicomputer might be configured to fit the algorithm without idling processors. For example, if data in a large array must be compared, reduced, combined, or in other ways interact, they can be passed to suitably smaller arrays or other structures of processors, and the original processors freed either to continue pipelining continuing streams of data or to turn to other tasks. This broadcasting and combining of data can now be effected with no cost at all if, as in the formulation we are about to examine, transforming operations are also effected on these data, which were moved together for that purpose also. Thus fan-in can serve processing at the same time it serves what might be thought of as the bookkeeping functions of broadcasting, sorting, choosing, and outputting.

This suggests restructuring the problem and formulating the program to fit such a structure. The following is an example of such a combined program—architecture formulation.

An Example Formulation of a Program for a Parallel Pyramid Network

I shall sketch through an example that is one I have thought about a great deal. I should mention that my approach to the problem, the kinds of programs I have formulated and coded, the kinds of arrays I have programmed, and the kinds of network designs I have explored (and discuss in this book) have all had a strong influence on one another.

I want to input sequences of images (as from a movie or a TV camera) and give a running description of the changing scene. (Other parts of the network might then drive motors that moved wheels or/and arms and hands to go to and manipulate objects the system recognizes and decides are worthy of such actions, or/and drive printers or voice synthesizers to output the description; but I shall leave those issues for another time.)

I shall take the role of a programmer using a (semifictional) programming language for a network. (Note how I attempt to design the program, the computer network, and the flow of information through that network all together, and how the design of each affects the others.)

First, I choose an input array for the images. If I can assume a large network from which I can carve out whatever I need (costs be damned), I shall think in terms of the expected sizes of the objects that need to be recognized (not every speck of dust, but occasional leaves and mosquitoes as well as trees and elephants). Then my personal preference is to choose an array size that fully resolves (for a human eye) the smallest of those objects, but without much additional detail, in a total field large enough that they will remain in the field for a reasonable time no matter how fast they are moving and that there will be a sufficient number of other objects to serve as a background/framework.

This guarantees that the needed information is present, but as simply and as locally as possible, thus reducing the computing burden as much as possible. (Many people would prefer higher resolution. That is one factor leading to the many different programs a system must handle—I here give only one relatively concrete example.)

I have raised a number of very complex issues, and for exotic scenes one might have to give them a great deal of thought. But if our scenes are of the ordinary everyday sort—streets, farms, living rooms, gardens—the field of vision that evolution has developed for human eyes, and the fields we have developed to (roughly) reflect that, in TV cameras, strongly suggest numbers like 256×256 to 4096×4096.

Let us assume that I am satisfied with 512×512 for my input array. [I choose such a standard number partly because I suspect that the total system offered me will have several magic numbers like 64, 512, and 1024, which if chosen will enormously speed up processing, and partly to suggest that I do not think it worth spending much effort trying to decide between 512×512 and, say, 390×416 for a "best" resolving power—it would not make much difference, and I hardly know how to go about choosing (except by using a program that tries to zoom and adjust to an appropriate size; but then I would have to start with some initial size, clearly now without that a priori guess being very important).]

I want the system to load the raw image into an input buffer store. The store should be large enough to handle full-color images. Again, my personal preference is to not resolve the image at a finer level of detail of intensity than is needed to contain the useful information, so I would be happy with $12-24$ bits of storage for each pixel, $4-8$ bits for each of the three primary colors. Thus I want a $512 \times 512 \times 24$-bit input buffer store.

Next I want an array of simple SIMD processors arranged so that each can input from a very small, compactly connected subarray of this largest array. Here I opt for a 256×256 array of processors, each with good access to a 3×3 subarray of the input buffer store and a convenient store into which it can output intermediate results (either for its own subsequent input, or so that it will be conveniently accessed by the next processor down the line).

This suggests a pyramid. It would best have each 3×3 array of stores input (through distributing registers) to the processor lying directly behind its center store, but with processors lying directly behind every other store. That is, a 256×256 array of processors lies behind the 512×512 array of input buffer memories. Thus the raw input image will be reduced to one-fourth its size after this array of processors has finished working it over, and there will be a reasonable amount of overlap of the local 3×3 windows during this set of transformations.

Therefore we now have a structure of 256×256 processors sandwiched between a 512×512 input buffer store and a 256×256 (output) store. The stores for the raw input image might well need only 12 bits of memory, but 64 bits would be convenient, if cheap. The higher-level stores might usefully be employed to store several different pieces of intermediate results/information, e.g., about color, local texture, and several differently oriented local edges. So I would be inclined to have at minimum 128 bits or, preferably, 1024 or 4096 (or more!). I would be happy with the today-traditional 1-bit processors. They are perfect for logical feature compounding, and adequate for (variable precision) bit-serial arithmetic. (But 4-, 8-, 16-, or 32-bit processors, if economic, would certainly work.)

I have now described the first layer of a converging pyramidal system. This should continue for several more layers: Sandwich a 128×128 layer of processors behind the 256×256 array of memory stores, so that their information is input to these processors, just as the original 512×512 input buffer stores input to the first layer of processors. This layer will output to a 128×128 array of stores. Continue in this way.

(I have a slight desire to see each successive store a bit larger and, possibly, each processor a bit more powerful. So I might, if I as programmer could really design my subsystem completely, as would be the case with a radical reconfiguring capability, choose 32-bit, then 128-bit, then 512-bit, then 2048-bit memories, moving inward through converging layers. But my final choices would depend upon extensive test runs and much tuning of the program, so it is probably simplest to ask for

2000 or 4000 bits of store at every level, except for the raw input array.)

Several other issues should be considered here. We do not merely have to use all the processors we ask for; we have to keep them busy and use them efficiently, but without overloading either processors or memories. In the program I am sketching out the overriding factor (the "real-time" aspect) is the speed at which the images are input to the system.

Today's TV cameras (that use U.S. standards) take one picture every 33 msec. To resolve the motion of ordinary real-world objects moving at normal speeds, think of a range between 20 and 200 msec per image. We can easily input the image to the input buffer stores in far less than 33 msec. We can also input the image in parallel with the processing of that image by the first layer of processors (and all the other layers, all of which will be working in parallel). So each processor has 33 msec in which to execute the sequence of processes programmed for its layer. If the sequence takes less than 33 msec, the processors at that layer will then sit idle until the next 33 msec period begins.

It is therefore important for efficiency to distribute the computing load equally over the different layers of processors, in as close to 33 msec bursts as possible. That may well be the most important consideration in deciding exactly how to decompose the total sequence of processes to be effected by the whole program, and exactly which layer of processors to assign to any particular process.

Here we are in a strange predicament. Thirty-three msec is a very long time when a useful and powerful process can be executed in only 1 or 3 μsec, or even less, over the entire image! It may well be that we have jumped from having far too little time to having far too much! For we literally have enough time to execute from 1000 to 100,000 instructions at each layer on each image input.

We shall not really know whether this is true until we have extensively coded, tested, and tuned programs of this sort. Nor is this anything to be worried about; on the contrary, it will be great if we find we have achieved computer systems powerful enough to handle this very difficult and very important class of problems.

But if indeed such a network has more than enough time we can do several things to save money by making it cheaper, and therefore be able to use it for a wider range of applications:

(1) We can make it smaller, and assign subarrays to each processor (this means each of the [smaller number of] memories will have to be made bigger).

(2) We can use slower technologies (if that saves money).

(3) We can definitely use 1-bit rather than more powerful processors.

(4) We can multiprogram the processors, having them run other programs during their idle time.

At some point I should want to assign MIMD processors to the much-reduced array of relatively high-level information that the prior layers of processors had achieved. This again would depend heavily on my knowledge of the size of the total system, the capabilities of each type of processor available, the overall structure connecting and configuring the different types and regions of processors, and the costs of using the system with respect to my ability to pay these costs.

The MIMD processors would be very useful for a variety of purposes; I shall mention only the problems in following up the clues already uncovered at earlier layers; their suggestions as to specific types of objects that appear to possibly be present in the different subregions of the scene; and the computation of trajectories and development of predictions of where objects will be next, and what these objects might be doing.

So I specify that once a layer with only an 8×8 array of 64 processors is reached, these should be MIMD processors. Each should be a 32-bit processor: Each will need a separate chip most of whose area will be devoted to memory, so the savings from 16-bit or 8-bit processors would be negligible. If I know that the network I am using offers such processors, but only at a layer one or two deeper than the layer my program has reached, I may well decide either to reallocate processes to use that extra layer or two, or to try to reformulate the set of processes I had intended to execute in the more flexible MIMD mode so that they can be executed instead by an SIMD layer.

If I knew that the MIMD layer had only nearest-neighbor connections among processors, I would continue to try to code processes that worked with more distant interactions using the same kinds of strategies of moving them successively closer together that I routinely try to use at the SIMD levels. But we can probably assume that richer interconnections are offered, as in the I-O control pyramid of Chapters 16−18.

Finally, I want to be able to code processes that feed back information uncovered at higher levels, using this information to make adjustments and branches at lower levels. This, again, would be expedited by the rich two-way connections at a distance offered by the I-O control pyramid.

Summarizing Observation

To carry out such an enterprise efficiently, programming, reconfiguring, and even specifying hardware, it seems necessary that I be given good information about the interconnections in the physical network that might be used for these purposes (or/and information about the reconfiguring possibilities, and how to make use of them).

Note that this whole exercise is designed to develop a program in which everything is specified. Every procedure is a "slave" or a "drone." There are no "smart" nodes where a flurry of messages get passed back and forth among procedures, and these procedures (or, possibly, a decision procedure) engage in consensus reaching and decision making. On the contrary, this program (and I strongly suspect that, at least for a long time to come, all programs should do the same) not only specifies the procedures in complete detail but also tries to specify as many as possible of the details of mapping these procedures into the physical processors.

The crux of efficiency will be how well this mapping is effected, so that unnecessary message passing and moving around of code are minimized. At least until we know a great deal more than we do today, all sources of help should be used, and that certainly includes the one intelligent person involved, who is also the person who knows by far the most about the program — the programmer.

It seems desirable to have the compiler give the programmer feedback at many steps, showing how it is mapping the program onto the architecture, mentioning problems and points of inefficiency, suggesting that different alternative approaches to the same algorithm might be explored, and mentioning ones that in its (very rough) categorizations it infers may be appropriate. This can, of course, become very verbose, irrelevant, and a large bother rather than a help; so the programmer should be able to turn it down, or off.

But until we know that we do not need the programmer's help, and that she/he does not need all the help the compiler can give, we should use these two sources of information as fully as possible. At the same time the computer should collect as much data as possible about which resources are being used and which are most useful to help tune and redesign the system.

CHAPTER 10

Some Short Examples Mapping Algorithms onto Networks

I here examine a variety of problems for which networks would appear to be attractive. But they do not have an obvious mapping onto some simple structure, in the way that near-neighbor perceptual processes map onto SIMD arrays. And some are problems that people have scarcely begun to formulate for networks. So these should be read for what they are—rough sketches that almost certainly would change beyond recognition during the process of precise formulation and programming for a network.

Breadth-First and Heuristic Searches

Breadth-First Search through a Growing Set of Possibilities

A tree seems a good structure for a breadth-first search, when the following procedure is used: Assign each node from which the search begins to a different processor. Each processor applies the set of rules that create buds. Each bud is examined to see if it is the desired goal and, if it is, a message is broadcast to that effect (this will reach the output and controller processors, and also tell the processors active on this program to stop).

If a processor has fewer offspring than the number of buds generated, these extra buds must be passed to some vacant processor elsewhere (it is desirable that it be close, and also that it root a spreading tree of lower-level nodes). Alternatively, pruning might be effected, to cut the number of buds down to size.

If the bottom layer is reached, the node must send buds to its parent(s) rather than its offspring. Now the process might continue indefinitely, bouncing back and forth between the two layers.

This suggests an attractive alternative mapping, onto a two-layered system.

Heuristic Search

A heuristic search algorithm might proceed much like the breadth-first search just described, with the difference that "heuristic" procedures (that is, additional subroutines) would be applied to determine the exact subset of buds to assign to the node's offspring. A pruning algorithm is used to choose the required number of (most highly weighted) buds. These are passed to the node's offspring, and each offspring (of each parent) repeats the process.

Typical heuristic search algorithms use the best or the n best nodes to bud from next. This would seem to be difficult and time-consuming for most networks, since it would involve a lot of message passing to poll all the processors involved in the search, then choose the best. Here is a place where one might want to reformulate the procedure, to continue along the locally best paths. Processors might only occasionally send messages about what they are finding. They could then choose to drop unpromising parts of the net and reassign processors that needed more.

A different formulation would have each processor responsible for its own subtree. This slows things down, but it greatly reduces the need for message passing, and is probably the most suitable when only small numbers of processors are available. This alternative is often also preferable for breadth-first search.

Fishburn (1979) developed an algorithm for a best-first alpha–beta search into a game tree, using a tree of processors in this fashion, that gives at least $k^{1/2}$ speedup with k processors. Fishburn (1981a) has improved this algorithm to get linear speedups, approaching k as the degree of the search tree increases. Checkers gives $k^{.6}$, chess $k^{.8}$, and Go close to k^{1} speedups. This algorithm incorporates a reordering of tasks, where all necessary processes are executed first, along lines suggested by Akl *et al.* (1980). See also Knuth and Moore (1975) for an examination of alpha–beta search.

It seems plausible that both a breadth-first and a heuristic-search program might be executed on an SIMD network as well as by an MIMD network. But searching through a tree, although it appears to have a very high degree of parallelness, is by no means easy to formulate for efficient execution in SIMD mode.

Searches for Question-Answering
and Database Access

Searches through Semantic Memory Nets

A "semantic memory net" typically has the following structure: A large store of information is broken up into small pieces (nodes) that are connected by pointers (directed links). Often there are several different types of pointer (perhaps only two or three, rarely more than five or ten), though usually there is only one type. Paths along pointers are usually followed to see whether (a) some piece of information is stored within the network, (b) a reasonably short path can be found between two (or, occasionally, three or four) pieces of information, or (c) what information (occasionally of a certain type) is closely associated with (that is, is stored in the neighborhood of) some stored piece of information.

Usually a breadth-first search is made in such a net, starting from one or more nodes that have been directly accessed (by some previous algorithm, e.g., a system that parses a question asked of the semantic memory network). Occasionally a directed heuristic search is made, the system attempting to choose the most promising paths. Often the heuristics will take into consideration information about one of the other nodes from which the search is spreading out, or found along some of the paths being followed (this is sometimes the case in the problem-solver's heuristic searches of the sort discussed above, but it was ignored there to keep things simple and because it is relatively less common).

The straightforward way to handle semantic memory searches would be to assign each node (that is, each piece of information) in the semantic memory network to a different processor, with a link between processors wherever there is a link between semantic memory nodes. This would require very large numbers of processors, and is quite unreasonable in terms of the technology of the immediate future. On the other hand, each processor could be extremely simple, e.g., a synapse-like threshold gate. So each 1990s VLSI chip might contain many thousands.

A more realistic alternative would assign large subnets to each processor by decomposing the entire net into a set of subgraphs surrounding the nodes from which the search is to occur. When the search moves beyond a subgraph, new processors are assigned to new subgraphs that project out in the directions of movement.

But now all nodes that are traversed must be tagged (probably in a copy) and these tagged nodes must be made available to all processors, so they can check at each move whether an intersection has been achieved. So this problem may necessitate a large amount of message passing—unless large numbers of processors can be given access to a single common memory. Or tagging each node as traversed may be sufficient, if the program can be formulated so that this information is needed only locally, at that node. The problems of notifying processors when a goal has been attained, and outputting the results, appear to be much the same as for the problem-solving searches.

Database Management Systems

Database management systems use much more highly structured and much simpler data. For example, a company will store its department names and under them the products made, materials used, and employees' names. Under each employee will be stored payroll and other personnel information. Often a tree is used; sometimes a "relational database" stores this information in the form of tables of triples—two items connected by one or another (of a specified set of) relation(s).

If the database can be segregated into tracks, then a different processor can be assigned to each track, and queries can be given to appropriate processors (possibly by a single master processor that computes this assignment). But usually this kind of nice segregation is not considered possible or desirable (though opinions may well change as large parallel sets of processors become available and parallel database systems are designed).

Therefore there appears to be a need to assign and reassign processors to new (sub)searches, with the attendant problems of finding free processors and passing messages. Nor is there much possibility of using SIMD sets of processors if different queries might be expected to arrive at different times. But it would appear that the simultaneous design of the database's structure and the network of hardware processors might achieve a level of organized structure that the database manager could live with but would make processing by a network extremely efficient. Whether such an architecture would also be useful for other types of programs, or would best be dedicated to database problems, is yet another interesting problem for exploration.

Su *et al.* (1979) and Schuster *et al.* (1979) have designed parallel associative arrays for SIMD database computers. DeWitt (1979) has designed a parallel database computer that links up to sixteen MIMD computers via a cross-point switch to a large bulk memory. Boral (1981) used what he calls an "algorithm-directed" design approach to specify an architecture (computers on a ring over which tasks are broadcast and assigned). He studied the relatively simple set of eight algorithms needed to interrogate a relational database (e.g., "delete," "append," "select," "join") and concluded that a ring-based data-flow approach was most feasible, because the relatively large amount of processing needed to handle a query would dominate the necessary message passing.

Goodman and Despain (1980) have made a careful analysis of how buses, trees, n-cubes, and several different augmented X-tree architectures might be used to handle a closely related, and quite important, problem in managing a large database—that of eliminating multiple entries of the same item. They compared several algorithms, including ones that have each of the buds of a tree work on a subset of the total database, matching and eliminating duplicates, then sending the results to the parents. They found that parallel processing can appreciably improve performance, and that a hyper-tree with rings linking internal nodes and perfect shuffle links at the leaves is best.

Modeling Systems of Muscles and of Neurons

Modeling Robot Effector Systems

Consider problems of coordinating a set of muscles (or other motors) that move a set of bones (or other appendages) around in space—for example, the legs—arms—trunk system that is coordinated and balanced in walking, or the fingers—hand—arm system that must work together when picking up and manipulating objects). Huen *et al.* (1977) are developing a ring network for systems of this sort, where a hierarchical structure with many simple low-level "virtual arms" is partitioned heuristically for efficient mapping onto the processors.

An alternative approach would assign each muscle and each controlling ensemble of neurons to a different processor. In contrast to the semantic memory net, the muscle net should be small enough (on the order of from three to at most a few hundred muscles) for this to be feasible. And the interconnections between muscles and their controllers will be relatively sparse, and hierarchically structured. Now each processor can do its job of muscle-effector control, passing messages back and forth only to its directly connected nearest-neighbor controllers and other muscles. Essentially, this suggests a straightforward mapping of the neuromuscular system onto the network. The hierarchical aspects of the organization suggest an augmented tree. But possibly any network whose edge/node ratio was commensurate with the needs of the system being programmed would work about as well.

Networks of Neurons

Neuron nets can be similarly mapped onto a hardware network. But if each synapse is assigned to a different processor, the net will quickly grow too large for today's systems.

It is important to mention this possibility, however, since it will shortly become feasible and is an extremely attractive and direct way to model intelligent systems simultaneously at the physiological and psychological level, imposing the constraints that the real world imposes, yet allowing for the real possibility of fast and efficient processing in real time (because of the appropriate mapping onto the parallel−serial network structure).

For now, a subnet of neurons might be mapped onto each processor.

Modeling the Whole System,
in Interaction with Other Systems

A discouragingly small amount of work (considering the importance of this problem) has been done on modeling systems that attempt to combine perception, memory, thinking, and motor behavior into a single "wholistic" system. The computing burden is still far too great to handle everything without many oversimplifications. But parallel computers may well make this feasible.

The game of life is one of the simplest possible examples. Its rules of birth and death are trivial, deterministic, gamelike. Its organisms simply get born and die; they have no other characteristics. Yet it generates surprisingly complex behavior. It is a standard exercise in beginning programming courses: The program in a language like Pascal takes roughly 100 instructions. The program for Kruse's (1978) PICAP takes 1 instruction! Burks (1966) contains several interesting chapters on early proposed simulations of systems of organisms that extend the game of life to games between competing species. Rosenstiehl *et al.*

(1972) have explored "myopic" (that is, local) algorithms executed by "intelligent" graphs of finite automata in similar situations.

Toda (1962) explored (but never programmed) the situation of a "fungus eater" gathering information and valuables from an alien planet where only fungus might grow. Doran (1968a,b), attempted to simulate a need-driven robot in a simple environment. Uhr and Kochen (1969) coded a "MIKROKOSM" where a specified number of very simple objects and organisms were generated at the start of a run, scattered through the environment. Then they interacted over time according to simple abstractions from the need-driven interactions of real-world organisms.

Uhr (1975) coded a more sophisticated "wholistic" system, one that included a relatively powerful pattern recognizer, and also procedures for simple problem solving, semantic memory search, and motor action. But this system was never embedded in an environment, either simulated or (as in the case of the robot systems) real.

I mention these attempts to handle the extremely large and complex problems of modeling a whole perceptual−cognitive−motor system, and several such systems cooperating and competing with one another within a larger environment, because they so obviously contain a great deal of parallelness, much of which could certainly be exploited if suitably structured networks were available. Each of the many organisms and regions in the environment could be handled separately, in parallel. The perceptual procedures, the motor coordination of the several limbs or muscles, and the simultaneous acts of perceiving, thinking, and moving by each simulated organism also all take place in parallel. They could quite naturally and straightforwardly be programmed for a parallel network that handled these parallel procedures explicitly.

Parallel Algorithms for Processes That May Be Basically Serial

Some problems are simply very serial. These include problems where a number of pieces of information interact with one another in such a way that any piece is as likely as any other to affect and restructure the situation. Often such a problem can be handled, by using divide-and-conquer techniques, in a parallel−serial fashion, with different processors assigned to the different subdivided subproblems. But sometimes this seems hard or impossible to do.

The n -Queen Puzzle

An interesting example is the n-queen puzzle that is often used to demonstrate the extensive amount of serial processing needed to search large trees of possibilities. The problem is to place as many chess queens as possible on a chess board (of any size), such that none can be captured by any other.

The typical algorithm will place each queen in turn, moving it until all queens are safe, placing another, until failure, then backtracking to move the last-placed queen to its next safe position.

The procedure can be made parallel in several ways:

The test whether a new placement is safe can be made by a parallel search in all the directions the queen can move. All moves of all queens can be tested in parallel.

David Hunt (personal communication) has coded an n-queen program for DAP (an SIMD array). It takes four hours to find all safe configurations on a 32×32 board. This is a very large search, and four hours is far shorter than the time that any serial machine would need.

A Parallel Algorithm for the Traveling-Salesman Problem

Another interesting example is the traveling-salesman problem, which is traditionally handled by serial searches. It is known that in two-dimensional space there are no polynomial-time algorithms for achieving an exact solution. Heuristic techniques (e.g., Lin and Kernighan, 1973; Dionne and Florian, 1979) have been used for up to 300 points, with what appear to be good results, obtained in an acceptable amount of time.

Halton and Terada (1978, 1982) have developed a highly parallel algorithm that takes an extreme divide-and-conquer approach, decomposing any bounded region in n-dimensional Euclidean space into a large number of equal cells, each containing roughly $\log f(n)$ points, for any choice of an increasing function f. A shortest path is found within each of these cells, and these are then combined into the grand tour. This algorithm takes $O(n f(n))$ time. It can be used to find tours of thousands of points. The resulting tour length will be asymptotic to the correct answer as n approaches infinity. Thus even on a serial computer this algorithm is suited for tours several orders of magnitude larger than can be handled by customary serial algorithms.

It would appear that this algorithm might map very nicely onto an SIMD array. For example into each of DAP's 4000 processors could be input one cell of a very large space with thousands of points. Parallel code would then connect points in each cell in parallel, probably by expanding from each point, again in parallel. It is not clear how efficiently an SIMD system could then combine the individual cells, or whether this might better be done by a network, or by a single host. Possibly better would be a mixed SIMD−MIMD system like the array/net (Uhr *et al.*, 1980) or the kinds of structures described in Chapters 15−18.

CHAPTER 11

A Language for Parallel Processing, Embedded in Pascal

This chapter examines parallel programming languages, chiefly for arrays. It briefly surveys the major languages that have been developed. Then, to give the reader a detailed look at one such language, it focuses on PascalPL1, a language I have been working on. [See Appendix E for examples of code in several of these languages.]

PascalPL1 extends Pascal with constructs to process parallel arrays and pyramids. It allows the programmer to code procedures that effect parallel operations over arrays of integer (or, optionally, Boolean) values of the sort commonly used in image processing, pattern recognition, and scene description programs. Statements can specify sets of relative locations (with respect to each cell of the array) and compounds of such sets, embedded in assignment statements or in conditional statements. It also has constructs that handle convergence of information that is taken from one array, then transformed and merged into another smaller array, giving a pyramid structure.

PascalPL1 does not purport to handle efficiently a wide range of numerical operations. Nor does it attempt to handle the problems of parallel-serial processing by networks in general. Rather, arrays of information [e.g., as input by a television camera, and transformed by arrays of hardware (or virtual) processors] are the major type of parallel structures that it addresses.

But since it is embedded in Pascal all the facilities and power of Pascal are available to the programmer.

PascalPL1 presently exists as a program (coded in Pascal and running under the UNIX operating system on the VAX) that inputs a legal PascalPL1 program and outputs a legal Pascal program that contains the code the PascalPL1 preprocessor output (which will now effect the parallel operations called for by the PascalPL1 code). Its special parallel constructs (to be described below) include parallel

```
assignment:     ||set ... := ...,
conditional:    ||if ... ||then ... ||else ...,
input:          ||read(...)
output:         ||write(...)
```

statement types. These all execute operations, in true parallel fashion, on arrays of information. This is emulated on the VAX, as on any other serial computer with only one processor, by setting up temporary arrays (to keep processes parallel) and embedding the processes within nested DO loops.

It is not at all clear whether present-day languages that have been developed for serial computers should be extended to handle arrays and networks of parallel processors, or whether entirely new languages should be developed. In the long run we shall inevitably see entirely new languages that reflect the new hardware architectures of computers as well as our advances in understanding language design.

But for the next three or five (or ten?) years it seems likely that the most powerful systems will use a serial computer (to serve as "host," file manipulator, compiler, simulator, etc., as well as to execute the serial portions of programs) along with any hardware arrays and networks of processors. It therefore seems appropriate to take advantage of the serial computer, and of the power of the newer languages that have been developed for serial computers.

Pascal (Wirth, 1971; Jensen and Wirth, 1973) seems a reasonable choice for the language to be extended. It is already widely used and promises to be used increasingly, especially on microprocessors of the sort often used to build networks and arrays. It is both powerful and relatively efficient. It has been extended, in Concurrent Pascal (Brinch-Hansen, 1973a,b, 1975) and Modula (Wirth, 1977b), to begin to handle networks of processors. It is being extended, in Telos (Travis *et al.*, 1977), to handle a wide range of artificial-intelligence and database management tasks. At the first of a continuing series of annual workshops on parallel architectures and higher-level languages for image processing (see Duff and Levialdi, 1981; Preston and Uhr, 1982) the consensus of those who felt a modern language was needed was that it should be embedded in Pascal (until a completely new language emerges), and a working group was formed with that goal (see Duff, 1979). Finally, since it seemed desirable to implement an experimental version of a parallel language, to explore its problems and its uses, a decision had to be made as to the specific language to extend.

A Brief Examination of Parallel
Programming Languages

A large number of languages have been coded to handle parallel array processes (see Preston, 1979, 1981). Until recently these have tended to be at the assembly-language level, or to be subroutine calls in Fortran. But they include a number of languages that are quite simple and straightforward to use, and offer the user great power (to a large extent because the parallel arrays, with their very large numbers of processors, are so powerful). Many of these languages execute exclusively on a particular machine with parallel hardware (or on a simulation of parallel hardware), and are directed toward image processing.

Recently a number of higher-level languages have been proposed or implemented. We shall briefly examine these now.

Languages Designed for Numerical Problems on SIMD Arrays

Several languages have been developed for array processing of numerical problems. These include APL (Iverson, 1962), Glypnir (Lawrie *et al.*, 1975), and Actus (Perrott and Stevenson, 1978) for the ILLIAC IV supercomputer, and the parallel DAP—Fortran language for the DAP (Flanders, 1979).

Glypnir is an Algol-like language that forces the user to think in terms of a fixed number of processors. Actus attempts to be more general. For example, an array is declared: VAR a: ARRAY[$1..m$; $1..n$; $1:p$] OF integer, where the "parallel dots" (the colon) show that this index is to be processed in parallel, by either an array or a network vector processor.

The following fragment shows how a simple solution of Laplace's equation on an $N \times N$ grid, where we replace each element by the average of its four nearest neighbors, could be coded in Fortran:

```
DO 3 I = 2,N-1
DO 3 J = 2,N-1
3    Y(I,J)=(X(I+1,J)+X(I-1,J)
      +X(I,J+1)+X(I,J-1))/4.0
```

(more code is needed to handle the border elements).

The equivalent DAP-Fortran code is

```
X = (X(+,)+X(-,)+X(,+)+X(,-))/4.0
```

The constructs $X(+,)$, $X(,-)$, etc. shift-access the four neighbors. Since every cell is computed in parallel there is no need to store the results in a different array.

A Fortran-based language is now under development for the MPP (Potter, 1982). In contrast to DAP—Fortran, it is more specifically directed toward image processing. Code equivalent to the above follows:

```
X($,$) = (X(+1,$)+X(-1,$)+X($,+1)+X($,-1))/4
```

Languages Designed for Image Processing on SIMD Arrays

Kruse and his associates (see Gudmundsson, 1979) have developed a very nice Algol-like language, PPL, in which programs can be coded that call both the PICAP parallel processor and the serial host computer. PPL contains two major parallel constructs, for (1) logical and (2) numerical operations, with the format (type = logic or numeric):

$$
\text{(type)} \quad
\begin{matrix}
1 & 2 & 3 \\
4 & 5 & 6 \\
7 & 8 & 9
\end{matrix}
\quad = \text{result}
$$

That is, the programmer is asked to specify a 3×3 array of nine values (these may be Boolean values, or integers specifying weights or inequalities), plus operation and a result. This closely reflects the nearest-neighbor structure of the PICAP I window (and of most of the arrays and pipelines specialized for image processing that emulate hardware arrays).

Uhr (1979; see also Chapter 12) has developed a higher-level language, PascalPL0, for the CLIP parallel array that makes use of "compounds" and "implications" similar to, and precursors of, some of the constructs in the present Pascal-based PascalPL1.

Sternberg (1980) is developing an elegant and potentially very powerful language that basically uses set-theory operations.

Reeves (1979) has developed a system that extends APL to handle image processing as well as numerical processing of arrays. He argues (personal communication) that APL, since it handles arrays quite naturally, may well be preferable to Pascal as the "host language" into which parallel constructs should be embedded. But since Pascal is far more widely available, and far more consonant with the feelings that most people have today about what is "good structure," it seems the better choice, and Reeves *et al.* (1980) are now developing a new Parallel Pascal language for the MPP.

Parallel Pascal extends Pascal chiefly as follows:

(a) Parallel arrays for MPP processing may be declared, e.g.,
image1, image2:
   PARALLEL ARRAY [1..128,1..128] OF integer;

(b) Expressions may involve whole arrays, and functions return arrays:

a := b + sin(c) + 1   {means}
a[i,j] := b[i,j]+sin(c[i,j])+1 {for all i,j};

(c) Control statements may have arrays for control variables:

IF a>b THEN c := 99   {means}
IF a[i,j]>b[i,j] THEN c[i,j]:= 99 {for all i,j};

(d) New standard functions (e.g., shift, rotate, sum, max) are added;

(e) Read and Write will input and output whole arrays.

Levialdi and his associates (1980, 1981) are developing Pixal, which consists of parallel extensions embedded in Algol-60 (which they used chiefly because Pascal was not available on their computer). Pixal uses a "frame" construct that allows the programmer to specify a set of relative locations (not only nearest neighbors) to be looked at everywhere (that is, relative to each cell in the array), and a "mask" construct with which the programmer can specify a set of weights to be applied to the relative locations specified in the coordinate structure.

Languages That Handle MIMD as Well as SIMD Processes

Three major systems have been developed to handle concurrent asynchronous parallel processing, chiefly motivated by operating system problems, Concurrent Pascal (Brinch-Hansen, 1975), Modula (Wirth, 1977b), and StarMod (Cook, 1979). Conway (1963) first developed their key "fork" and "join" constructs, for spawning new processes that execute in parallel, and for starting up a process when all its necessary inputs have been joined together. [See Andrews (1981) for an examination of the major constructs in languages of this sort.] Wittie and van Tilborg (1980) are developing a language for MICRONET written in and based on Concurrent Pascal.

Douglass (1980) has proposed extensions to Pascal to handle parallel arrays that build on earlier proposed extensions by Pratt and Ison (see Ison, 1977) to handle networks. These include a "fork" operation to set up new processors that are executing different sets of instructions in parallel, and a "split" operation, to invoke whole sets (e.g., arrays) of processors to execute the same set of instructions, but each on different data (e.g., different local regions of a visual image). The programmer can specify "windows" [sets of (relative) coordinate locations]. Douglass (1981) has also given the specifications of a new language, MAC, for programs with both SIMD and MIMD processes, along with example programs.

Flynn and Hennessy (1980) are exploring the use of functional languages of the sort proposed by Backus (1978). Ackerman (1979), Ackerman and Dennis (1979; see McGraw, 1980), Arvind *et al.* (1978), and Plas *et al.* (1976) have developed Petri net—like languages for data-flow computers that might have relatively large numbers of processors.

Schmitt (1979) has described extensions to Pascal to define and then execute operations on any arbitrary network structure, rather than limiting the programmer to arrays. Radhakrishnan *et al.* (1979) are developing a language for image processing using networks of computers.

Many of these languages use as their key construct an operation on a set of relative locations. The languages designed for a specific hardware array typically build in the interconnection pattern of that array (usually, as in the case of PICAP I, the 3×3 subarray of the nearest neighbors). Pixal's "mask" and Douglass's "window" generalize this to any set of arbitrarily distant neighbors. (But when such a language is actually executed on a hardware—parallel array this gives excessively long sequences of nearest-neighbor shifting operations.) Uhr added constructs for the convenient compounding and implying of information over several different arrays. Douglass and Schmitt make suggestions for constructs to handle more general networks, with MIMD processors as well as SIMD arrays.

A Description, with Examples, of PascalPL1

PascalPL1 is designed to handle arrays, and sets of arrays, that contain binary, gray-scale, and/or numerical values. It allows one to code procedures that look at and compound sets of relative locations (around each "pixel" cell) at every location in a single array, and to compound sets of these sets across several arrays, using either arithmetic or Boolean operations.

It seems best to introduce PascalPL1 by starting with very simple examples, gradually introducing its full set of constructs and features.

An Overview of PascalPL1 Constructs

A PascalPL1 program looks like a Pascal program, except that it contains several new constructs [all announced and made visible by two vertical (parallel) bars, e.g., ‖procedure..; ‖set..; ‖if..; ‖read..; ‖write..;]. This means that the programmer must declare all the necessary constants, data types, and variables, and strictly follow all the conventions of Pascal [e.g., a program must begin with program {programname} (input, output{...}); and end with end] The programmer can code as much as she/he desires in ordinary Pascal. Only when parallel constructs are desired do any deviations occur. These parallel constructs are handled as follows:

The procedure that will contain these (one or more) parallel constructs must be declared:

‖procedure {procedurename};

Any time after this procedure's "begin" statement, a "dimension declaration" statement must be placed, e.g.,

‖dim   [0..127,0..127];

Then come, interspersed with ordinary Pascal statements, parallel constructs of the sort

```
‖read{..};
‖write{..};
‖set   {array(s) assigned to}
    := {compound of array(s)..};
‖if {compound of arrays) (ineq) }
  ‖then {array(s) modified}
    ‖else {arrays modified on failure - optional};
```

A Very Simple Example Program

The following program makes only the simplest use of PascalPL1 constructs. [See Uhr (1979b) for the Pascal program into which the PascalPL1 preprocessor translates it, along with other examples of PascalPL1. See Appendix E for a longer, but still very simple, program, and for an example of Pascal code into which a PascalPL statement translates.)

```
program simple(input,output);
{the programmer must declare the array
    data structures used}
{  for this program they are:
   image, negative, edgedimage  }

procedure sayhello;
begin
  writeln('hello');
end;

||procedure demonstrate;
begin
   ||dim    [0..2,0..2];
     ||read(image);
       ||write(image);
     ||read(negative);
   writeln('the image and the negative image',
      ' have been input');
     ||set edgedimage := image - negative;
     ||write(edgedimage);
     ||set image, negative := 0;
end;

begin {program}
  sayhello;
  demonstrate;
end.
```

An Example of Actual Computer Output
from the Simple Program Above

```
hello
TYPE IN INTEGERS FOR  image[ 0.. 2, 0.. 2]
   1   2   3       were input to row  0
   4   5   6       were input to row  1
   7   8   9       were input to row  2
ARRAY =  image  contains:
   1   2   3 ;
   4   5   6 ;
   7   8   9 ;
END OF ARRAY.
```

```
TYPE IN INTEGERS FOR  negative[ 0.. 2, 0.. 2]
    5   5   5         were input to row  0
    5   5   5         were input to row  1
    5   5   5         were input to row  2

the image and the negative image have been input

ARRAY = edgedimage  contains:
   -4  -3  -2 ;
   -1   0   1 ;
    2   3   4 ;
END OF ARRAY.
```

This program first outputs "hello" and then inputs a 3 × 3 array, naming it "image." (As it is presently implemented to run interactively, it outputs the message to TYPE an array, and outputs the inputs, to verify.)

The next statement | | write (image) outputs the array stored in image, in array form. | | read (negative) inputs the array that it names "negative." Now the program outputs that the two arrays have been input. [Note that this is a regular Pascal writeln(..); command. It illustrates how regular Pascal statements can be interspersed.]

Now, for each cell in the array "edgedimage" the program subtracts what negative contains from what image contains. Next it reinitializes the two arrays image and negative, so that each of their cells contains a zero. Finally, it outputs the array named edgedimage.

This is a trivial example, and does not begin to indicate any of the more powerful ways that the parallel—assignment construct (| | set) can be used. But it shows how PascalPL1's parallel constructs can be intermixed with standard Pascal, and it begins to show how input and output are extended, in a straightforward but what appears to be satisfactory way, to handle arrays.

The Assignment Statement

An assignment statement contains a set of "arrays to be assigned results" (called "assignees") to the left of the assignment operator (":=") and a "compound of structures" (called "compounds") to the right of the ":=."

Now let us examine a sequence of more powerful assignment statements:

```
||set vert := image[+(0:1,0:0,0:-1)];
```

(This statement looks, for each cell in image, at the three relative locations specified, sums what it finds, and stores the result in the corresponding cell of vert.)

```
||set cross := vert[+(0:0,0:2,0:-2)] * hor[+(0:0,2:0,-2:0)];
```

(Sums the three relative locations in vert, does the same for hor, then multiplies these sums and stores the result in cross.) Note that

```
||set array1 := array2 (op) array3;
```

is equivalent to

```
||set array1 := array2[0:0] (op) array3[0:0];
||set vert, hor, image := 2 * image[+(0:1,1:0)] - #average;
```

[This illustrates how several arrays can be assigned the values computed by the compound, and how the compound can include constants and variable identifiers (each must be preceded by '#').]

```
||set gradient := image[+(0:0*12,1:0*-2,
    0:1*-2,-1:-1*-1,...)];
```

[This will get a weighted difference between the center cell (weighted $+12$) and the four square neighbor cells (weighted -2 each) and 4 diagonal neighbors (weighted -1 each) (only the center and three of the eight neighbors are shown).]

```
||set featurei, featurej, labelk
    := featurem[*(4:-3*2>5,-5:7*3>14,0:0*21>112)];
```

(Multiplies what is found in each relative location by the specified weight and accepts the result only if it exceeds the specified threshold, then stores the sum of these results in the corresponding cells of featurei, featurej and labelk.)

To summarize: An assignment statement consists of an arbitrarily long compound (to the right of the assignment operator ":=") of array specifications. An array specification consists of an array name followed (optionally) by a "structure." A structure consists of an (integer or Boolean) operator followed by a set of relative locations. Each relative location in an integer array can, optionally, be followed by an operator and a weight and/or an inequality and a constant.

One or more arrays can be named to the left of the assignment operator. Each will be assigned the result of the set of operations on arrays specified in the compound. [An option that is probably not very useful also allows (for integer arrays) an arithmetic operator followed by an integer, to modify each result by the specified constant before storing it in its corresponding cell.]

The Conditional Statement

A conditional statement can also be constructed, with the form

```
||if {compound} {optional inequality}
   ||then {modifications};
```
(or)
```
||if {compound} {opt. ineq.}
   ||then {modif.} ||else {modif.};
```
For example,
```
||if featurei[+(0:1,0:-1)]
   * featurej[+(1:0,-1:0)]  >  37
        ||then  labelk*9
        ||else  labell*2, labelm+27;
```

The conditional statement first computes the compound (for each cell in the arrays). Then, if an inequality is specified, it tests it and (only when it is satisfied or, when using Boolean arrays, if the compound is true) makes modifications to the arrays following the ||then If there is also a ||else, what follows it is modified (only when the conditional fails, or is false).

The modifications can specify an operator and an integer (e.g., labeli + 19, indicating "add 19 to labeli," or **vert** * **36**, to multiply vert by 36).

Reading and Writing Arrays

The two simple constructs

```
||read({arrayname});
```
and
```
||write({arrayname});
```
input and output the specified array.

Constructs That Declare Parallel Procedures

The

‖procedure {procedurename};

construct must declare each procedure that contains one or more
‖if... or ‖set... statements. (Its purpose is to signal the Pas-
calPL1 preprocessor that a parallel procedure will have to be set up, so
it can declare that procedure here, with a "forward." This avoids any
problems that might arise if the programmer were to use the same
name in the declarations that PascalPL1 uses in declaring the temporary
lists and data structures that it must use. If these extensions were embedded in the Pascal compiler itself this declaration could be eliminated
or, probably better, combined with the ‖dimension declaration.)

The ‖dim statement declares the array type and dimensions of the
arrays to be used in the parallel constructs that follow in this procedure.
Its shortest and simplest form is

‖dim;

(which declares the previously declared, or default, dimensions and integer arrays).

Its full form is

‖dim {arraytype}merge
    from {from-setname}[{arraydimensions}]
    to {to-setname} [{shrink-convergence}];

The arraytype can be & (for Boolean) or * (for integer). The from-
setname and to-setname are optional. If used, they are concatenated in
front of the names of arrays in (a) compounds and in (b) assignees or
modifieds, respectively. (This is an option that may be useful when
sets of arrays are collected into layers or characteristics, and if and
when a feature that lets the programmer declare and use them as arrays
of records, using the Pascal "with" construct, is implemented.)

Arraydimensions must be given in the form: [minx..maxx,
miny..maxy], optionally declared Boolean or integer—e.g., [0..127, 0..63
: integer]. (Note that this means the arraytype can either be declared
before or within the arraydimensions. Whichever seems more convenient and more natural will be retained in the future.) Thus still
another alternative declaration is

‖dim [0..7,0..15 : boolean] {optional shrink};

The shrink—convergence indicates how information is converged
from a from-array to a to-array. For example, if it is 2,3 then the x
coordinate of a from-layer cell is divided by 2, and the y coordinate is

divided by 3, to compute the coordinates of the to-layer's cell. (If no shrink−convergence is specified the program will use whatever was last specified. The default, if shrink−convergence was never specified, is 1,1−that is, coordinates are divided by 1, so that no convergence occurs.)

An end statement

```
| |end;   {or end;}
```

ends the parallel procedure.

The type of border to be used when a relative location to be looked at lies outside the array is specified by

```
| |border := {bordertype};
```

where bordertype can equal 0 (typically signifying "white") or 1 ("black,"), or 2 (for "what is contained by the nearest cell within the array"). (The default condition is border = 2.)

Discussion

Should a language for parallel arrays and, more generally, for networks, remain as an extension embedded in Pascal? The present and projected extensions appear to cohabit rather congenially within Pascal. PascalPL1 seems surprisingly simple in the extensions that were needed, and it makes use of standard Pascal in many ways. A programmer should be able to code in this mixture with little interference between the two systems. And Pascal is useful when the programmer codes at least some processes for a serial computer (as she/he certainly will, at least until all the very difficult problems of parallel arrays and networks are solved).

But it seems likely that the parallel aspects of computing are of overriding importance, that we are entering a completely new era of parallel networks of computers. We shall need to develop completely new parallel algorithms and programs; we shall find that whole new types of approaches to problems, and of problems themselves, are now amenable to and invite attack.

At some point, possibly quite soon, entirely new languages will be called for. But it seems best to move toward new languages by first extending those that exist, and also trying to combine the features of different types of languages that appear to be relevant [e.g., APL, array languages like PICAP's PPL (Gudmundsson, 1979) and CLIP's CAP4 (Wood, 1977), and multiprocessor languages like Concurrent Pascal and Modula]. These endeavors, along with the continuing design of and experience with arrays and networks, should give us the understanding needed to develop entirely new and more appropriate languages.

CHAPTER 12

A Programming Language
for a Very Large Array

This chapter describes PascalPL0, a programming language for the CLIP parallel arrays. It should help give a feeling for how to use large SIMD arrays, along with a fuller idea of what such arrays, and their languages, are like.

PascalPL0 outputs code for either CLIP3, a 12×16 array of 1-bit processors, each with 16 bits of memory, or CLIP4, a 96×96 array of 1-bit processors, each with 32 bits of memory. Every CLIP processor executes the same instruction, in SIMD mode, on information fetched from its own and/or one to eight near-neighbor memories. Each processor is thus linked as the center of its local eight-neighbor array (or, when the program specifies it, a six-neighbor hexagonal array). When processors fetch and store information, each addresses the bit in its own and its specified neighbors' memory specified by the instruction address. Therefore the entire array of processors works with the entire array of bits (called a "bit—plane") at that address.

PascalPL0 was a first attempt to develop a "higher-level" language for CLIP. It was directed toward pattern recognition and scene description (as opposed to image processing), and tried to free the programmer from having to contend with the details of the machine's architecture and the machine language code.

PascalPL0 was developed because the author's attempts to code pattern recognition programs for CLIP (in contrast to the much shorter image processing programs that have typically been coded for arrays) were resulting in the need for long sequences of hundreds, or thousands, of machine language instructions to handle relatively mundane feature detection and compound characterizing tasks.

That it served this purpose reasonably well is attested to by the fact that a 48-statement PascalPL0 program that was coded in a few hours compiled to 1085 CLIP statements that would have taken several days or weeks to code and debug.

PascalPL0 exists as a SNOBOL program that inputs PascalPL0 code and outputs either CLIP3 or CLIP4 code, as desired and specified at the start of the program. It was coded and run on the University College London IBM 360/67. It was used to output a number of paper tapes with CLIP3 code that were then assembled and executed, without any errors, on CLIP3. Example fragments from these programs are shown in Appendix E.

An Introductory Description of PascalPL0

The flavor of the language can best be got by examining the following basic types of PascalPL0 statements:

(a) Nearest-neighbor masking statements effect processes that are functions of the (eight or six) directly connected nearest neighbors.

(b) Compounding statements effect processes that are functions of more distant neighbors. (The programmer can specify whatever number of such neighbors he chooses, from whatever distance.) Thus, e.g., several different features, such as the strokes, joins, and ends of a letter, can be compounded together, to determine whether a higher-level feature is in the input. (The CLIP code to which such statements are compiled executes sequences of nearest-neighbor masking statements that shift more distant cells to be compounded until they are adjacent.)

(c) Implication statements specify sets of possible output names and weights that should be associated with them if the cell being tested is positive (e.g., if some specified or just-computed compound feature is present).

PascalPL0 has a number of other constructs, including several macros for commonly used sequences of instructions, that are not described here.

The Compounding Statement

The compounding statement is coded in the general form

Compound =

$$\text{Part}_1(\text{Mask}_1)\text{OP}_1\text{Part}_2(\text{Mask}_2)\text{OP}_2 \cdots \text{OP}_{n-1}\text{PART}_n(\text{Mask}_n)$$

e.g.,

   TRIANGLE = SLOPE1(3,4) * SLOPE2(1,4) * SLOPE3(6,4);

SLOPE1, SLOPE2, and SLOPE3 each refer to the particular word in each processor's memory that contains information about whether an edge with that particular slope is present at that processor's location. (This is information that would have been got by previous nearest-neighbor masking and compounding instructions.)

   The asterisk is the logical "AND" (other operations are + for "OR," @ for "EXCLUSIVE−OR," and − for "NOT").

   The two numbers in the nearest-neighbor mask (separated by a comma) indicate (a) the direction from which the information is to be fetched and passed (Figure 1 shows the directions used by CLIP and PascalPL0), and (b) the distance.) Thus, in the example above SLOPE1 is passed four steps to the southwest.

```
   1 2 3
    \|/
   8-P-4
    /|\
   7 6 5
```

Fig. 1 The directions in which CLIP and PascalPL0 pass information (these are the standard name/numbers assigned to CLIP). (Thus 1 = NW, 2 = N, 3 = NE, 4 = E, 5 = SE, 6 = S, 7 = SW, 8 = W.)

The Nearest-Neighbor Masking Statement

   The nearest-neighbor masking statement is the basic CLIP statement, and can be quite complex. It fetches information from one or two memory locations (loading it into one or two registers), passes one of these pieces of information to the eight or fewer nearest neighbors specified in the instruction, computes a Boolean function of all this information, and stores the result in the specified memory location (and in one register). It can have a number of forms, but will typically be of the following sort:

$$\text{Output} = \text{Info}_1((\text{Mask})(\text{op})(\text{Info}_2))$$

e.g.,

```
IMAGE = IMAGE(2468)*-IMAGE
```

equivalent CLIP3 code is

```
LD 3 C 6  {load 3 into A-register,
             clear B, store result in 6}
PR 0(2468)A,-A*P, S
     {process A through directions 2468,
      call it P, "AND" it with A,
      use square array}
```

(the object code must specify the memory location—in this case 3—rather than simply specifying IMAGE).

Here IMAGE refers to the word in each processor's memory. IMAGE is passed on (by the nearest-neighbor mask) to the four horizontal and vertical nearest neighbors (e.g., the 2 passes it from North to South). A logical "OR" (which need not be specified within the nearest-neighbor mask) is effected over all these four passed-on pieces of information about IMAGE. The result is then logically AND-ed with the negation of the information (from the cell itself only).

A number of variations on such a command are possible. For example, the output might be stored in a third memory word, e.g.,

```
SLOPE1 = IMAGE(15)+FEATURE1
NEWIMAGE = IMAGE*NEGATIVE
```

[here no information is passed to neighbors; rather, a simple logical function is computed over two pieces of information fetched from the (each) cell's own memory].

The Implication Statement

Whenever a feature, compound, characteristic, or any other type of information implies possible alternatives an Implication statement of the following sort can be written:

Feature == Implied$_1$(Wt$_1$)Implied(Wt$_2$) $\cdots$ Implied$_n$(Wt$_n$)

e.g.,

```
VERTICAL = POLE(5)TREE(4)LETTER-D(3)
     CHAIR(2)LETTER-I(4).
```

The weight (in parentheses) associated with each of the names is added into the total weight that the program is accumulating for that object.

Several legal variants on this statement, that make for more efficient programming and simpler code, are described in Uhr (1979a); one variant is shown in Fig. 2. The programs that have been coded to date typically begin with nearest-neighbor masking operations that effect the simple low-level preprocessing and local edge detection functions. Then a sequence of compounding statements shifts and compounds lower-level features and characteristics into successively higher-level ones. Each of these implies one or more possible names, each with its associated weight. If the feature is found, anywhere, these weights are combined into the total weights for these names. Finally, the program chooses the most highly implied name. In this be way an approximate cone/pyramid structure can be simulated on a CLIP-like cellular array.

Coding an Array like CLIP
in a Language like PascalPL0

It was discovered that an especially convenient way to set up a code sheet is in the form of a transition matrix whose column labels are the possible names of objects and whose rows are labeled with the features and characteristics to be looked for (see Fig. 2 for an example of actual PascalPL0 code in this matrix format).

PascalPL0 Code   Feature   Names of Objects to be Recognized:
A B C O U chair tree table flag

; (";" means comment—so this is a comment!)
; Get simple contours in all 8 directions at once.
0 = 0(1>8)∗P   ;contour

; Get 4 local edges.

;		A	B	C	O	U	chair	tree	table	flag
diag1 = 0 (∗15X) {edge\}		1	1	1			-1	-1	-1	-1
vert = 0 (∗26X) { " |}		1	2				2	2	2	3
diag2 = 0 (∗37X) { " / }				1	1	1		1	1	-1
hor = 0 (∗48X) { " — }		2	1				2		2	1

; Get the simple compounds *L*-angle and *V*-angle.

| 5 = 2(#34,2)∗4(#56,2) {*L*-angle} | | | | | | | 2 | | | 2 |
| 6 = +1(#7,3)∗3(#5,3) {*V*-angle} | -1 | | -1 | 1 | | | | | | -2 |

; (More compounds...of compounds are needed here.)
; The numbers 1 = ,...,6 = refer to the memory locations
;   for storing results.
; The numbers under the object names indicate the weights
;   to be added if the instruction succeeds.
; The weights are kept low because these are very poor
;   features. At higher levels features improve; but many are
;   needed to combine into reasonably good decisions.

Fig. 2   A fragment of a transition matrix-like code sheet.

First, one would write a feature name (e.g., "vert" or "*L*-angle" or "window") as the row label, or simply draw the feature. Then the weights with which this feature implied each possible object would be put in the object's column for this feature row. Finally, the actual PascalPL0 code, for near-neighbor masking or compounding, would be written for that row. The code for Implication statements would already be present, in the cells of the array!

This turned out to be extremely simple for the author to use, and reasonably easy to teach and get somebody else to use. (Only one other person beside the author was asked to code a program in PascalPL0. He spent about two hours learning the language and one hour coding a program, one with about 20 statements that compiled into over 200 CLIP statements, and executed without bugs.)

This kind of transition-matrix format is probably too narrow and constraining a discipline, but it is a simple way to start programming. And it is not as constraining as it might appear, since a very wide variety of programs can be coded in this way. In any case, PascalPL0 offers many additional constructs, and can be used in a number of different ways.

Summary of Experience with PascalPL0

About eight programs were coded in PascalPL0, ranging from roughly 15 to roughly 50 statements in length, but generating CLIP3 object code ranging from roughly 200 to 1100 statements in length. These programs attempted to recognize from among four up to fifteen different objects. They used roughly ten, twenty, or thirty different features and characterizers. Shifting characterizers into near-neighbor position and merging weights into position and adding them took the bulk of the (object code) instructions and computing time. Note that long, tedious, time-consuming sequences of near-neighbor shifts must be used to converge features together, when simulating a pyramid to detect compound features.

Finding the object name with the highest weight took strikingly little time, since it could take advantage of parallel processing, as follows: Each of up to sixteen possible object weights was stored in a different column of one bit−plane of the 12×16 CLIP3 memory. A bar of 1s is shifted down the columns that stored the combined weights, and AND-ed at each shift. This gives a 1 in the column(s) with the highest nonzero bit(s). The procedure continues recursively if more than one 1 is found.

PascalPL0 is too close in spirit to CLIP assembly language, and too closely tailored to a converging pyramid approach to pattern recognition and scene description, to be successful as a general higher-level language. But it is a significant step above CLIP assembly language, and it tangibly eases the programmer's burden.

This language was motivated by an approach to scene analysis that is highly parallel, and can take advantage of the potential great increases in speed offered by a parallel computer like CLIP. PascalPL0 was designed to expedite programs that take this approach, especially in its major compounding and implication statements. But it is a general-purpose language for a general-purpose computer, so any kind of scene analysis or pattern recognition program could be coded in it.

On the other hand, there is little point in using a parallel computer for a serial program. The important questions are, What is the range of parallel programs that people would like to code? How commonly can serial programs be replaced by equivalent parallel programs? and What kind(s) of language(s) would best serve the kinds of programs people would like to code?

PART IV

Toward General and Powerful
Future Network Architectures

CHAPTER 13

A Quick Introduction to
Future Possibilities

This chapter surveys some of the plans and predictions of the people who have actually been developing and building networks and arrays.

Predictions 10, 20, or 50 years into the future are extremely dangerous in a field that is advancing as rapidly as computer science, with so many potential new developments and applications of major, sometimes revolutionary, importance not yet even contemplated. But we must think about the future. That is the only way we can shape it to our goals, rather than let it blow us where it will.

In the case of parallel networks there are several basic factors that are known with a good deal of assurance:

We shall need very large arrays and networks, both to handle very large problems and to handle problems with severe real-time constraints that must be solved far more quickly than is possible with only one, or even a few, processors.

We can be almost certain that the present steep and linear rate at which we have been increasing the packing density of devices on LSI and VLSI chips for the past 10 years will continue for the next 10 or 20 years. This is so because several approaches are being taken; each has already proved to be tractable and been highly successful; each is still only in its early stages of maturity; from all appearances systems can continually be improved by orderly steps that combine them with still other already-improved technologies.

So the extrapolation from today's 100,000-device chip to the 450,000-device chip (which is already being used in a commercial computer) to 1,000,000- and then 10,000,000-device chips by 1985 to 199?

is a bit on the conservative side. It is impossible to assign real probabilities to such unique future possibilities, but I shall use the kind of fake probabilities one uses when predicting "80% chance of rain": The 1,000,000-device chip will be sold in quantity by 1990 with 97% probability; the 10,000,000-device chip will be sold by 1995 with 60% probability and by 2001 with 90% probability. Then still larger chips will become available, in 20??.

Very Very Large Networks

Larry Wittie (1976) appears to be the person who coined the rather suggestive and euphonious term "mega—micro networks." He, as his long-term goal (personal communication; only now and then hinted at in his papers), would like to have a network large enough that he could then model the human nervous system, or a significant part of the nervous system.

He prefers to model at the neuronal level. But he would probably assign a whole structure of neurons to each processor, rather than wait until he could build ten billion processor systems. He has, however, suggested that a system with a million processors is possible, and desirable.

I should mention that Wittie seems to be of mixed mind about this. His present intent is to put one processor, plus its attendant message-passing processor, on each chip. The limited number of pins on each chip, along with his desire to use powerful processors with large memories, lead him to this choice. Since a computer built from a million chips is too unwieldy to be feasible, he seems to feel that 10,000 or 50,000 processors, each relatively powerful, is a more plausible goal.

Herbert Sullivan (Sullivan *et al.*, 1977) has also written about systems with millions of processors. Both Sullivan and Wittie have been interested in the problems of feasible designs for such very large networks. Both have come up with designs that would appear to be workable—if the technology allowed and our programming capabilities could handle it.

Vincent Rideout's original hope for WISPAC (Cyre *et al.*, 1977) was that it could grow almost without bounds, as future technologies made that feasible. The regular three-dimensional lattice structure would appear to make that appropriate. But a small prototype lattice, on the order of a $5 \times 5 \times 1$, is all that is now planned.

Very Large Networks

Despain and Patterson (1978) are designing the X-tree with VLSI technology central to their plans. Rather than build anything now, they appear to be designing a suitable processor for a custom-built VLSI chip, for the period around 1985−1988 (see Patterson *et al.*, 1979). This will be a chip containing one relatively powerful processor, as large a memory as possible, separate faster small memory caches for instructions, data, and addressing, and a "writable control store" that allows one to change the microcode that defines the instruction set. They point out that chip area is now a more valid measure than device-count, but they give the device-count estimates for their ~1,000,000-device chip shown in Table I. (They assume one device per bit for slower dynamic RAMs and four devices per bit for faster static RAMS.)

Table I

Projected VLSI Chip for One Computer Node in the X-Tree[a]

	Bits/word	Number of words	Number of blocks	k bytes	k transistors
Memory					
Dyn.RAM	32−64	128−256	32−64	16−128	128−1024
Stat.RAM u−code	32−64	256−1024	1	1−8	32−256
Instr.cache	16−32	256	1	0.5−1	16−32
Data cache	16−32	256	1−2	0.5−2	16−64
Addrs. cache	32−64	64	1	0.25−0.5	8−16
Descriptors	16−64	64	1	0.125−0.5	4−16
Process cache	32−64	16	1	0.063−0.125	2−4
Stack	16−64	64	1	0.125−0.5	4−16
ROM u−code	32	1−2	1	4−8	32−64
Total Memory					242−1492
Processor					
Arith.Logic Unit	16−32		2		20−40
Control					20−80
Input−Output					10−20
Grand Total					292−1622

[a] Bytes and transistor counts are in thousands.

There will also be additional larger memories off the chips. It should be possible to build X-trees with a few hundred, or even a few thousand, of these relatively powerful computers with relative ease.

Siegel (1979) appears to be thinking in terms of an ultimate goal of 1024 computers for his reconfigurable PASM system. The computers he is considering using today for a possible 16-computer prototype (32-bit processors built from fast bit—slice chips) are a good deal more powerful than the LSI-11 computers Wittie has used. So it seems reasonable to extrapolate to future technologies and conjecture that Siegel's system would make up in greater power of individual processors for smaller total numbers of processors.

The increasingly large hardware costs of reconfiguring probably mean that such a system could not be made much larger than 1000. But if a less-than-complete reconfiguring capability were used, this limiting factor might be overcome.

Briggs *et al.* (1979) are apparently also thinking about a reconfigurable system that might grow as large as 1024 processors. Since their reconfiguring network appears to be more economical and less complete than the one envisioned by Siegel, their system might have more possibilities for further growth.

A Small, Powerful Network
of Very Powerful Processors

Widdoes and his colleagues (1980) have plans to build successively larger S-1 systems with each major new advance in technology. That they expect to be able to do this with relative ease, using their package of design programs with which they can recompile the processor's logical design into each new technology, is one of the major strengths of their project.

Present plans are to put 16 processors into one system, connected with a cross-point switch. Each of these processors would be about one-fourth the power of a Cray-1, and would cost about $250,000. (One has been built; apparently the plans for the rest of the network are by no means firm.)

They are already working on the next-generation processor. This will reduce the processor's chip count, hence its cost but not its power, so that expected costs are only $25,000 per processor. They plan to link 64 such processors using a (rather expensive, but in terms of the system's total power not excessively so) cross-point switch.

Very Very Very Large Arrays
and 3D Lattices of Stacked Arrays

Michael Duff (personal communication) has talked (but never written) about the possibility of putting the entire CLIP array, along with the sensing/input device, on a single chip or wafer (the circular disk of several hundreds of chips from which chips are diced). This is something he explored, using optical techniques, when he first began to develop the CLIP systems. But then the technology was not ready. Indeed even today optical technologies are not yet ready, and are not predicted firmly enough for one to have much reliance on them.

Duff now points out that the entire 96×96 array of processors can be put on a single wafer, by the following argument: The size of the present CLIP4 chips (each holding eight computers) is such that with only a 40% reduction (well within even today's technology) the total chip area covered by all 10,000 processors would fit on the single wafer. Whole wafers can be used in this way. But this is by no means certain, for wafers are mechanically rather weak and breakable, and the yield would have to be much higher than it usually is at present for an entire wafer to be free of defects. Or, alternatively, arrays could be made fault-tolerant. (They lend themselves fairly well to this, since they are micromodular and already have many redundant paths.)

Arguing this point from a different perspective, I want to point out that since each CLIP4 processor + memory uses fewer than 600 devices we can envision packing 10,000 such processor + memories onto one single 10,000,000-device chip (although that dense a packing will almost certainly not be achieved).

Many chips are light-sensitive (but most in an uncontrolled way that precludes using this quality as a direct input to the chip's devices themselves). Solid-state CCD (charge-coupled device) chips are now being used for high-resolution television cameras. So we can begin to think seriously about light-sensitive CCD chips that handle not only the input of the image but also the storing and the processing of that image.

David Schaeffer and James Strong (1977) explored, in their first designs of the "Tse" (Chinese ideogram for vision) computer, the possibilities of using large bundles of 10,000 optical fibers, one such bundle for each computer, and then connecting thousands of these bundles together into a large network. They soon came to the conclusion that optical technology was not yet ready for such a development, and turned back to electronics. The MPP grew out of this work.

Schaeffer's more recent (1979) vision is the following (for the 1990 era): Only 128 chips should be needed for the entire 128×128 MPP array. His goal is to make a multicomputer small enough to fit into a satellite, to be used as a complete stand-alone image processing and pattern recognition system.

Chips Containing Arrays Stacked into 3D Lattices

Next, Schaeffer wants to stack a number of such MPP layers into an array lattice. To this end NASA has let several contracts, to GE, Yale, and other groups, to explore fabricating such arrays, developing input sensors, etc. For example, GE is looking into using lasers to burn holes in the chip into which a whole two-dimensional array of pins can be inserted, to connect directly to the processors in the next layer. This, then, might be a $1000 \times 1000 \times 100$ system, all contained in a 6-in. cube.

Etchells *et al.* (1981) describe a three-dimensional system of stacked 512×512 arrays of very basic highly modular specialized processors, each on a VLSI wafer, and expect to build a "fully functional 32×32 demonstration computer consisting of 10 processing planes of memory and accumulator circuits." They appear to have successfully developed and tested a procedure for thermomigration of aluminum droplets (with the wafer heated to 1000° C) to build the feedthroughs connecting the two opposite surfaces of a wafer. Spacers separate wafers and "microspring interconnects" that stand roughly 1 mil away from the wafer's surface and are spaced 10 to 20 mils apart form what appears to be a resilient connection between wafers, with very low parasitic impedances, propagation delays, and power dissipation. They argue that "for equivalent-circuit technologies, the 3D architecture/packaging combination will always yield an order of magnitude higher performance in each of the three areas of speed, power, and volume, when compared to more conventionally designed and packaged systems."

A successful technology for stacking wafers, or chips, one that allows many interconnects between them, will open up enormously exciting possibilities for very small portable yet extremely powerful computer systems. A computer with 100 stacked 512×512 arrays would appear to be sufficient for very sophisticated pattern recognition, scene description, and modeling programs. Even 20 stacked 128×128, or 10 stacked 32×32 arrays would open up important new possibilities for real-time perception, especially for dynamically changing scenes, robot control, control of missiles, planes, and autos, and mobile vision systems for the blind. Appreciably smaller pyramids may give further increases in power.

Very Powerful Pipelines and Arrays
and Trees of Pipelines

Stanley Sternberg (personal communication) is planning pipelines that are a good bit more powerful than the already-very-powerful Cytocomputer. The new Cytocomputer chip will allow one to extend the pipeline to potentially any length. So one can conceive of pipelines with 1000, or even more, processors.

H. T. Kung (1980a) and his co-workers have been designing systems of special-purpose processors for VLSI chips in which data flow "systolically" through the chip in a manner reminiscent of a pipeline. Bentley and Kung's (1979) tree for database processing, although highly specialized, is no longer special purpose. The tree architecture they envision is reasonably dense, as the recent results on packing processor nodes near one another described in Appendix A suggest. And trees can be embedded relatively compactly into grids on chips. Another intriguing possibility is a general-purpose systolic computer, built from many chips configured in a relatively general tree, or from several specialized algorithm-suggested structures with limited reconfiguring, forming part of a general-purpose supercomputer or network.

Sternberg (personal communication) has also been thinking about circular pipelines, and about sets or arrays of pipelines working in parallel. This seems to me especially interesting because it begins to reconcile arrays and pipelines into a single system that combines the virtues of each. Note that Schaeffer's layer-stacked MPP and Etchells, Grinberg, and Nudd's three-dimensional processor, can be thought of as pipelines, the whole SIMD array at each layer being one stage in the pipe. Similarly, the "pyramids" and "whorl" systems discussed in subsequent chapters have some of the features of pipelines, as well as of arrays and of networks.

A Summarizing Comment

Several of the systems described in this chapter are under active development, with very promising results to date; others are more or less conjectural. Some of the people I have mentioned may prefer to disavow the systems I have described. But it seems to me important to think about such possibilities if we are to develop computers to their full potential.

Millions of processors (as Wittie and Sullivan have suggested) would enormously increase the power and speed of computers (assuming the problems of efficiently programming and using them are solved). A thousand very powerful processors of the sort Siegel and Briggs are considering would give tangible increases, and would be very attractive systems for medium-large computer facilities. If Cray-sized supercomputers are used, as by the S-1 group, even 64 will give enormous power.

The entire vision system in a single package, whether a single chip, wafer, or sandwich of layered chips, as Duff and Schaeffer have envisioned and Etchells, Grinberg, and Nudd are pursuing, would open up image processing and pattern recognition to a wide variety of applications. A very large $1000 \times 1000 \times 100$ MPP could handle esoteric satellite images on-board. Much smaller 100×100 arrays might well be cheap enough ($5 or $50 each seems plausible) to serve in automobiles, and certainly to serve in robots.

Some of these tending-toward-the-blue-sky ideas are ones I have heard in conversations with people I know. There are probably many others around. And I shall be presenting still others in subsequent chapters, with, I hope, some reasonable arguments for their interest and plausibility.

CHAPTER 14

Construction Principles
for Efficient
Parallel Structures

This chapter continues our examination and comparison of network structures. It first explores the variety of possible types of parallelism, possible sources of speedup from parallelism and from hardware and software design, and the range of possibilities between "tightly" and "loosely" coupled networks. Next it surveys recent applications of graph theory (see Appendix A) that have produced a number of interesting new possibilities for networks.

These graph theory results are still too narrowly focused on one or two formal global characteristics (chiefly the density of packing of nodes). But local structural characteristics (in particular a close fit of the "anatomy" of the network to the "physiology" of the program's flow of procedures) are almost certainly of overriding importance. Therefore the rest of this book concentrates on architectures that attempt to embody programs' flow in appropriate local structures, and to combine and mold these local structures into appropriately interconnected global structures. Such architectures appear to be appropriate for (a) a decomposition of the program graph into relatively separate subgraphs, each mapped onto and executed by a subcluster of the multicomputer network, and (b) a pipe-network whose structure mirrors the program's structure, with problems' information flow pipelined through it.

The Great Variety
of Possible Sources
of Parallelism

This is a good point at which to gather together the various sources and types of parallelism that can be incorporated into a network, since we have finished examining a wide variety of systems, and are about to explore extensions and new possibilities for future very large multicomputers. In theory, a single system might incorporate virtually every type of parallelism. But that would be too expensive in practice, and overall efficiency would probably not be increased by simply adding all together. Architecture design is a judicious engineering art, where the various costs and trade-offs of technologies and alternative ways of attacking and partially solving problems must be carefully assessed.

A computer that is 10%−40% more powerful than its competitors will, everything else equal, sweep the market. Users will applaud the computer that executes programs twice as fast, or at half the cost. The ranges of potential speedups and economies from the different sources of parallelism are often far greater. A designer often can afford to trade great sources of power for even greater sources. Today it is not at all clear how many of these potential sources of speed and power can be exploited. We can consider with equanimity losing an order of magnitude here, but gaining three there, and losing a factor of 2 over there. The net results might be large, but even small gains can be very worthwhile.

At this point, it is useful to distinguish between several types of parallelism: (a) actual, (b) serial, and (c) virtual.

(a) Actual parallelism actually does several things in parallel, as when several people, together, push a car, or several computers execute instructions at the same moment. It is roughly equivalent to a logical AND operation.

(b) Serial parallelism refers to a capability to effect any one of a variety of alternatives. For example, we can pick up the phone and dial any number in the phone book (but only one at a time); a computer can fetch information from any word in its main memory (but only one at a time). It is roughly equivalent to an EXCLUSIVE−OR operation.

(c) Virtual parallelism looks parallel on the surface, but may actually be effected serially, or partly serially, at the hardware level.

Conventional serial computers incorporate occasional bits of parallelism. These include (rough estimates of possible speedups in brackets)

(1) Random access to a large memory (serial).

(2) Parallel fetch or store of a word (actual) [8, 16, 32].

(3) Random access to the appropriate special processor (serial) [3−10].

(4) Parallel operations on several words (sometimes actual, sometimes virtual) [2−5].

(5) Parallel operation of several processors (e.g., I-O processors working in parallel with the CPU) (actual) [2].

(6) Instruction fetch and decoding in parallel with execution of the previous instruction(s) (actual) [2].

(7) Parallel transfer of word(s), e.g., to main memory from disks, to cache, to high-speed registers, several or all at the same time (actual) [2].

(8) Execution of several instructions (by several different processors) at the same time, as in the pipeline of the Cray-1's (actual) [2−50].

It seems illuminating to examine these in terms of graphs. Random access to memory means that one of the processor's registers is linked, as the center of a star, to each of the cells in memory. These are EXCLUSIVE-OR (serial) links, since information can pass to or from the register and only one memory cell at a time. Similarly, random access to the appropriate special processor means the set of registers into which the operands are fetched are linked, as the center of a star, to each of the processors. This is best drawn as two stars with their centers linked together. Such a graph mirrors the actual flow of information. But a conceptual graph might be drawn to emphasize the linkages; this would be a complete bipartite graph, with links from each memory word to each special processor.

Parallel computers introduce additional types of parallelism:

(9) Several processors execute the same instruction at the same time (actual) [4000−1,000,000+]. This may mean that one controller sends signals specifying the instruction to be executed and the location of its operands and results that must flow in parallel to all processors, or that each processor has stored, fetches, and executes the instruction in parallel. That is, one controller can send parallel signals to all processors (the controller linked to those processors actually or virtually as the center of a star); or the program can be loaded in each processor's memory; or each processor can access a common memory for the

program (but this would give impossible contentions, since many processors would be trying to access the same word).

(10) A window of information is fetched in parallel (actual or virtual, as when that window is actually processed serially) [4−25]. This is much like the fetch of a word. But it typically means that several connected words are fetched from other computers' memories, and that these words are fetched across direct links to the adjacent neighbor nodes. Therefore the window implies (but does not necessitate) a regular local near-neighbor linking scheme, as in an array.

(11) Pipelining, much as in the Cray-1, but with much longer, possibly specialized, pipes of processors (actual across processors, virtual across data) [2−1000+].

(12) Parallel processors, more or less specialized, several (if not all executing in parallel (actual) [3−20].

The variety of sources of parallel speedups is large. The total potential speedup is so large that few individual sources are essential.

Note that most types of parallelism can already be found in traditional serial computers, both within the single CPU, in word-parallel operations, and in parallel I-O and other types of data transfer (which is equivalent to message passing in networks). The parallel arrays and networks chiefly increase the amount of parallelism. The only place where increases can be continued potentially without limits is in the addition of separate computers, as when arrays, networks, and pipelines are made bigger.

The arrays chiefly add the very large numbers of parallel processors, plus the local window. These can be viewed as new kinds of chunks, or data-processing structures. Thus we have the 32-bit word, the 3×3 9-element window, the 64×64 to 4096×4096 processor-memory array.

The networks also add parallel controllers, which allow for parallel execution of different asynchronous instruction streams. They offer the possibility of massive parallel execution and transfer of data. But it is far more difficult to conceptualize and program an algorithm to achieve those possibilities.

Pipelines add processors that work in parallel, but each executing a different instruction (therefore each must have its own controller). From the point of view of data streaming through the pipe, each instruction is executed over all data, in a virtual parallel mode.

Sets of specialized processors on a bus or some other interconnection hardware act much as do the special-purpose processors in a traditional CPU, except that they can often be used several at a time, rather than one at a time.

Additional Speedups and Efficiencies
from Hardware and Software

Parallelism (when it is used effectively) potentially gives enormous speedups. But we should briefly examine a number of additional sources.

First comes faster hardware, in terms of gate delays, bandwidth of links for data transfers, and speed of clocking and driving the total system [2−100]. Asynchronous operation eliminates the delays inherent in clocking the system to cycle at the speed of its slowest process [2−4]. More efficient logical design (which often includes parallelism at the level of the graph of gates) is another major source of speedups and efficiency [2−10].

Directly linking computers, rather than connecting several computers via a common bus or ring can make message passing appreciably faster [2−100]. Message-passing protocols and possibilities of conflict are virtually eliminated [2−5000 or more, depending upon how elaborate is the protocol, bandwidth of buses, and message-passing requirements of the program].

Code can be executed fastest when it is built into hardware. Almost as fast, but inherently slower, is microcode; then comes machine language code, then carefully compiled code from higher-level languages; slowest is interpreted code [2−10 at each level]. One of the biggest expenses with code can simply be the loading of that code into the computer [2−10].

If a network gets slowed down by messages that reconfigure switches or gets bogged down by messages that oversee and coordinate the execution of programs, it may well be swamped by the need to pass a program's procedures as well as data to processors dynamically added to the set executing that program. A carefully designed systolic array will thus be most efficient on these grounds, since code is embedded in elegant special-purpose hardware, and gates continue to work steadily when a long stream of data flows through. Pipelines share in not needing to load new instructions. This suggests providing large streams of data, possibly by batching, to keep systems busy so that they do not have to set up and start up a new program.

SIMD arrays avoid almost all the problems and delays resulting from the general message-passing and instruction−execution capabilities of MIMD networks, and therefore eliminate the operating system's software overheads. Here, for example, we see messages passed to every computer in the array with no delays at all [2−5000+, depending upon the amount of message passing].

Reconfiguring networks that shuffle large amounts of data around in parallel can give similar efficiencies, usually on a smaller scale [2−50].

Conventional serial computers suffer from several major inefficiencies (see Uhr, 1981c):

(1) Very large memories, I-O processors, and other hardware are added to keep the single CPU busy.

(2) The CPU itself is made of several more or less special-purpose processors, but only one is actually executing code at each moment in time.

(3) When iterating through sets of data, a number of additional instructions are needed to index and test for borders.

Potentially arrays and pipelines can avoid all these inefficiencies. But note that to avoid indexing, the array must be large enough so that each processor need handle only one stream of data. And the pipeline can greatly reduce the time needed to index the stream of data by handling this with hardware, but it cannot eliminate it completely.

In sharp (and discouraging) contrast, MIMD networks of large traditional computers suffer from all these problems, plus additional sources of inefficiencies in message passing and the attendant additional overheads, and the increased likelihood of being forced to remain idle.

In general, to the extent a system is specialized to a particular job for which efficient solutions can be found, it becomes more efficient and cost-effective. But (as Shore, 1973, argues) it is also less usable, and therefore may sit largely idle a higher percentage of the time. And it will fill a smaller niche, and therefore is less likely to be bought in large quantities. Thus a beautifully designed special-purpose system may gobble up data when it is run but almost always sit unused, or cost too much because it can be sold only in small quantities, or not be built in the first place. For example, the floating-point array processors are dramatically faster than a traditional computer, and fill an important niche. But they appear to sit idle a good bit of the time, and often their cost effectiveness is questionable. Simply adding special-purpose systolic chips to a traditional CPU may have the same problems.

The ultimate measure for comparing different kinds of computers will be the volume of information they transform into solutions per unit of cost. Cost can be measured as a function of gate count, chip area, wiring and packaging, time, etc. But it is also a function of market and volume of sale, especially when very high development costs, as in the design of a large new computer, or chip, must be amortized. So the finding of new niches and important applications, the development of good programs and effective total systems, and the building up of markets all play an appreciable role in making a computer system more cost effective and more efficient.

From Tightly Coupled to Loosely Coupled and Possibilities in Between

Every computer on earth is in a network, since we can phone one another about the results our computer output. They are just very very ... very loosely coupled.

The term "tightly coupled" network is typically used today to refer to an SIMD array like CLIP or DAP, or to a network where processors share a common memory or are linked over a cross-point switch. The term "loosely coupled" usually refers to an ETHERNET-like connection of completely separate computers via a bus or ring, or to an MIMD network with processors connected via buses. Both terms have been used rather loosely, but the underlying contrast being suggested is important.

There are a number of aspects to coupling computers (or any complex systems).

The SIMD array, where each processor executes the same instruction on the same addresses (each in its own memory), all slaves of a single controller, is the most tightly coupled in most respects. Here the single controller is universal master. (Note that the sharing of memory is a different type of coupling, not feasible and not used in SIMD networks.)

But the functions of the controller can be split apart. We can envision, for example, each processor executing the same instruction, but each on different locations in (its own) memory; or each processor executing a different instruction, but on the same locations in memory; or each processor fetching from different directions (but this would introduce contention problems that the standard SIMD array architecture largely avoids).

Today's SIMD arrays are all synchronous, since all processors execute the same instruction. They could be kept synchronous even if different instructions are executed, simply by having the system clocked to the slowest instruction (but that would waste time, possibly large amounts of time, if some processes took much longer than others).

The physical linkages can vary in many ways. Processors can connect directly to processors, or through registers, or through shared (local or global) memories, or through buses. Each of these types of connections can be direct or through reconfiguring networks of switches.

Processor — processor connections can easily waste a lot of the processors' time. A bus with many processors can easily generate bottlenecks and delays.

The bandwidth of the connection can vary greatly. Within-chip links can be fast and simple; from chip to outer environment they can be slow and cumbersome. A chip's pin or wire can transmit 1 bit at the rate of 10^4–10^6 bits/sec; coaxial cables and sophisticated parallel interfaces can handle 10^7 bits/sec or more; optical fiber links should be able to handle 10^{11} bits/sec or more.

Possibly the most important determinant of coupling is the operating system. An ETHERNET is loosely coupled because many processors are interfaced to a common bus with mediocre bandwidth (typically on the order of 50 processors on a 10^7 bits/sec bus). But it becomes very much more loosely coupled because the operating system is built and tuned to handle electronic mail and document handling (sharing, editing, displaying, printing, etc.), rather than efficient execution of a single program by many processors. Most of today's operating systems ignore the structure of the network, and therefore take as long to send messages between near-neighbor processors as between the most distant processors.

Possibly the best measure of the degree of coupling is the amount of time needed to execute a typical instruction on data not stored in the processor's memory. Here we see a DAP processor taking 250 nsec to fetch and process data in its own memory or any of its four neighbors' memories. In contrast, as we saw in Chapter 8, an Arachne (Finkel and Solomon, 1980b) or MICRONET (Wittie and van Tilborg, 1980) processor takes 10 μsec to fetch one word of data in its own memory, but 10 − 35 msec to get that word (as part of a 10- to 40-byte packet) from some other processor's memory.

This is an appropriate spot to reemphasize the crucial importance of designing algorithm, program, network structure, and operating system all together. The efficient execution of the program is the ultimate goal.

Promising Construction Methods:
General Principles

Recent applications of graph theory to the problem of building large graphs suitable for computer networks seem to indicate that several major types of construction procedures may be emerging.

This work has concentrated almost entirely on issues of diameter and girth, and in particular on finding "globally denser" graphs (that is, with more nodes for a given diameter and degree). But a variety of techniques for finding and constructing new denser graphs have recently been devised (see Arden and Lee, 1978; Imase and Itoh, 1981, using de Bruijn, 1946, networks; Goodman and Sequin, 1981; Uhr, 1980; Leland *et al.*, 1980; Leland, 1982; Bermond *et al.*, 1982), and most appear to give graphs improved on other desirable characteristics as well, including connectivity and symmetry.

Hoffman and Singleton (1960) proved that the only nontrivial graphs that achieve the upper bound that E.F. Moore pointed out to them for the number of nodes in a graph of given degree and diameter are the 10-node Petersen graph (with degree 3, diameter 2) the 50-node Singleton graph (with degree 7, diameter 2), and, if it exists, a not-yet-found graph of degree 57, diameter 2. These "Moore graphs" have a number of apparently desirable characteristics, including maximal girth, very good connectivity and cohesion, and outstanding symmetries.

Many new, denser graphs have recently been found. They include de Bruijn networks, some of the compounded graphs achieved by the construction operations of Uhr, Li, Leland, and Bermond *et al.* and several graphs constructed by a powerful heuristic search program written by Leland.

One might conjecture that to be very dense a graph must be highly symmetric, and that symmetry underlies a variety of other desirable characteristics as well. But the whole problem of what is the most appropriate set of characteristics for a large computer network is one about which discouragingly little is known. Density, connectivity, and especially symmetry are certainly desirable. But when programs are mapped onto architectures so as to maximize information flow, the most appropriate are probably measures of local density and other local properties. *And a structure that best fits the data flow of a particular program or class of programs is almost certainly of even greater importance.*

We now know how to build networks appreciably denser than trees, by augmenting the tree, to reduce its diameter; yet the Moore bound shows we can hardly ever reduce the diameter to even one-half that of the original tree. So the relatively small increases in density that can occasionally be achieved may be less important than good local structural characteristics.

It seems more important to construct networks *with good local properties,* including local density and symmetries, but also with *a structure that mirrors the structures of the programs they will execute.* This implies developing operating systems that try hard to assign each individual program to a closely connected set of nodes. The overall structure should use as many links to tie clusters together as are needed to handle the (if all goes well relatively rare) message passing between clusters.

We can now build structures by addressing two major issues: (a) local cluster characteristics and (b) global characteristics. Several preferable procedures appear to be emerging:

(1) *To optimize global characteristics, use a properly augmented tree.* In particular, if global density is the overriding criterion, use de Bruijn (1946) network shift registers, since Imase and Itoh (1981) have proved they give the densest graphs found so far, starting at roughly 15,000 nodes.

(2) *To optimize local characteristics, choose one, two, or a few local clusters, using whatever criteria seem most appropriate.* Then:

(i) *compound these clusters,* using whichever of the recently discovered compounding operations seems most appropriate; or

(ii) *choose an augmented tree,* in terms of whatever global properties are most appropriate, and replace each of its nodes with a cluster; or

(iii) *tile these clusters,* using a tessellation pattern that best fits the needs for data flow between clusters.

De Bruijn networks are (for very large but not for small graphs) the densest found to date. They are also relatively symmetrical, since they are basically trees that have been augmented by adding links to their buds (including links that join buds to internal nodes, thus raising degree by 1). But if a program is mapped into a relatively small local subgraph of a large de Bruijn network, the local density of that subnetwork will be relatively poor, since it is simply a tree, locally. Even a large program will almost always decompose into several closely coupled structures of procedures, and each of these structures might best be mapped into local subgraph clusters. So unless one is to run a very large program on the entire de Bruijn network without bothering to map related procedures into neighbor nodes, its good global density may not result in good message passing (which will almost always rely on local density).

Therefore the two-stage procedure, of choosing appropriate clusters and then compounding them into an appropriate overall architecture, may be the best approach. It also simplifies our problems, since they need not be handled all at once. And the original cluster graphs are relatively small, so that much more is known, or can be learned, about them. Only relatively small graphs (up to 100 nodes or so) approach the Moore bound at all closely. And a heuristic search program like Leland's can be used only for graphs with up to 500 or so nodes.

Clusters can be chosen for structural reasons as well. This follows the general philosophy that the algorithm and program should determine the anatomy of the network that executes it. The array and lattice are obviously widely usable basic structures. They will often be used in the example designs that follow. Other types of clusters, chosen for other criteria, might similarly be combined into other types of overall structures.

Load-Balancing Processor, Input, Output, and Memory Resources

There are several crucial reasons for the strong emphasis on serial techniques that has dominated computing. Serializing

(a) gives underlying theoretical models like Turing machines their analytic powers;

(b) makes algorithm development far simpler using our serial-conscious minds;

(c) makes building computers far simpler, since we need only one processor.

After almost 40 years we feel we have become expert at serial techniques. Iteration, recursion, random-access memory, depth first and heuristic search, divide and conquer, top-down programming, and structured programming are often pointed to as serial.

But many of these are actually parallel or neutral. The random-access memory makes large amounts of data potentially available in parallel; only the hardware limit of one word fetched at a time makes it serial. Divide and conquer and iteration are clearly tools of parallel processing (today usually simulated by a serial machine). Structured programming ultimately means structured to the computer architecture, itself structured to the algorithm, with whatever mixture of parallel and serial the problem needs.

Parallel versus serial is a pseudoargument. Almost every system combines the two. The real issue is to optimize the amount and types of parallelism with respect to the costs of space, time, and effort and the human—years and difficulty in developing the needed formulations, algorithms, programs, and hardware structures.

Balancing input, output, and processing is an interesting way of viewing our problem. Parallel processing increases speed (bandwidth), and therefore needs faster input and output.

A transistor is at bottom an input—output wire that does some work in between to effect some transformation, and therefore takes a bit of extra time. The bandwidth of a processor is typically the bandwidth of its fetch and store operations (which themselves are functions of that processor and the memories in which its data are stored). Fetch and store are input and output for that processor. Just because they are quite slow, most processors have high-speed registers for indexing and for other repeated uses of the same data. Typically, fetch and store might each take 1 time unit, addition 2−10, multiplication 10−100, and transfers from and to high-speed registers 0.1. Typically, 16,000, 64,000, or 1,000,000 or more words are available in high-speed memory for immediate fetching. This large memory buffer is needed to store the program (which must also be fetched) and to make sure all needed data will be available without the far longer delays of input from slower memories.

A processor that fetches or stores data in 1 μsec has a one-million-word input—output bandwidth. It will typically take longer to process those data, e.g., 5 μsec to add, 20 to multiply.

When I-O becomes a bottleneck we can introduce new input processors that work in parallel, plus more memory to store and buffer input data until needed. This should be no problem, since I-O is faster, simpler, and cheaper than actually transforming. But pin fan-out bottlenecks and, if and as increasingly dense VLSI chips are achieved, the relatively smaller surface for input—output surrounding the region containing the processor can impose limits. It is not clear how generally one can design around them.

For a 1-μsec computer a megaword bus is usually adequate. But the more computers we put on the same I-O device like a common bus, the lower the bandwidth to each (unless we can ensure they need I-O only intermittently and we can spread their needs out in ways that cause no delays). Thus even four or eight computers on a common bus can make I-O a bottleneck unless that bus is given four or eight times the bandwidth. The relative cost benefit of the bus becomes a relevant factor.

But today's operating systems take milliseconds for I-O. This is comparable to the imbalances of disk or tape access in contrast to high-speed memory, and calls for strategies for loading data before they are needed. Increasing the physical bandwidth of the I-O channels will not speed up such a software-limited system. Only by eliminating this software overhead, either by using more appropriate structures (e.g., point-to-point links rather than common buses), simplifying the message-passing protocol, building some of the specific code into hardware, and/or allowing and asking the programmer to handle communication by organizing procedures, data, and message passing so as to load balance and to minimize contention and I-O problems, can we expect networks to be efficient.

Parallel processors increase the system's bandwidth. Simpler processors (like the 1-bit processors in the large arrays) increase bandwidth—if they can really be used more efficiently. The I-O bandwidths must therefore be increased commensurately. Increased power and efficiency from parallel multiprocessors will be realized only when we have learned how to augment I-O, manage memory to give commensurate bandwidth, and orchestrate message passing to eliminate overheads and delays.

CHAPTER 15

Designing VLSI Chip-Embodied
Information-Flow Multicomputers

The basic VLSI chip is roughly $4-8$ mm square. Circuits of devices and wires are packed in near-planar fashion as densely as possible. In the early 1980s this means transistors roughly $2-10 \times 2-10$ μm are linked by 1-μm wires, and separated by $1-4$-μm spacings. These sizes are continually being decreased. The resulting VLSI packing densities of 10^5 (today), 10^6 (by 1986), 10^7 (by 1990?), and 10^8 (by 199?) offer the real possibility of designing highly modularized information-flow multicomputers, where each chip executes a relatively self-contained process on the data being pumped through it, and the whole multicomputer organizes these separate (highly nontrivial) processes into whole (very large) programs.

Programming may well become the design of algorithms in silicon. Architecture design and program development may well combine into the design of properly structured data flow multicomputers. The overall constraints -10^0-10^5 chips, each with 10^5-10^8 gates and with 128 or 256 off-chip pins—offer many interesting problems, and enormous possibilities.

Some designs will be highly efficient special-purpose embodiments of specific algorithms. Others will be general-purpose systems whose chip components are modularized, as appropriate, into more or less special or general components.

The overall problem now becomes the development of algorithms that can be mapped into and executed efficiently by such multicomputers. This means the algorithm should decompose into modules that map compactly into one chip (or a module of chips), and modules should be interconnected so that the whole flow of information through processes is also handled efficiently.

"Efficiency" means minimum area on the chip for the circuitry (transistors and wires), adequate pins (I-O) from the chip, so that there is no slowing because pins must be multiplexed or processors must wait for inputs, and minimum wires (and also reconfiguring switches) between chips. Underlying these is the maximization of the "volume" of data flow through the system, as measured by the total physical resources filled with and doing useful work on data at each moment.

Underlying these, efficiency means minimum costs of formulation, design, and construction (including work on the algorithms that the system will execute). Ultimately, far more important than efficiency is the value of the programs that the system executes. That is, the most important multicomputer (hardware—software—program development) system we can develop is one that stimulates and obtains new programs that handle new problems at higher levels of performance.

This suggests a modularized design procedure, each chip treated separately but within the larger multichip system, with its pins, switches, and larger I-O and power problems. Such a design should cycle through the various levels until reasonably optimized. It should start with a firm idea of the kinds of algorithms to be handled. This should lead to decisions whether to handle them with one general system, with a reconfigurable (or partially reconfigurable system), or with several specialized systems.

Embedding Networks on VLSI Chips

A network consists of nodes and links. Each link has a certain width (dependent upon the wire width and spacing requirements for a particular technology). Each node has an area (dependent upon the device size dictated by the particular technology) and a perimeter (this can be made a function of the total width requirements of the links joining it, as well as the area requirements of the device size). Each technology sets its rules for embedding the graph of device—nodes and wire—links into the chip's several-layered landscape, in terms of area needed for devices and wires, heat dissipation, pin fan-out, etc. But in general, chip perimeters determine input—output capacities, and interior gates can be linked, and programs written, so that the times they spend computing are commensurate with the input—output possibilities and limits.

The problem of embedding a multicomputer network of processors and memories into a set of VLSI chips is today being examined in the form of the problem of embedding graphs whose nodes are computers, or gates, into grids so as to minimize area (usually of a bounding rectangle). Here, then, is still another criterion for packing nodes, rather than the density and average distance measures used by researchers trying to develop good topologies without considering the constraints of chip technologies.

Today most people exploring the problem of mapping networks onto VLSI chips, (e.g., Thompson, 1979, 1980; Leiserson, 1980) assume that

(a) Wires run and devices are oriented in only horizontal and vertical directions.

(b) Everything is embedded in a square grid.

(c) Devices are square.

(d) All device nodes are at the same layer (this is often called "planar," but note that since wires lie in two or more additional layers, nonplanar graphs can be embedded).

(e) Wires lie on two (or a constant number of) parallel layers; therefore there can be at most a constant number of crossovers at any point.

(f) Wires have a minimum width and minimum spacing.

VLSI chip design rules, especially when simplifying assumptions are made to allow for quick design, as by Mead and Conway (1980), typically specify a grid in which the graph structure of linked gates must be embedded. Vuillemin (1980), Chazelle and Monier (1981), and others have begun to explore the space and time complexity of different simple algorithms when effected by optimal embeddings of the computation in such a grid. These grids have to date typically been checkerboardlike, with links allowed only in the vertical and horizontal directions. This problem is reminiscent of the problem of laying out circuits in logic arrays (Akers, 1971), IC and LSI chips (Uehara and Van Cleemput, 1981), and other types of design regimens.

Whatever the raw materials from which a network must be built, there will be inherent constraints and optimal packing possibilities (Dertouzos, 1971). Often it will be desirable to specify a user interface that allows for easier design, at the expense of less-than-densest packings. (Here we see a typical trade-off between user development time and the cost of building and running the final system. So costs must be considered for the total development—use cycle.) At that point the user is confronted with a graph embedding problem, where one graph must be matched with a subgraph of another.

Possible Worthwhile Relaxations of VLSI Design Assumptions

Several relaxations of today's assumptions would appear to be desirable and worth exploring.

First, there may well be an iterated basic mesh to the system that is other than a square array. In particular, arrays where each node has eight links, connecting to four diagonal as well as to four square neighbors, are frequently of interest for image processing and number-crunching tasks.

Similarly, hexagonal arrays (Golay, 1969; Kung, 1980a), where each node has six direct connections to its hexagonal neighbors, offer attractive, as yet little explored, potential power. And they are quite appropriate for VLSI technology.

Trees map reasonably well into the chip, as typically done via square grids. But diagonal wires might increase their mapping density appreciably, by a factor of $2-10$.

The basic rules of silicon appear to be the following:

(1) Electrical characteristics determine transistor area, wire diameter, and minimal spacing between components.

(2) Transistor area determines the possible range of transistor perimeters, which determines the number of wires that can be joined to it.

(3) Chip yields (which are a function of many variables, including packing and heat dissipation of the circuits etched onto that chip and the number of pins from that chip) determine the total size of the chip.

(4) Pins from the chip determine the maximum number of links (without delays from serial data transfer over the same pin) from the subgraph etched onto that chip to the outside world of other chips.

(5) The possibility of crossovers (giving nonplanar graphs) is introduced by allowing special junctions for wires that burrow into a substrate layer.

Within the limits of these constraints a variety of regularizing, simplifying assumptions can be made, in order to expedite the design of real systems, and to analyze pertinent idealized problems by using graph theory and other appropriate tools.

Today's simplifying assumptions for computer-aided design (CAD) impose horizontal and vertical links in the square grid. But the basic VLSI technology does not favor horizontals, verticals, and rectangles. So whenever it is convenient to draw a network with wires and borders of different orientation this might be encouraged. Any saving in area from elegant customized designs of the lowest-level circuits, like memories and logic, will, because they are used so widely, have a major effect on efficiency. This is much like the issue of careful machine language coding of inner loops in programs compiled from a higher-level language.

Other issues, of crossovers (which, depending upon the technology, are more or less likely to decrease yield), of time to traverse wires and gates, and of energy dissipation, might also be taken into account.

Thus a number of specific variants might be examined, e.g.:

(1) Time of propagation along a wire might be a function of length (Chazelle and Monier, 1981).

(2) Gates might take any orientation.

(3) Wires might take any orientation.

(4) Gates might be rectangular.

(5) Gates might have areas and shapes (perimeters) of any sort.

(6) Crossover junctions might be specified when wires are needed in layers other than the two basic horizontal and vertical layers. (This would accommodate more ports.)

Embedding Different Graph Topologies into the VLSI Chip

Thompson (1980) gives a lower bound for A (area) and T (time) for n-input, n-output functions like integer and matrix multiplication, polynomial evaluation, discrete Fourier transform, and sorting:

$$AT^2 > O(n^2)$$

(this is a lower bound). Several key networks embed naturally in the plane. These include square, diagonal, and hexagonal arrays, and, indeed, all arrays and tilings of locally planar networks. Trees embed in chip grids (Mead and Rem, 1979) of area proportional to the number of nodes. Paterson *et al.* (1981) explore several possible embeddings, and Yao (1981) proves that a tree (to compute binary addition or any Boolean function) can achieve the lower bounds of $A = O(n)$ and $T = (\log n)$.

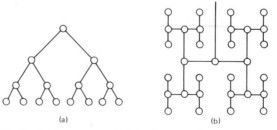

(a)   (b)

Fig. 1 A binary tree embedded in a two-dimensional grid.

Valiant (1981) shows that embedding arbitrary graphs, and also planar graphs, into a planar graph is quadratic (whereas embedding a tree into a planar graph is linear). Valiant uses the planar separator theorem of Lipton and Tarjan (1977) (in any n-node planar graph there is a set of $2 * 2^{1/2} n^{1/2}$ nodes whose removal leaves no connected graph with more than $\frac{2}{3}n$ nodes) to recursively split planar graphs and embed them into nonplanar graphs in $O(n \log n^2)$ space.

Augmenting trees with links at their buds, or with internal links, can give problems. Probably the best augmentations in terms of the recent graph-theoretic results in increasing density, connectivity, and symmetries (see, e.g, Imase and Itol, 1981; Leland *et al.*, 1980; Bermond *et al.*, 1982) are more difficult to realize (since they are nonplanar) on a near-planar chip than are a number of other, possibly almost as good.

The shuffle-exchange networks (Stone, 1971) are also quite irregular (which may be a function of their ability to permute data). Thompson (1980) has shown that an n-node shuffle-exchange network needs at least $O(n^2/\log^2 n)$ area on a chip, and Kleitman *et al.* (1981) give a layout that achieves this limit. This means that small nets shuffling a few thousand nodes will become feasible, but not larger ones.

De Bruijn networks (de Bruijn; 1946; Imase and Itoh, 1981), hypertrees (Goodman and Sequin, 1981), heuristically augmented trees (Leland *et al.*, 1980), and other relatively dense augmented trees appear to be poor candidates for embedding in a chip grid, just because of the links that draw distant parts closer together. N-cubes and other nonplanar graphs are probably similarly poor.

Theoretical Bounds and Actual Achievements

It is not clear how relevant these kinds of results will be for the realization of designs. When actual graph layouts are achieved they become powerful candidates for modules in the larger system. They must be integrated into that total multicomputer in an efficient yet flexible way.

Galil and Paul (1981) prove that there exists a parallel computer [using Preparata and Vuillemin's (1979) cube-connected cycles to interconnect nodes] that when given a description of a t-time-bounded p-processor-bounded computer C with m nodes simulates C in time $O(t \log p \log m)$. Valiant and Brebner (1981) show that there exists a "realistic" computer (a bipartite graph with links between processor nodes and memory nodes, of some realistic degree d) that can [using an $O(\log t)$ "d-way shuffle" randomized sorting algorithm] simulate any "idealistic" p-processor system (where all processors share a common memory, thus forming a complete bipartite graph), with only $O(\log p)$ loss in efficiency.

On the one hand these results suggest that in theory we might use a variety of interconnection topologies with only logarithmic degradations in performance. On the other hand, when actual systems are built, tenfold, twofold, and even 20% losses in efficiency can be of crucial economic importance. Note that augmented trees and compounds of the sort examined in Chapters 4, 7, and 14 would appear to be more efficient topologies, but embed less efficiently.

Constructing VLSI-Based Interconnection Topologies

There appear to be no known general principles for drawing graphs, much less for drawing graphs to minimize area. So for the moment it seems best to explore some rule-of-thumb construction procedures. For example, one might start by laying out some known subgraph of special interest—e.g., a subgraph that is itself dense; or that appears to be at the "center," "corner," or "edge" of the graph (placing it at the center, corner, or edge of the chip); or that serves as a connector, e.g., as a star with spokes that connect many graphs. Then the additional parts of the graph might be added as needed.

The above assumes that the graph has been designed first (whether or not with the layout problem in mind); then the layout of graph on chip is a subsequent, completely separate problem. It seems far more reasonable to make the design process itself one whose central problem is to thoroughly populate the chip with as many powerful components as possible, placed and interconnected to effect as much useful computation as possible.

This suggests designing relatively small modules, then tiling, compounding, or linking them into larger modules. Small modules can be made more efficient. For example, a tree of degree 5 and diameter 2 with four offspring can be embedded in a 5×5 grid (if diagonal links are used) leaving 4 of the 25 cells to serve for I-O.

Now this tree can be linked to other trees, or/and to other kinds of small modules. Any empty cells might as well be filled, giving extra processors (or I-O processors). Reconfiguring switches between modules would allow for a variety of topologies. For example, larger trees could be built from these small tree modules, and also trees with arrays at their buds, or placed wherever the program dictates that they would be of use.

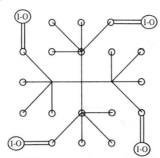

Fig. 2 A small module of a tree embedded in a 5 × 5 grid.

It may well be that the best design practices (at least until we develop a much better understanding of the problem of embedding networks in limited areas) are of the following sort:

(1) Construct a tree and then add those links that can be added, given the constraints of the layout achieved so far.

(2) Construct small local trees that can be embedded efficiently, then tile and link them together as desired.

(3) Construct small local clusters of any desirable sort, then tile and link them together.

(4) Emboss small local clusters of any desired sort into an appropriately augmented tree or other structure that spans a larger area of the chip.

Examples of Related Analogous Design Problems

Designing a computer has interesting analogies to designing a building, block, neighborhood, or city; or designing a kitchen, restaurant, or food processing plant (among other loci for performing sequences of operations).

Consider the design of a house in the 1980s. The available technologies offer a variety of standardized, modularized components. Almost all of them are square cornered, and (in the U.S.) have dimensions that are multiples of 6, 12, or 18 in. Doors and windows (among other components) fit only in certain places, in certain ways; and they are expensive and hard to make energytight. Most people (in the U.S.) want their houses to have a kitchen, bathrooms, closets, and private bedrooms. Three bedrooms, two baths (and/or showers), one kitchen, living room, etc., might be called the "standard-purpose" house. But each family has (or thinks it has) its own needs and preferences.

An individual hires an architect (at extra expense) who designs a special house for that family (or adapts one of the architect's previous designs). Such an "architect-built house" is usually far more expensive since it attempts a custom design and pushes against the economics of mass merchandizing and fabrication. It often leaks, needs redoing, and is in general less bug-free than the mass-produced house. It often appeals more to its owners, who then may find it a bit harder to sell. But the people who finally buy it (if they pay the owners' original price) have probably been chosen by the house; it appeals to them also.

When a new house technology emerges (which is very infrequent), people try to put the old plans into the new package, usually with relative waste. Assume we can build only round domes (which can be built quite cheaply, but are not, largely because they do not fit with square ideas). How do we fit in flat doors, walls, pictures, furniture backs?

The house constraints must be lived with, and exploited, in building the total design. Assume the set of room dimensions and linking doors between rooms as given (as by a builder who ran out of money, then sold the shell to somebody else). This is the closest to today's situation with square chips with few pins and computers of hundreds or a few thousands of chips. One would naturally redesign the way sinks, refrigerator, and furniture were placed, the assignment of rooms, the flow of traffic from room to room. That is, the anatomy of each room, and of the whole house, would be designed with the dynamic uses to which the rooms are put and with the flow of traffic through these uses, and also with the overall fixed and given structure of the rooms and the whole house in mind.

The operating system and the implementation language would also be designed at the same time (e.g., the fuse boxes, lighting controls, knobs, and keys). But they would be subservient to the purposes of living efficiently and comfortably that they served.

If this were a factory, it would be increasingly important to maximize the utilization of space in order to maximize the total throughput and output of produce. That is, the design would be considered successful to the extent that the cheapest factory (including its machines and workers) produced the most, best, most profitable (that is, buildable at low cost with respect to sales value) goods.

Consider the factory that is placed in a building bought when discarded for its original purpose, or a building or set of buildings that must be laid out according to the restrictions imposed by the land's terrain and the city's zoning rules. One would not simply take the design of the previous successful factory and duplicate it, ignoring that half of the building was thus left unused, or blithely adding extra extensions so the plan would fit, ignoring that even more space were being left idle. Instead, one would redesign the factory's conveyors, belts, and flow of processes to fit that building as well as possible.

Similarly, with the radically different boxes the VLSI chips will offer and the strong constraints they impose on input and output and the interconnections between them, it would be very poor design practice to take as given previous designs of CPUs and computers and try to fit them within this new and very different straitjacket. The people actually designing VLSI chips do not think of doing such a thing for the individual small modules, like memory or simple logic or arithmetic processors. But the larger design of the CPU often tends to be imposed upon them, by tradition and inertia as well as a lack of time to explore the potentialities now open. There is a great need for a large amount of open-minded exploration and rethinking of how computing is done, and how it might best be done in the new terrain of thousands of VLSI chips. The building materials and blocks on the one hand, and the algorithms, programs, and problems on the other hand, should determine the actual layout and control of the ensemble of workers that forms the hardware — multicomputer — software operating system/language.

Arrays of Processors Surrounding
Local Shared Memories

Arrays are among the densest graphs with which to populate the chip [see Sequin (1981) for an interesting design]. Consider arrays whose nodes are the following different sorts of processors:

(a) $1-5$ gates, effecting a single primitive operation.

(b) $5-10$ gates, effecting a simple logical or arithmetic operation (almost certainly a special-purpose processor in the spirit of the systolic arrays).

(c) $50-500$ gates, giving a general-purpose albeit simple 1-bit processor like STARAN, DAP, or CLIP.

(d) 2000 gates, giving a simple 8-bit or 4-bit processor.

(e) 10,000 gates, giving a conventional 16-bit or 32-bit computer (albeit with very limited memory from the traditional perspective).

(f) 100,000 gates, giving a relatively powerful traditional computer, albeit still with very limited memory from the traditional perspective.

In each of the above cases, necessary memory is part of the node. With 100,000 gates, memory might be as much as 90% of the total gates; but in the other cases memory must be limited to 50%, or even less. But as the node grows larger the space interior to each local cycle of cell in the array grows larger. Here we can assume that memory spills into whatever empty space is available, and it seems attractive to have this memory shared among all the nodes surrounding it.

Such designs enormously simplify the chip layout problem. First, one must develop a good dense design for the individual node. But, since most of the nodes are small and simple, this is relatively easy (compared to the traditional problems in designing larger processors). The density of processors will be determined by the size of each processor, which gives the basic array spacing. This also determines the space contained by each cycle of cells, and hence the size of the (shared or separated) memory(s).

A variety of considerations all point to arrays that are grids with processors at the crossings and memory spreading into each cell. The cycle of processors surrounding each memory shares that memory, and therefore interconnects via and sends messages via that memory. This seems desirable from the point of view of VLSI technology, where iterated small modules give densest packing. The memories are highly modular, and the (at least sometimes) relatively small and simple processors can be very carefully designed, and will be iterated many times. Arrays have a relatively good density, connectivity, cohesion, and symmetry. Local connectivity is especially good.

Designing a large graph by compounding structurally desirable and appropriate small graphs and tessellating them out into the region available appears to be a good procedure, from the point of view of ease of design and control of local properties. And arrays are especially well fitted to at least several very important classes of problems, including perception (image processing, pattern recognition, scene description) and number crunching with large matrices (e.g., for weather prediction, modeling masses of matter, wind-tunnel simulation).

The most straightforward, and most efficient, way to use such a system is by having data input via a row of pins to the first row of processors, and then to continue to pump the transformed data through subsequent rows. Thus the whole chip can be used as a two-dimensional pipeline through which one-dimensional arrays of data are pumped.

The major bottleneck is probably the number of pins that can be placed on a chip. Assuming 64, 128, or 256 pins *in toto,* only 32, 64, or 128 bits input at each cycle seems plausible. This suggests rows of from 32 to 128 1-bit processors, or 4 to 16 8-bit, 2 to 8 1-bit, or 1 to 4 32-bit processors. Or, alternatively, the processor might be designed to be reconfigurable to do either, e.g., 32 1-bit logic operations or 1 32-bit numeric operation, or 4 8-bit numeric or byte-matching operations, or combinations of the above.

Assume that a gate needs an $l \times d$ (length by depth) rectangle (including surrounding separating space) and a wire is w units wide (including surrounding space).

A variant would mix memory gates with processor gates, e.g., in checkerboard fashion with red squares processors and black cells memory, or with every other row memories, or with memory surrounded by 8 processors in a 3×3 square or surrounded by 6 processors in a hexagonal array.

The next step is to have each processor made from a small number of gates, on the order of 5 or 10. The processor will be quite special-purpose, and therefore a number of different types of processors might be interspersed throughout the array.

The only fundamental restriction on shapes is that a processor should be adjacent to the memories and processors to which it links (to keep wires minimal). The array might be conceptualized as "square" but the actual processors might be long, even curved and convoluted, in general the processors accessing a common memory surrounding them. (Bubble memories might be placed like mortar between rows of processors, so each could access that memory by waiting for the circulating bubbles.)

Assume each gate needs 5×5 units on a 5000×5000 unit chip. This would allow up to a 1000×1000 array. But the 32 to 128 pin input would be a rather stringent bottleneck. Large buffer memories would be needed near the pins unless pipelines could be designed to keep processors busy with only the restricted flow of information allowed by the pins. An attractive layout might be a tree that fans out from the pins, with adjacent buffer memories filling the corners of the chip.

A 10-gate processor would minimally need 15×15 units, but more realistically 30×30. A 100-gate processor would minimally need 100×100, but more realistically 300×300. The amount of memory needed would determine the spacing between gates, since memory can fill spaces within cycles.

The 100-gate to 300-gate processor fits rather nicely within the pin constraints and chip size. Depending upon whether they could be designed, 10,000-gate 32-bit processors that needed 200 to 500 units square could be put 10 or 20 to a row, which would be reasonably compatible with $64 - 128$ pins.

Chips built from regularly spaced micromodular components also have intriguing fault-tolerance potentialities. Rather than reject a whole chip because of one or two tiny defects, the system might better shunt out the one or two tiny modules that contain that defect.

Designing in Terms of Chip Area

Assume chips with roughly 1-μm resolution, which might mean wires 1 or 2 μm apart and transistors $2-10$ μm square. This is the kind of chip that is thought of as holding 1,000,000 or so reasonably packed devices. But with unusually high degrees of packing, increases of up to an order of magnitude might be approached.

A number of alternative chips of great interest are now possible, and each may well be best for its set of purposes. Total systems that use a variety of these chips are also of great interest, and should be explored.

(a) At the traditional extreme, each chip would have one processor, much like today's CPU, plus as much memory as could be crammed into the remaining space. The processor might well be made as powerful as possible, since there is enough space for a super Cray-1, yet not enough space for all the needed memory. Such a system would use several additional chips for additional memory . Therefore a total large system of, say, 10,000 chips might give a network of 1000 or fewer large computers, or 2000 or 3000 computers at the most.

(b) Next, a small number of processors (4, 8, 16, or, possibly, 32) might be put on a single chip, all interdigitating a common memory that fills whatever space remains. Here we can assume processors each with $10,000-50,000$ gates, e.g. each needing a 1000×1000 μm area on a 6000×6000 μm chip. Thus eight processors would use 8/36 of the chip area, and memory would predominate. This would also give an interesting kind of design, with the shared memory serving for message passing, data flow, and convenient access to a common program. Two types of systems might be developed:

(1) All memory is of this shared sort, so that the total amount of memory per processor is now drastically less than in today's traditional computers. Such systems might have $40,000-320,000$ computers on a 10,000-chip system, with the on-chip local clusters interconnected in any way desired.

(2) Additional chips might be used to give each computer its own traditional memory. This would cut the total network quite drastically, to $1000-3000$—since now the traditional philosophy would prevail and each processor would need the same amount of memory as in (a) above.

(c) A relatively large number of traditional 8-bit computers might be placed on each chip. Assume 200-μm-square areas for processors, which would now be placed in arrays (or/and trees). A 16×16 array now becomes viable, albeit with interspersed memories that totaled to far less than found in traditional computers, instead approaching the typical 1-bit array computer's memory. Such a system might achieve 1,000,000 or more processors on 10,000 chips. And a number of alternatives would make attractive combinations of such chips with chips of types (a) and (b) above. For example, an array of type (c) might be connected to a tree or pyramid of types (a) and (b).

(d) A very large number of 1-bit computers might be used, arranged in arrays or trees. Here it would be especially desirable to have links in the substrate for crossings to connect at a distance, and/or circulating memories that could serve to pass messages and data. Assuming 50 to 500 gate processors, with each processor an especially modular design (because it is small and simple, and can be optimized), and only one or two controllers on each chip, a 20- or 30-μm-square processor seems plausible. This would allow 32×32, 64×64, and even 128×128 processors per chip, with at least an equal amount of memory. Thus 10,000,000 or more computers of this sort might be achieved! Again, alternatives that mixed an array built of chips of this sort with chips containing more powerful computers would appear worth exploring.

Using Circulating Memories (e.g., Bubbles, Shift Registers)

Interesting possibilities arise when memory circulates past processors, as in shift registers and bubble memories. Assume some pattern of processors, with memory filling enclosed but otherwise empty regions, and extruding through small gaps between processors. To simplify, consider a row of processors touching a continuous memory row that touches another row of processors. If information circulates, then any processor can store into the memory, and any processor can fetch that information.

This effectively gives a complete graph of links between all processors, of the sort that a separate bus would give, and also without any waste, since the links serve also as buffer, exchange, and main memories. Such a setup depends upon the appropriate technologies, and also a design of the sort we have been pursuing, where processors and memory tile and extrude into and around one another much as living systems multiply to fill empty spaces.

Summary Discussion

The theoretical examination of embeddings of graphs into grids or other graphs that in their turn appear to embed well into VLSI chips is increasing our understanding of our problems, and of techniques for coping with them. A real computer's programs will have some mix of input, output, and computing, and the fan-in, fan-out, logic, and memory can be allocated accordingly, trading off computing time versus chip area as desired. But it seems preferable to design specific architectures for their structural and formal properties. We design a house, a factory, or a town by composing a suitable structure for the particular real estate, not by mapping graphs of arbitrary possible structures into the plane.

The enormous numbers of devices involved, and the modular structure of our algorithms and programs, strongly suggest designing large chip layouts in terms of small local modules, then building modules of modules. This plus the ability to allow for at least small numbers of crossings, thus requiring (at worst) almost-planar rather than fully planar graphs, and the ability to fill empty spaces with local memory, may make it relatively straightforward to pack VLSI chip real estate reasonably well.

We can even design specific algorithms onto the chip. For example, since shuffle exchanges do not embed efficiently on a chip, we might search for better data-permuting graphs in terms of chip area. Arrays and trees, which are efficient as well as structurally attractive, appear to be good starting points from which to search for good augmentations in terms of chip area.

Relatively little is known about the range of algorithms that multicomputers will, potentially, be able to execute. So their whole development should itself be treated as a cycling design process, as new algorithms are searched for and found. Therefore it should be made as simple as possible for algorithm- and problem-oriented people to design such chips and systems, jointly with algorithm design, with an absolute minimum of special knowledge of hardware needed. This poses large problems for developing good support systems of program-design language, interactive CAD environment, and on-line algorithm development, modification, and testing packages.

The role of such a multicomputer's operating system should be to make sure the flow of data through the hardware is efficient, in that it expresses the structure of the algorithm as best mapped into the hardware. All the effort of maximizing the number of devices on a chip and interconnecting these devices so that they will execute algorithms efficiently is wasted if programs are not loaded properly into the network and allowed to execute as planned.

Kung's (1980a) "systolic" chips are designed to execute specific algorithms in a specific way, and are (potentially) highly efficient as a result—for those algorithms (only). Specialized arrays like Duff's (1978) CLIP4 are much more general, since they are programmable multicomputers, each with its own memory and a general-purpose instruction repertoire. But they assume that data will be mapped into them in a certain "fitting" manner (e.g., a matrix of numbers or a picture in a television raster should have its neighborhoods preserved, and be properly sized).

Information-flow VLSI networks, since they must be designed in a highly integrated way, must similarly execute programs and handle the flow of data so as to preserve and use that integration. The achievement of the operating system is to analyze and specify (or suggest) ways of improving this homomorphism, and then to use these structural correspondences efficiently. Thus the operating system serves as the larger environment's adviser/helper in minimizing area, pins, volume of information flow, and even algorithm and formulation effort.

CHAPTER 16

Compounding Clusters
and Augmenting Trees

This chapter begins to suggest some of the desirable characteristics of networks, and to sketch out combinations and generalizations of what appear to be the more powerful network structures.

Arrays, Tree/Pyramids, N-Cubes

Arrays and trees appear to be basic structures of overriding importance. Both are widely used. Trees fan in and fan out, converge and diverge, and therefore serve to distribute and collect (as in fanning into and out of a chip or a memory RAM, or in sorting, broadcasting, and combining information). Arrays execute highly parallel relatively local processes. Arrays plus trees supplement each other well, as the workers plus the distributing−collecting−combining−communicators. And, as we see in structures like pyramids, arrays and trees can be combined to simultaneously process and communicate. They are simple and embed well in VLSI chips. Each can profit from augmentation, the array at its edges, the tree at its buds.

The tree structure can be used as the primitive skeleton with which to start to build a variety of interesting, potentially much more powerful systems. Think of a straight-limbed tree all of whose branches continue as rigid straight lines. Take a two-dimensional array and cover it with such a straight-limbed tree as follows: Connect all nodes in each (nonoverlapping) $R \times C$ subarray to one "parent" node. Similarly connect these parents, until a single parent (the apex of the "tree/pyramid") is reached.

For simplicity, let us assume $R = C = 2$, giving a 4-tree/pyramid (4-pyramid for short), each parent has four offspring, and each layer of the pyramid has one-fourth the nodes of the layer just "below" it (in the direction of the original array). Each node in the base array gets one new diagonal link to its parent node (each parent getting four new diagonal links to its offspring).

Now cells in the original array connect along paths through the higher layers of the pyramid. Thus 4-pyramids reduce maximum distances to $\log_4 N$ links. For example, the two cells at opposite corners of a 1024 row by 1024 column array will be only 20 links from one another, 10 moving up the pyramid, and another 10 moving back down, in contrast to 2024 in the array itself.

Binary N-cubes have 2^n nodes, each node with N links, the longest path being N. So a 20-cube would be needed for 1,000,000 processors. It would therefore have some pairs of processors (at opposite corners of the cube) that were at the maximum distance of $\log N$, or in this case 20, links apart. It would also need 20 ports per processor (an uncomfortably large number); whereas the 4-pyramid would need only 5 (4 to offspring, plus 1 to parent).

Table I gives a few comparisons between N-cubes and tree/pyramids. Only the trivial 2-cube and 3-cube pack more nodes within a given diameter. As diameter grows, pyramids and trees become increasingly efficient in their use of links.

A pyramid with 11 links to each node thus has the same maximum distance between nodes (diameter) as an n-cube that is as large as the pyramid's base alone and has 20 links to each node. A pyramid with each parent connected to a 4×4 array of 16 offspring has a maximum distance of 10 and uses only 17 port—links per node.

Table I

A Comparison of Some Selected Pyramids and N-Cubes

Links/node		Diameter		Nodes	
N-cube	Pyramid	N-cube	Pyramid	N-cube	Pyramid
2	2	2	2	4	3
	2		4		7
3	4	3	3	8	7
	4		4		13
4	5	4	4	16	21
10	11	10	10	1,024	111,111
11	11	11	10	2,048	111,111
12	11	12	12	4,096	1,111,111

Probably more important when it comes to actually mapping programs into the network and executing these programs, the pyramid appears to have more useful links between just those nearby processors that will most often need to communicate. These links appear to be in a more regular and more useful pattern, that of the array. And it seems much easier to make meaningful and appropriate changes to the interconnection pattern in a pyramid, if only because its structure can be visualized, and can be mapped into (often three-dimensional) data flows for meaningful sets of problems.

Pyramids of Arrays

Pyramids are like X-trees, except that they are rarely rather than usually binary, the bottom layer is an array with square or hexagonal nearest-neighbor connections and the two-dimensional geometry of the array is implied in the pyramid. Rather than use a diagonal tree/pyramid structure everywhere, except where stitched to a base that has the square or hexagonal array structure, we can generalize either structure into the other.

All or any of the plies above the base can be turned into arrays. That is, we can add square or hexagonal links to nearest neighbors in the same layer.

Links can be added to nearest neighbors in adjacent layers. Diagonal links can be added within the base array, or any other new array. One can also add links at the base, and also at higher layers of the pyramid, to connect opposite edge/surfaces.

Pyramids and Lattices

The pyramid appears preferable to the n-cube for several reasons:

(a) Its diameter (worst possible distance for message passing) is smaller.

(b) It needs fewer links.

(c) It has more links between nearby processors, the ones that are more likely to need to intercommunicate.

(d) There are more feasible sizes (not just the powers of 2).

(e) For multiprogramming, it would appear easier to embed a program in a small local section of a pyramid, because of its uniform, relatively rich, interconnection pattern.

The following are the simplest near-neighbor interconnection patterns that might be used:

(a) "Square" for square links, two in each dimension.
(b) "Diagonal" for diagonal links, two in each dimension.
(c) "Square/diagonal" for both.
(d) "Hexagonal" for hexagonal links to six nearest neighbors.

It would be attractive to generalize a pyramid into a "lattice-pyramid" of the same dimensionality as the underlying pyramid. Such a lattice would have N^2 nodes in every layer, rather than in the base layer only. The diameter of the lattice would remain unchanged, but at the cost of adding more links—that is, at the cost of raising its degree.

But it appears that nobody has investigated this possibility for multicomputer networks, and it is not clear whether this can be done in a reasonably powerful way. One approach might be to use a reconfiguring network like the shuffle network, replacing each of its bank of switches by a bank of computers and replacing each layer in the pyramid by the corresponding bank in the shuffle network.

A simple way to achieve many of the desirable characteristics of a pyramid while increasing the number of nodes in the multicomputer network and filling out and regularizing its shape is to generate an $N \times N \times N$ "lattice-cube" from the $N \times N$ array that forms the base of a pyramid. Now diagonal links will give the converging characteristics of pyramids. These can be in one, two, ..., or in all six directions. Each interior node now forms the center of a 27-node $3 \times 3 \times 3$ cube. Since 26 ports to 26 links is an uncomfortably large number, it may be preferable to link only to some of these neighbors. Possibly the best symmetric scheme is for each node to link to its 6 square plus its 8 corner-diagonal neighbors. An attractive directional scheme would link each node to the 4 square neighbors in one direction plus the 4 square neighbors in the opposite direction, thus achieving convergence and divergence in these 2 directions (but not in the other 4 directions).

Convergence is far slower (and diameter greater) in a lattice than in a pyramid, since it takes N rather than $\log(N)$ steps. But the most desirable number of transformation steps from the raw image to perceptual recognition probably lies between these two bounds. Therefore another attractive possibility would alternate several layers of tree/pyramid converging structure with several layers of a lattice.

A lattice will often be better in terms of finding a reasonably compact and well-connected subsection into which to input a program to be executed. (I would argue that most programs tend to reduce the amount of data, giving a pyramidlike flow from base to apex; some programs tend to disperse data, giving a pyramidlike flow from apex to base; and only a few programs impose varying shapes/volumes on their data. By this argument a pyramid might indeed be the most desirable shape. But if we assume that a large network should be multiprogrammed, carving a number of subnetworks out of the total system, each the proper shape for the particular program it is executing, then a lattice or sphere is a fuller and simpler shape from which to carve.)

Toroidal Lattices and Spheres

We can improve upon the lattice by connecting the network at (one or more pairs of) its opposite faces (just as in an array like DAP or the array/net the opposite edges are connected). This gives a "toroidal lattice." This simple addition cuts the longest distance between any two processors almost in half.

We can simplify the lattice further, and make it a bit more general and more efficient, by increasing the number of its faces until it becomes a "sphere." A "toroidal sphere" would connect each surface cell to the cell at the opposite side of the sphere, that is, the cell most distant to it. (By "sphere" I mean a network approximation to a sphere).

Consider a lattice with each node linked to its 6 square and 8 corner neighbors (of the 26 neighbors in the $3 \times 3 \times 3$ cube that surrounds it). There will be 7 paths connecting opposite points on the surface through the center node that are all the same length, plus other paths that are slightly longer. If links went to all 26 neighbors, the lattice would be pulled inward (as by "tufting" in a quilt) by 13 paths. More and more link—paths could be added (by using basic cubes whose edges are longer than 3 nodes), pulling it closer and closer to a sphere.

Concentrically Nested Spheres within Spheres

Alternatively, a sphere can be built like an onion, nesting concentric spheres (each an approximation) "laced" together in a variety of possible ways (some of which are described in Chapter 17).

To visualize this, generate a sphere as follows: build a tree/pyramid with many limbs, with lateral links at every layer. Spread these limbs apart, widening the base, until the buds are uniformly dispersed in all directions. This gives a sphere centered around the root of the pyramid. Alternatively, lay more tree/pyramids next to the first, their roots touching, their surface branches side by side, and thus use sectional pyramids to build toward a sphere.

Each of the inner layers of the sphere can be made an array, just as in the lattice. The original links from the tree structures can be augmented, as desired, to increase the number of diagonal inward-flowing links.

Here it seems desirable to think of "facets" that are small subarrays oriented at slightly different angles, to build the curved surface of the sphere. (Each facet might be the set of nodes that can be put on a single chip.) For example, with VLSI technology a 16×16 array of computers (of the simple 1-bit kind) might make 1 facet, most appropriately placed on the surface layer.

Hexagonal facets, building hexagonal arrays, seem an attractive choice, since they have more of the feeling of the sphere's smooth curved surface. [To build a hexagonal facet take a hexagonal "block" of six neighbors surrounding the seventh central computer, using seven of these to build a hexagonal "module" of blocks (see Appendix D).]

But any type of building blocks, modules, and facets can be used. It is the pattern of links that imposes a geometry over the network, and gives it a distance metric. The physical shape of block, module, facet, or network is a function of the materials used; the "shape" and distance metric of the network's flow of information is a function of its interconnection pattern.

Spheres Conditioned and Carved Out to Suit Their Purposes

The chief virtue of a sphere is that it is the simplest, most regular, and most compact shape. So there is no reason to go to great lengths to get a "more nearly perfect" sphere if that is less simple. On the contrary, it seems more fruitful to try to carve the generally bulging spherelike structure to more closely fit the "shapes" of programs and data flows—for example, by using flattened spheres, ovals or eggs, cones, or whatever.

A Summarizing Tentative Choice
and Design of Attractive Structures

This chapter and previous chapters have taken a first look at possible network designs. They indicate how similar are several of the most powerful, as well as the most simple structures—trees, arrays, lattices, pyramids, and spheres. The ones that seem the most promising will be developed further in the chapters that follow. A few preliminary summarizing observations are in order:

Trees are an attractive place to start. They are compact, with simple and regular interconnection patterns and reasonably good average and longest distances. In contrast, stars, flakes, n-cubes, and most other interconnection patterns are rather like trees that have been pared and stitched in more or less strange ways.

But a strict binary (or n-ary) tree is not satisfactory, since it has no alternative paths for fault tolerance or alternative routings to level the load of messages, and its diameter can be improved. Unless input, output, and mass memory are connected only to the bud nodes at the end of the tree, some nodes must be given 4 ports. And since in most technologies either 3 or 4 ports means 2 binary switches are needed, along with 2-bit addresses, the extra cost for 4 links will often be insignificant. So it seems more reasonable to move to augmented trees with 4 links and a judicious pattern of augmenting stitchery. The hyper-tree and the de Bruijn network give the densest structures. Note that among the augmented 4-trees is the cellular array.

It will often be desirable to increase the number of links, using a 3-bit switch, for (up to) 8 links, or a 4-bit switch for (up to) 16 links. This, as we shall see, opens up many powerful new possibilities for interconnection patterns, can dramatically decrease diameter and average distances, and makes for extremely robust fault tolerance.

Two-dimensional arrays with links to 4, 6, or 8 neighbors are not only appropriately connected for many kinds of programs, they also have often acceptable network properties in terms of longest and average distances, ease of routing, etc. Their many lateral connections make for robust fault tolerance and possibilities of spreading messages over alternative paths if any overload occurs toward the root.

Still better connectivity can be achieved at the relatively small additional cost of stitching a tree over the array. This establishes many new, much shorter paths for messages, moving up toward the apex and back down to the array. A good example is a 9-pyramid, with 9 links to a 3×3 subarray of offspring and 1 (10th) link to the parent. This superimposes a 6-layer pyramid (giving a diameter of 12) over a 729×729 array of processors (which might look at a 2187×2187 array of image-containing memory stores).

Such an array—tree/pyramid has lateral connections only within the array at its base (each node now 5-connected, to its 4 sibling neighbors and to its parent), and only diagonal connections within the 10-connected pyramid nodes. Diagonal connections can be added, if desired, within the base array (giving 9 links per node). Vertical connections can be added within each subsequent layer (giving a total of 14 links per node).

More lateral connections (which means cycles, alternative paths, and fault tolerance) can be added by "overlapping" nodes' offspring, e.g., 2 "adjacent" parent nodes linking to 3 (of the 9) offspring in common. This adds more nodes in the higher layers, increasing the height of the pyramid and therefore its longest distance. For example, with a 3×3 window of offspring this means that convergence is now 4 to 1 rather than 9 to 1. A 729×729 array needs a 10-layer pyramid (which could also be used to cover any array from 513×513 to 1024×1024) to cover it.

When a converging (or diverging) flow of data is to be processed, pyramids may be most appropriate. But a more general structure, especially for a large network that can multiprogram many users, would appear to be a lattice. Its near-neighbor connections can be varied in a number of ways, and its overall shape can be carved, as deemed appropriate, e.g., to bulging lattice, egg, or sphere. (These extensions will be examined more thoroughly in subsequent chapters.)

One or more pairs of opposite faces can be linked together, just as arrays often use wraparound links between opposite edges. This substantially reduces distances between nodes (at minimal cost since these surface nodes have fewer links).

Three major types of interconnections appear to be useful:

(1) Near-neighbor local connections that determine the network's microstructure and most of the actual flow of data through processors.

(2) Very distant connections (e.g., among nodes at the buds of a tree, or between opposite faces of a lattice).

(3) Reconfiguring networks that physically change the interconnection pattern under program control.

A number of alternative structures will be described in later chapters. But one of special importance is mentioned here:

(4) Connections at intermediate distances (e.g., the "input−output control pyramid" introduced in Chapter 17) that lace together still too distant regions, as appropriate.

Pyramid, Lattice, Discus, Torus, Sphere

We shall now begin to examine several architectures that have a number of desirable characteristics when used for very large networks of thousands, or millions, of processors.

All combine large numbers of "worker" processors into densely packed networks with minimal diameters, very good near-neighbor linkages, robust and usable redundancies, and rich possibilities for fault tolerance. All can be expanded easily. And all offer an overall shape and modular microstructure, plus the possibility of adding judiciously planted reconfiguring capabilities, that should make for efficient packing and execution of many programs.

Many of the basic building blocks and modules suggested for the microstructure of the architectures are described in Appendix D. In general, I prefer simple regular near-neighbor connections, as in square and hexagonal arrays. But denser, more symmetric (and also more complex) connection patterns, as found in the Moore graphs (see Appendix A) might be preferable, at least for certain types of problems, and toward the innermost parts of the network.

The layered pyramid is the simplest macrostructure. It is designed to handle a flow of data through (programmed) processors from base toward apex, or from apex toward base.

The lattice extends and fills out the pyramid, making its macroshape more regular and more amenable to multiprogramming.

The directional layers can now (optionally) be replaced by modules that allow convenient program flow in any direction, including simultaneous multiprogrammed flow in opposite directions.

Several still more efficient shapes are now possible, including discus, egg, and sphere for multiprogramming, and pipe and torus for programs with special, or repetitive, flow. Each of these can further be connected back onto itself, giving toroidal lattices and spheres.

An IOC (input−output−control) pyramid is introduced to handle problems of input−output, access to large memories, message passing at great distances, fault tolerance, and local/global program control.

The Basic Pyramid Structure

The pyramid consists, roughly, of arrays of processors sandwiched between arrays of memory stores. A "layer" contains one processor array and one memory array. Each "cell" in a layer contains one processor and its "companion" memory store.

The pyramid starts with the "base" array of memory stores (it is often convenient to think of this as the "retina"). Sandwiched right behind the retina lies a (usually smaller) layer (with its one processor array and one memory array). This layered structure continues until the "apex," a layer containing only one cell, is reached (Fig. 1). (See Table I for ranges of plausible parameter values for pyramids.)

Retina	Layer1	Layer2	Layer3
M0	P1−M1	P2−M2	P3−M3
M			
M	P−M		
M	P−M	P−M	
M	P−M	P−M	P−M
M	P−M	P−M	
M	P−M		
M			

Fig. 1 Overall structure of a pyramid.

An input device (e.g., a TV camera) inputs information (e.g., a picture image) to the retinal stores. If the image is larger than the retina, a subarray can be input to each store (e.g., 8×8 subarrays of a 1024×1024 image can be input to each cell of a 128×128 retina and processed serially.)

Each processor is linked to, and "looks at," a subarray of retinal memory storage cells, that processor's "window" (see Fig. 2).

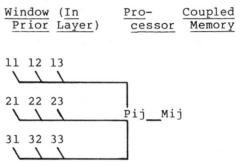

Fig. 2 Links between a processor, its window, and its coupled memory. [Note: Only one processor is shown, linking to a full 3 × 3 window, and to one coupled memory at the same (*i,j*) location in the same layer.]

Full Windows

```
MM    MMM    MMMMM    MMMMMMMMM
MM    MMM    MMMMM    MMMMMMMMM
      MMM    MMMMM    MMMMMMMMM
             MMMMM    MMMMMMMMM
             MMMMM    MMMMMMMMM
```

Cross-Shaped Windows

```
    M        M              M
  MMM        M              M
    M      MMMMM    MMMMMMMMM
           M              M
           M              M
```

Augmented Cross-Shaped Windows

```
      M              M
    MMM            MMMMM
  MMMMM        MMMMMMMMM
    MMM            MMMMM
      M              M
```

  2 × 2   3 × 3   5 × 5   5 × 9

Fig. 3 Some examples of possible window sizes and shapes.

Assume a window is a 3 × 3 subarray. Such a window can have any of several different detailed forms: The processor can be linked to all nine stores in the "full" window (as in Fig. 2); the processor can be linked only to the five "cross" stores (but not the four "diagonal" stores). In larger windows (e.g., 5 × 5, 7 × 7, 11 × 11) there are a larger number of possibilities, some of which are shown in Fig. 3.

The most reasonable windows would appear to be 3×3, 2×2, and 5×5. Larger windows seem attractive, but they quickly become expensive in terms of links and processor time needed, and whatever transformations they can achieve can always be achieved by a short sequence of smaller-window transformations.

Each processor has one other memory store, its "coupled" cell in its layer's memory array. (Each of these memory stores also links to processors in the next processor array.)

Each layer has some "shrink" factor that converges the structure toward the apex, forming the pyramid shape. (Figure 1 shows a shrink of 2; additional examples are given in Fig. 4.) This pattern continues, with the same amount of shrink/convergence at every layer, until a layer with a single processor is reached. This forms the "apex" of and completes the pyramid.

Fig. 4 Some examples of shrink-convergence from layer to layer. [Note: (P) processor, (.) no processor at that location, (|) border (memory cell, or a programmable input).]

The relation between the size of the window each processor looks at and the shrink factor gives the "overlap," that is, the percentage of the memory stores that more than one processor can access directly. (But note that information from arbitrarily more distant memories can be accessed by any processor with a sequence of instructions. So this is simply the immediately accessible overlap.)

Table I gives examples of pyramids that all have the same overlap and shrink characteristics at every layer. There might be reasons to make these different at different layers, to reflect the kinds of processes executed at each layer. For example, the first layer of processors might be the same size as the retina, so that details can be explored quickly.

Or a layer that executes processes that look at a bigger window (e.g., to get textures over a 7×7 or 11×11 window) could have such a larger window (rather than execute this process with a sequence of smaller processes, either in the single layer or in this case more conveniently spread out serially over several layers). And the convergence might be made different, to be compatible with this different kind of process and window size.

Table I

Some Interesting and Plausible Sizes for Pyramids

Size of Array (length = breadth) Retina	Layer1	Window	Shrink	Depth	Overlap
32	16	2	1/2	5	0
32	16	3	1/2	5	4/9
32	16	5	1/2	5	4/25
64	32	2	1/2	6	0
64	32	3	1/2	6	4/9
64	32	3	1/2	6	4/9
64	22	3	1/3	4	0
64	22	5	1/3	4	9/25
96	48	3	1/2	7	4/9
96	48	5	1/2	7	4/25
96	32	3	1/3	5	0
96	32	5	1/3	5	9/25
256	128	3	1/2	8	4/9
256	128	5	1/2	8	4/25
256	128	7	1/2	8	4/49
256	64	5	1/4	4	16/25
1024	512	3	1/2	10	4/9
1024	512	7	1/2	10	4/49
2048	1024	3	1/2	11	4/9
2048	683	5	1/3	7	9/25
4096	2048	3	1/2	12	4/9
4096	1366	6	1/3	8	9/36

Achieving Good Time and Cost Trade-offs with Links and Registers

Registers can be interspersed between memory stores and processors where needed to keep the number of links and ports down to acceptable size. This in effect turns one single fetch or store instruction into a sequence of two instructions. It also introduces the additional hardware needed for the register stores and the extra links. But it reduces the number of links to each processor node.

Most of the pyramid designs require reasonable numbers of links. (With today's technologies, up to 8 links seems simple; up to 16 seems acceptable. This may grow, but slowly.) For example, a system with 3 × 3 full windows needs 9 links from each window to the processor that handles it, plus one more link between that processor and its store (which is one of the cells in the next layer). The preferred 3 × 3 cross-window needs only 5 links for the window, plus 1 link to the coupled store.

A 5×5 or 7×7 full window needs unacceptably large numbers of ports (25 and 49) for links from each processor. Cross-windows need only 9 or 13 links, and would often be an attractive alternative. But full windows can be handled quite straightforwardly by interspersing registers. That is, a word will be fetched from a store to a register.

A simple design has one such register for each row in the window. This means a 5×5 window has five registers, each with five links to its window — row and one link to the processor; the processor has five links to these registers (plus one to its next-layer store). Thus if n registers are added, n new links are added also, but the number of ports and links for each processor is divided by n. Any other division of the window into subwindows of acceptable size would be possible, and straightforward.

DAP and the MPP use 3×3 cross-windows. CLIP and the Cytocomputer use 3×3 full windows. PICAP II has several processors that use larger windows. The economics of a very large pyramid suggest that the simple 3×3 cross-window be used.

Several or Dual-Port Memories to Speed Up Processing

Such a system is most effectively used by pipelining successively processed images through the layers. This means that all processors can at the same time be fetching information from the stores in their window in the previous layer, or storing information into their own memory stores. Processing can be speeded up if one layer of processors can store at the same time that the next layer fetches. This could be effected with only a little extra expense by using two or more memories instead of one, or dual-port memories.

Serial versus Parallel Processor Access to Memory Stores

Each processor has direct access to the memory stores in its window and to its one coupled store in its cell of its layer. This access is most simply made, as in DAP and MPP, one word at a time. That is, the processor can fetch one word from any one of its window or cell stores in one instruction cycle.

But this access can be speeded up, if desired, by adding a logic/switching network so that (in window parallel fashion) several words can be fetched, and also transformed (e.g., by using logical or arithmetical operations) at the same time. This is relatively expensive (in

CLIP it accounts for almost half the devices in each processor) and specialized. Depending upon trade-offs and the mix of instructions used by the system's programmers, it is probably better, as well as simpler, to have the program fetch neighborhood information from the window one word at a time, and have the processor effect all transformations itself.

Input and Output Devices

The retina is connected to input devices from which it receives successive pieces of information (e.g., television pictures and other images). The details of input connections can vary, depending upon the size of the retina and the costs of hardware, as follows:

Input might come to every cell in the retina, to every row, or in some scattered pattern throughout the array. In large arrays with present technology input to every cell does not seem feasible—it is too expensive and requires too many pins. But if light-sensitive sensors can be combined with processor and memory gates on a single chip this might become entirely feasible.

Today's very large arrays input to the 32, 64, 96, or 128 rows, shifting information from each of these input channels to 32, 64, 96, or 128 columns. But as the size of arrays increases the number of columns that each of these input channels must input to (by successive serial shifts) will increase. (Indeed people have already suggested that the 128×128 MPP may be I-O bound.)

Probably the best alternative (when it is too expensive to input to every cell) is to shotgun inputs uniformly dispersed throughout the array. For example, each 8×8 subarray of retinal memory stores could be connected to one input—output link. An array of any size could now be shifted into the retinal array in 64 steps. Obviously, the number of steps to input an image can be made independent of the array size by interspersing the proper numbers of input links.

Building Blocks and Modules

The processor—memory layers are themselves built from "modules" which are in turn built of "blocks" (see Appendix D). This serves both to move toward an easily constructed prefabricated system (where, e.g., each block or module is a single chip) and to establish nodes for communication, control, and input—output links.

The size of modules and blocks should be determined by

(1) required input—output speeds;

(2) desired maximum and average distances between processors for message passing;

(3) economics of packing processors, blocks, and modules onto chips and boards.

A block is a small array or lattice of processor groups (i.e., processor and associated memory, registers, and switches). A module is a small array or lattice of blocks. Input devices, output devices, mass stores, and any other shared resources are associated with a module. Input can either shift through registers into a layer's stores from the module or pass to the module's blocks and then shift into the stores.

There are many plausible sizes for pyramids, modules, building blocks, processors, and memory stores, on the background of projected sizes of chips (see Table II). (Two-dimensional modules are probably preferable on the single chip, since their near-neighbor links form, or almost form, a planar graph.)

Table II

Example Sizes of Processor-Store, Memory, Block, Module

		Possible sizes (Number of transistor-equivalent devices)			
Processor	Memory store	4×4 block	16×16 module	3×3×3 block	9×9×9 module
50	50	1,600	25,600	2,700	72,900
100	100	3,200	51,200	5,400	145,800
200	200	6,400	102,400	10,800	291,600
500	1,500	32,000	512,000	50,400	1,458,000
500	5,500	96,000	1,536,000	151,200	4,374,000
10,000	50,000	960,000	15,360,000	1,512,000	43,740,000

It is clear that only simple 1-bit processors and, even more important, memories that are very small (from the conventional perspective) can be packed many modules to a chip. (But from the perspective of today's SIMD arrays, memory can be relatively large, on the order of hundreds or many thousands, rather than dozens, of bits.)

This has very strong, possibly discouraging, consequences for people seriously interested in networks with millions of processors. If a chip contained an entire 16 × 16 module of 256 processors then only 4000 chips are needed for a 1,000,000 processor network. Indeed, we could pack almost 40 modules of very simple processors onto a 1,000,000-

device chip, and therefore need only 100 chips! Or only ten 10,000,000-device chips!

But unless we cut down the memory drastically there seems to be no hope of getting more than one conventional 16-bit or 32-bit computer on a single chip. Yet unless we can average 100 computers per chip we cannot expect to build a 1,000,000-computer network. But a conventional computer needs at least 256,000 bits of memory (and most people would dismiss that as far too small).

This suggests several possibilities:

(1) The memory stores linked directly to processors should be considered as cache memories, and a large network of larger memories should be added to the system.

(2) The conventional memory should be drastically reduced per processor, considered as a cache, and supplemented by a larger common memory on the chip.

(3) Simple processors and small memories should be used.

(4) Simple processors and small memories should be used wherever possible, but more powerful one-computer chips should also be interspersed as needed.

(5) Reconfigurable processors should be used, so that, e.g., either 32 1-bit, 4 8-bit, or 1 32-bit processor could be applied to data.

(6) Special-purpose systolic chips might be interspersed as needed.

These considerations suggest a network with regions that specialize, and therefore are sent appropriate programs by the operating system and use specialized hardware to process them.

The SIMD array of 1-bit processors is an obvious kind of specialist. A system that has a large enough percentage of such processors would indeed be able to average 100 processors per chip. In a pyramid, the retinal array is a good place to have SIMD processors (though possibly they should handle more than one bit at a time). Since most of the processors are in the retina, and 256 or even more can be packed on a chip, the one-million-computer network appears feasible, even if all the processors in every layer other than the retina need one chip per processor.

A module containing a 16×16 array of 1-bit processors, each with its own memory, that can be reconfigured into 8 32-bit, 16 16-bit, or 64 4-bit processors is an attractive alternative (if the extra costs, probably a reasonably small percentage, can be justified). Such a chip module could be used throughout the network, but specialized to one bit in sensory regions and to more powerful processors in "deeper" regions. It could also be reconfigured by the programmer as desired.

Handling Input–Output, Control, and Fault Tolerance with an IOC Pyramid

A special pyramid, for "input–output–control" (IOC), appears to be a relatively simple mechanism not only for input and output throughout a lattice or pyramid of "worker" processors, but also for message passing between distant processors and program sequencing and other control functions. This is simply a pyramid that connects modules, building them into larger and larger structures. (Alternatively, an IOC module of any desired shape might be used.) Since the module connects blocks, and the block connects individual workers, the IOC pyramid can be used to connect units of any type at any level.

Input–output requires links to distant input devices, like television cameras and tape readers, and output devices, like printers and plotters. For particular applications and architectures we know where these should be. For example, input should go into the retinal array for picture processing.

But when a general network is running a mix of programs, must load these programs and their data into almost any portion of the structure, and must output results from almost anywhere, it seems important to disperse input–output links widely. This can be expensive, since large packages of information are often involved and the link should have suitably high bandwidth. So it seems of value to consider whether the same links can conveniently be used for input–output and also for other related functions. There appear to be several rather attractive possibilities.

Input and output usually come in bursts, and, fortunately, at separate times. The loading of a (sometimes very large) program or procedure will also come in a burst, at still another time. And in a network it is also desirable to have the option of either loading a program into a processor that will then execute that program (in MIMD mode), or broadcasting that program's instructions to a whole set of processors that therefore execute it in SIMD mode. The latter, along with synchronizing signals and other kinds of control functions, tends to give a slow, steady patter of messages.

A pyramid seems an attractive structure for shipping input and output and messages, and also for handling control functions, for much the same reasons that it is an attractive structure for the network of workers themselves. Using an input–output–control pyramid structure also makes it possible to handle several other important problems with the same structure. The pyramid can be used, at its higher levels, to reconfigure, as desired. This gives the operating system a very powerful tool in its continuing endeavor to allocate programs to processors in as efficient an overall manner as possible.

Redundancy for Fault Tolerance

This system already has great potential for redundancy that can be used quite effectively for error correction purposes. If the processors' windows are large with respect to shrinkage, that is, to the extent there is overlap, there is redundancy. To take a simple example, if processors have full 4×4 windows, with 2×2 shrinkage, then, as processors (or memory stores) go bad, the system can (ignoring how it discovers these problems, deduces the needed modifications, and effects them) (a) move from full to cross-windows, and (b) reduce the window size until there is no overlap. That is, four processors are handling the 4×4 window, but only one might make do. Each processor has 16 links to memories, but, if these break, other several-link paths might be used instead. If memories die, other memories might be used to store their information (at the cost of smaller memory for each).

When one actually builds very large systems with millions of computers, it will be crucial that the system continue to work for many months or years, even though parts go bad. The reliability of the technology used will determine the amount of redundancy and fault tolerance it should be given. But to demonstrate how easily this can be incorporated into these systems, consider the following:

There is no need to shrink at all. Instead, an n-layer pyramid can be replaced by an n-layer lattice with all layers the same size. This gives complete overlap. But because each processor can look at all the memories within its window, one can program such a system to converge (rather than have the convergence built in). While the system is young and all is going well, these processors can be used in a variety of ways, for pyramidal convergence and for other kinds of programs. But as processors die, the system can be reconfigured more and more toward a pyramid, with successively less overlap.

In addition, we might doubt each processor to the point where we would use three or more to compute each function in parallel, using majority-rule decisions. This is quite feasible considering the very simple and small processors being used, and the ease with which we could embed this new example of parallel processing into the already highly parallel structure. Now we would have not only this great additional redundancy, but also, because of the way processors are arranged in groups—blocks and modules—where quite often not all processors are always busy, it might be possible to regroup processors, as needed, to keep each error-correcting group sufficiently large.

Rather than use larger common mass stores, it seems preferable to have the common store made up of small stores, with a small reconfigurable network between it and the processors in its block or module. This would make this memory more accessible to several processors at the same time, and it would also mean that pieces of the mass memory could be assigned, as needed, to replace processors' primary memories as they went bad.

To get millions of computers, memory must be kept small, at least from the conventional point of view. But now I shall argue that the memory is best judged in terms of the total problem, that when we have many processors, with a reasonably nice architecture structuring processors to data flow, each processor's memory requirement plummets. So I suggest that packing large numbers of computers on each chip is the most valuable thing we can do, and that we consider 100-device processors and 32-bit memories very seriously. But it would appear that 500-device processors and 256-bit or even 1024-bit memories may already fit, and will soon fit easily.

Examples of Attractive Pyramids

I shall give one example of a relatively large pyramid with a number of attractive characteristics, along with a table containing a few other interesting possibilities.

The retinal memory array is 1024×1024, so that large high-quality television images can be input one pixel (picture element) to each cell. A 512×512 array of processors looks at the retina, each processor using a 3×3 cross-window, giving four-ninths overlap and shrinking to one-fourth the size. This continues, each layer with 3×3 cross-windows and four-to-one shrinkage, giving a ten-layer pyramid.

Now superimpose the input−output−control pyramid, as follows: Every 4×4 subarray of processors forms a block, with one link to its module, which is a 4×4 subarray of blocks. The module (which is a 16×16 subarray of processor−stores) has an input−output channel (which also serves for control, including loading of program and data and other purposes) to the next smaller layer of the IOC pyramid.

Modules continue to converge, a 4×4 subarray to one parent. So the IOC pyramid for the retinal layer has six layers *in toto*: 1024×1024 (of stores, but no processors), 256×256, 64×64, 16×16, 4×4, 1×1. The subsequent layers of the main pyramid have fewer layers (see

Table II). A final node links to the root node of each layer's IOC pyramid.

The IOC pyramid connects sixteen offspring nodes to each parent. This means that the longest distance between any two processor nodes in a 1024 × 1024 array of over 1 million processors is only 10! (Using the main pyramid with a full 3 × 3 window it is 22.) The system could now choose between these alternative pyramids for message passing, for example taking the shorter one for vital messages only, or always choosing the one that was shorter between the two nodes involved.

A reconfiguring network can be added, if desired, at two particular points of interest:

(1) In the IOC pyramid, at whatever layer it becomes economically feasible (probably between the 16 × 16 layer nodes, but possibly at the 64 × 64 layer).

(2) On the individual chips, between the 16 blocks or the 256 processors in a module.

The basic building block has 16 processors; the basic module has 256. Either of these is a candidate for a single chip, as suggested in Table III.

Table III

A Feasible Minimal Building Block and Module Chip

	Unit		
Resource	One	16-block	64-module
Processor	200	3,200	12,800
Memory	100	2,000	8,000
Registers	10	160	640
Mass memory	1,000	1,000	4,000
Total	1,310	6,360	25,440

This uses minimal memory although it does include (rather too small) "mass" stores. But it fits reasonably well into today's technology, even for the entire module of 64 processors. If there is any extra room on a chip (as there shortly will be), that can be used to increase the processor slightly, up to 1000 or so devices, and to increase the memory as much as possible. Or one might add on-chip partial reconfiguring capability on some of the chip real estate.

A chip with only 16 processors will have relatively few input−output problems because of pin limitations. A chip with 16 × 16 processors will present problems, but, I suggest, fewer than one might fear: input — output can be effected through 64 or 32 pins, meaning only 4 or 8 shifts to load the entire chip — as opposed to the 64, 96, or 128 shifts for DAP, CLIP4, and MPP. These pins can also be be used for messages, programs, or control information via the IOC pyramid.

Lattices

A "lattice" can be generated from a pyramid of the sort just examined, as described in Chapter 16. It can be a lattice of pyramids of height $\log(N)$, or a lattice-cube of height N, or mixed, with an in-between height.

Such a system no longer has a shape that is in some senses specialized, for programs that converge/reduce or diverge/disperse information and, more specifically, for image processing and pattern perception. When a program does have such a pyramid shape, it can still be executed by a pyramid substructure carved out of such a lattice. The lattice is most appropriate for three-dimensional modeling. It is probably preferable for a general network that will usually be multiprogrammed, rather than be used for a single large program. For the lattice has a simpler, fuller shape with fewer narrow corners, so there is more likely to be room to load a new program.

The basic lattice is a very simple extension of the pyramid. There are several minor variants, and several major extensions, that are worth examining.

Input and Buffers to 0,1,2,..., or All Faces of the Lattice

The large input buffer store (the "retina") reflects the fact that pyramids were specialized to handle large amounts of data (pictures have been emphasized, but large databases might also be massaged or sifted through by such a structure). A lattice designed for a general mix of programs, or/and for three-dimensional modeling, will probably have little need for such a large input buffer. So simply eliminate it. If that increases the input and storage load on other parts of the system, simply augment input channels, by increasing the bandwidth to each module or by making modules smaller, and increase the size or number of the local mass memories.

Conversely, if the lattice will frequently be used for large-picture processing, database management, or other programs that need large inputs (whether or not they also map onto a pyramid), one might want to add a retina to the face opposite the original retina, and even to the other four faces.

These are good examples of a more general issue, one that seems worth mentioning here. The different kinds of resources—whether input, output, memory, or links for message passing—can all be dispersed, or/and concentrated where most needed. And they should be so dispersed, for efficiency and economy. Only when we have no a priori knowledge should they be dispersed uniformly. Nor need we take the attitude, "we're running a general computing center for an unknown mix of programs." For we can and should create and condition mixes that more efficiently utilize our resources, e.g., by shipping different programs to differently specialized subnetworks.

*Generalizing the Microstructure for Program Flow
in Several Directions*

The present lattice uses window—links oriented for programs that flow in one direction only, from the retina. This can be successively generalized, as follows.

To handle a flow either from or to the retina:

(a) Point the window in the direction away from the retina.

(b) Link each processor with all stores in this window.

The following is a system that would be able to simultaneously execute programs that flowed in both directions at the same time (probably using two retinas at opposite sides):

(a) Turn each processor into a "group" of two processor—store couples.

(b) Link one processor to the stores in its window facing the retina, and link its coupled store to the processor centered on this couple in the next layer (away from the retina); link the second processor to the stores in its window facing away from the retina, and link its coupled store to the processor centered on this couple in the next layer (toward the retina).

This adds a large number of new links—for the windowed stores plus the new coupled memory. So a reduction in the size of the window would often be indicated.

The system can be further generalized to handle (possibly simultaneous) flows in any subset of, or in all, directions. Start from scratch by eliminating the original processor, thus establishing a group that is empty:

(a) Point the window toward any of the faces of the lattice.

(b) Add one processor – store couple to the group.

(c) Link the processor to all the stores in its window, and to its couple store.

(d) Link its couple store to the processor centered on this group in the layer away from the window.

Lattice – Cubes, with Generalized IOC Pyramids (or Lattices)

A lattice-cube can grow very large. For example, a 1000×1000 retina generates a $512 \times 512 \times 9$ lattice (the retina stripped away), but a $512 \times 512 \times 512$ lattice – cube. Now the worst distance between pairs of processors is 1024, if the processor – memory links are traversed.

But with minor readjustments to the IOC pyramid we can achieve very short distances (note that this may not be satisfactory if there are large loads of messages between distant processors, since this second pyramid must serve many purposes). Rather than build a pyramid for each layer over two-dimensional modules, simply use three-dimensional modules and blocks.

Consider a $128 \times 128 \times 128$ cube of over 1,000,000 computers. Using modules that were only $2 \times 2 \times 2$ structures of blocks that were $2 \times 2 \times 2$ computers large, the greatest distance would be 16. (Note that a similar tree could be superimposed on a two-dimensional array with the same number of processors, or on a flattened lattice, simply by using blocks and modules of the same total volume but more suitably shaped.)

A further generalization would generate an IOC lattice from the IOC pyramid (just as the lattice was generated above from the pyramid of worker computers).

Specialized Processors at Various Regions on the Surface

Pyramids have bases that are often good candidates for specializing, as with an SIMD array of simple, general, 1-bit processors. One face of the lattice might similarly be specialized, probably in conjunction with a retinalike buffer store and suitable input links. Or the entire surface of the lattice might be so specialized.

The types of inputs the programs the lattice will process determine the types of input devices needed. This may suggest specializations of certain regions of the surface (or the interior) with clusters of specialized input devices, much as the skin surface of living systems has specialized regions for different kinds of sensory receptors.

N-Dimensional Lattices

Lattices of higher dimension than 3 could also be built, in exactly the same way, although there would seem to be little reason to do so. The more dimensions, the more faces, windows, and links. The major reason is to draw very large numbers of nodes closer together. But the IOC pyramid appears to be sufficient. And the problems in visualizing and working with systems with more than three dimensions make this seem undesirable when not necessary (e.g., for a direct mapping of *n*-dimensional problems).

Toroidal Lattices

Each face of an *n*-dimensional lattice can be linked to its opposite face, just as the leftmost column of a two-dimensional array can be linked to the rightmost column. This can be done with all pairs of faces, or with any subset of pairs. Thus we can have a 1-toroidal 4-lattice, etc. An *n*-toroidal *n*-lattice seems the most attractive. The term "lattice" alone will usually mean 3-toroidal 3-lattice (either flattened or cubical).

Discus, Torus, Egg, Cone, Sphere

A lattice with flat surfaces and sharp edges seems less alluring, and less simple, than several smoother and more symmetric shapes. More seriously, the structure of the program to be executed, and the needs of a whole set of programs to be packed and executed in a multiprogramming environment, suggest exploring still other shapes. But from now on the microstructure of layers, couples, groups, blocks, and modules will remain the same.

A discus is simply a slice of a lattice with more or less bulging centers and smoothed-over edges. It seems a nice shape for programs that spread out for a while, then contract, or for programs that need to expend more effort on certain parts of their data (which the input devices are therefore programmed to map into the bulging center).

A torus seems appropriate for programs that can be set up as pipelines that continue to iterate, so that they can be swirled continually through the torus.

An egg might be useful for the same reasons given for the discus.

Circular rather than square arrays can be used, for the retina and each of the processor—store layers of a system, to give cones instead of pyramids. Hexagonal facets and tiling seem especially compatible with such a structure.

It seems unlikely that any particular shape will be stable enough over enough computing to justify designing and building the particular shape in hardware. But it seems interesting to contemplate the possibility of tailoring the macrostructure to the program's overall structure. And one of the goals for an operating system (that would more likely work with a simpler structure like a lattice or a sphere) is to decompose the larger structure into program-shape-fitting structures.

Spheres

The sphere is the final variant shape I shall mention:

It seems the simplest and most desirable.

It has the most processors for a given diameter.

It seems the most accommodating, without narrow corners, in terms of being able to find room for programs.

It can be made into a toroidal sphere simply by connecting each cell on the surface to its opposite surface cell (that is, to the cell lying at the other end of a line drawn from that cell through the center of the sphere).

It can be made n-dimensional.

Spheres present several new problems. The direction(s) for window(s) can be established wherever one wishes, and a group, with the appropriate number of couples, established as for lattices.

It is not clear how many directions to establish. But the three orthogonal axes seem appropriate for a three-dimensional sphere. This can best be visualized by taking any of the lattice—cube designs and filling it out to form a sphere. (Now any retinal-input-like sensory areas bulge a bit, since they form a section of the sphere.) Another interesting possibility is to build the sphere from hexagonal modules. Now the nine axes might be used to establish directions of flow.

CHAPTER 18

Three-Dimensional
Array/Network
Whorls of Processors

This chapter describes several global structures that seem to me especially attractive, and gives a little bit of attention to how they might be used productively.

A sphere seems an especially compact, elegant, and simple structure. A three-dimensional sphere is the simplest to realize with actual physical processors built and housed within our three-dimensional world. Its surface is a convenient receptacle for storing and processing images. And its entire structure, or a substructure carved out from it, seems well structured for mapping a three-dimensional slice of atmosphere, earth, or other sections of the universe.

For a variety of reasons some variant shape (e.g., pyramid, lattice, dumbbell, torus, bulging cone, or flattened sphere) may be preferable for particular classes of problems. So I shall refer to such a fuzzily defined larger class as a "whorl."[*]

---

[*]Note: Chosen because whorls look and sound beautiful. The word also suggests, e.g., hand in hand, holistic, homogeneous, linked, organ, organized, oval, ovum, reconfigurable, swirl, symmetric, structure, wheel, whirlwind, whole, world, etc.

It might be an acronym for something like: "Whole Highly Organized Reconfigurable Linked Structures."

Spheres within Spheres
of Onion-Layered Faceted Arrays

Hexagonal arrays* on the two-dimensional surface of spheres arranged in onionlike layers would appear to give the most elegant structure for pyramidlike convergence toward the center of the whorl.

Petersen graphs or other good clusters might serve instead of hexagons as basic facets and building blocks. So they can be considered as alternatives whenever hexagons are mentioned. But since the simple near-neighbor connections of a hexagon facet are very useful for a variety of problems, they seem preferable.

The complete whorl imposes physical space constraints on the number and size of processors that can be packed into each internal whorl layer. There are several possible variant patterns of overlap and nonoverlap of links from one whorl layer to the next.

Consider a "hexagon facet" that consists of one central processor plus six processors surrounding it. Each processor might output to only one parent (placed directly behind the center of that seven-processor hexagon facet). This gives a reduction of seven to one moving inward, but with no overlap between hexagon facets.

If each processor has two inward links, reduction from layer shell to layer shell will be seven to two. Now each of the six surrounding processors links to two adjacent parents. The central processor's extra link might be used for other purposes: (a) for input—output; (b) to link to mass memories; (c) to link laterally at some distance in the same shell layer.

A reasonable balance between convergence and overlap would appear to be either two or three parents per offspring, giving seven to two or seven to three reductions in the number of processors from layer shell to layer shell linking inward.

The requirements for ports and links do not seem excessive: Each processor needs six links in its own layer (to each of its six hexagonal neighbors), plus seven links to its offspring, plus one, two, or three links to its parent(s). (The seven links to offspring could be replaced by one link to a processor that accumulated the information from the

---

*Note: A true sphere cannot be built from hexagons. The closest structure is a truncated icosahedron, made from 20 hexagons and 12 pentagons (see a soccer ball). By "sphere" I mean a hexagon-tiled curved surface more or less the shape of a sphere, but with some kind of irregularity or fissure at its back, that is in some region distant from its sensory/input regions. Whorls of other shapes would often have similar irregularities. The purpose of a highly symmetric shape like a sphere is not to give a complete surface array. A sensory region will only fill a relatively small part of the sphere's surface.

offspring. This might be a new processor, or the central processor in the hexagon facet.)

Processors on the outermost sphere layer need input—output links, but these replace the seven links to offspring. IOC links to input and output devices and to mass memories should also be scattered throughout the interior. But one-seventh of the processors (the central ones) have an extra link that can be used either to stitch or for input—output and memory. The number of links needed (fourteen to sixteen, or seven to nine if an extra processor is added) is not excessive, and economic trade-offs would determine the particular choices one would make for a specific problem.

Such a system would finally reduce to an innermost layer with only one, two, or three processors. Thus the system of two-dimensional spherical surfaces arranged in a three-dimensional onion-layered sphere collapses into its center in an irregular manner. This, plus the likelihood that problems at the center are quite different, suggests that a different kind of network, probably with more powerful and more densely packed processors, possibly with some reconfigurable interconnection patterns, be used for the central core. But almost certainly the near-neighbor connections will continue to be widely used.

Dense Three-Dimensional Whorls with Programs Pipeline-Swirling through Them

Rather than use two-dimensional hexagonal arrays arranged in concentric spheres with the expectation that programs will execute toward the center, we can build three-dimensional spherical sphere arrays like lattice-cubes. Now each processor has eighteen nearest adjacent neighbors. Subsequent layers now pack into this three-dimensional lattice, rather than into the two-dimensional surfaces of concentric spheres.

In such a system we might consider having programs "swirl around" under the surface, or, since the whole system (with the exception of its central core) is homogeneous, within some designated regions. This suggests a system that flows problems through processors, as though in a three-dimensional pipe network, each processor outputting the information the next processor downstream in the swirling whorl will use.

Now processors might best be linked through memories into which the intermediate results are placed, and from which these results are almost immediately fetched by the next processor. This also suggests using much smaller pipelinelike memories than are usually considered even for arrays, and that have been tacitly assumed. Rather than 1000 to 64,000 memory words, we might use only 8 to 128.

This gives a structure that might well be thought of as an active memory, with data flowing through it, and processors relatively densely scattered through it, to transform that data.

Limited Reconfiguring for Multiprogramming and Fault Tolerance

Reconfiguring capabilities are desirable both to map programs onto substructures of processors and to replace faulty processors (which will be inevitable in such systems when they grow very large). This suggests that the pool of reconfigurable processors include, in addition to substructures of varying sizes, a number of individual processors scattered throughout the whorl, in general in the cracks between substructures.

Such a whorl might best be used as a multiprogrammed computer, with different programs residing in different (often cone-shaped) substructures of the whorl. This would allow the operating system to carve out a cone or other indicated substructure whose dimensions were appropriate for each program, and to use any processors that lie idle at the margins of such substructures as traditional 1-CPU computers.

A limited reconfiguring capability would be especially useful when multiprogramming. The operating system would no longer have to resolve complex packing problems with each new program, and, possibly often, have to shuffle programs to new sets of processors. Instead, sets of processors could be added, or abandoned, by reconfiguring.

This suggests using a pool of processors that can be reconfigured by adding them to substructures to which the operating system has assigned a problem, and releasing them when the problem is solved. This is reminiscent of the way list-processing systems allocate and garbage-collect memory, except that such a complete reconfiguring capability would be too expensive. Rather, one might consider having a whorl that decomposes into substructures of the shapes and sizes that have proved to be most useful, plus a small pool of reconfigurable processors at the interstices between these structures. Reconfiguring should probably be further limited, so that only physically nearby processors and substructures could be reconfigured together.

This may well be a very difficult problem to pose for an operating system, if not one that is impossible to handle efficiently. But it is an interesting problem. Reconfiguring plus the ability to use a single processor in the old-fashioned serial background batch mode give the operating system several useful ploys. And it may turn out that many programs map onto a small number of regular structures, like cones, dumbbells, and toruses. For example, a large sphere decomposes nicely into cone segments with apexes toward the center of the sphere, with optional successively smaller toruses stacked around occasional cones.

Array/Networks That Are Sections of a Whorl

A simpler alternative would be an appropriately proportioned section carved out from such a whorl.

A cone or pyramid appears to be called for when problems involve two-dimensional arrays of information that must be processed and transformed into much smaller sets of information; or, conversely, a single starting point from which many paths must be explored. This includes most pattern recognition, scene analysis and description, and other perceptual tasks, as well as many numerical, artificial intelligence, and database management problems.

A two-dimensional array appears to be appropriate for, and is today used for, image enhancement and other image processing problems. But even for these it appears that at least a certain amount of multilayer convergence, as in a two- or three-layer truncated pyramid, would often be useful. Often a full pyramid is appropriate.

A lattice (with either square or hexagonal connections) is appropriate for three-dimensional modeling, e.g., of the atmosphere, a wind tunnel, or a whirlpool, and for numerical problems and image processing problems where information is *not* reduced.

A torus might be quite useful to contain a program that swirls through several revolutions. That is, it would contain the "swirling whorl" mentioned above. It seems attractive to consider giving such a system a central core, a simple starlike network emanating out from the host, its buds linked to, controlling, and monitoring the whorl.

Summary Discussion

It will not be clear which of the great variety of network architectures is "best" until large mixes of programs have been run on each. This will not happen for a long time. Nor can we ever expect a "definitive" answer, since new classes of problems and new algorithms will continue to be uncovered as new computer architectures move us closer to the possibility of handling them.

Given the expense of building large networks, and of coding programs with appropriately structured parallel algorithms that make good use of their potential power, it seems more reasonable to build rather general and regular structures, with enough reconfiguring capabilities so that data can be collected about the particular structures that are used enough to therefore build specifically.

Among the most promising structures would appear to be a three-dimensional whorl whose outermost surface layer is an array with hexagonally linked processors, with a (possibly similarly layered) internal network moving back to the center. Such a system would have input−output channels to each of its facets where (looking to the future) each facet might be a board, or a chip, containing hundreds or thousands of processors.

Concentric onionlike layers within layers, each processor linking inward to one, two, or three parent processors, would appear to be a useful design to handle (in a multiprogrammed fashion) programs that reduce, or expand, data.

A three-dimensional lattice, or a cone, pyramid, torus, or other shape whose structure suitably reflected the flow of information effected by some reasonably large set of programs, might also be considered. Such systems might not need to be multiprogrammed. But each would almost certainly be at least to some extent limited and not suited to all programs in a reasonable mix. So several more or less specialized architectures might be needed.

Lattices and spheres, in contrast, might be multiprogrammed, the operating system carving out and allocating to each program suitably shaped collections of processors. If given even a limited capability for reconfiguring under program control, the programmer could (if suitable algorithms for operating systems are developed) dynamically carve out differently shaped sets of processors, as needed. Those processors that cannot be shaped nicely into a needed structure can always be allocated in traditional serial-computer fashion, one processor to handle one program.

Such an array/network whorl poses a number of difficult problems and lies far in the future. But it would appear to be attractive from several points of view:

It fits well with future VLSI technologies, where many processors can be fabricated on a single chip.

It offers the hope of supersupercomputers that can grow as large as needed for increasingly larger problems.

It poses a fascinating array of research problems.

CHAPTER 19

Toward Very Large, Efficient, Usable Network Structures

The Great Variety of Potentially Powerful, Efficient Networks

This book has surveyed many of the different array and network architectures that have been proposed or, occasionally, constructed during the past few years. It examines their differences and similarities, and sketches out a variety of possibilities for future networks.

These include a relatively coherent set of extensions, from arrays and trees through tree/pyramids and pyramids to lattices, spheres, and whorls. In addition, a number of promising alternative possibilities have been described.

This book also examines some of the problems of programming networks, and of developing programming languages and good interactive software tools. And it explores the problem of developing parallel algorithms for parallel arrays and networks and of designing structures for these networks in a coherent, coordinated manner.

The Structure of Problems, Programs, and Networks

Each particular problem for which we develop an algorithm and a program has its own structure. Ideally, we should build (or reconfigure) a computer network that embodies this structure as closely as possible.

Occasionally that will be possible, when the problem is important enough and the program will run long enough on enough data to justify the expense. In such cases we may achieve a network structure that exhibits the "anatomy" of the problem, its nodes processing information flowing through it much as a human brain/nervous system processes information.

More frequently (at least until VLSI technologies mature to the point where cheap customized networks can be fabricated as easily as a program can be written today) the same network will have to process a variety of programs, with a variety of different underlying structures.

We therefore need more general networks.

Networks That Contain a Variety of Different Kinds of Resources

An interesting, relatively simple network is one that contains a variety of different kinds of resources, each specialized for particular kinds of processes (e.g., image enhancement, convolution, floating-point multiplication, text processing). Such a system is attractive because it also includes the conventional "host" 1-CPU computer, which can be used for program development, editing, compiling, and the other traditional support processes.

One such network might simply have different regions that are more or less specialized to different tasks. An attractive example is a "sensory" region on the parts of the surface of the network where large amounts of input are received. Other alternatives might have several different systolic and other special-purpose processors, along with pipelines and arrays linked to the host's high-speed bus, or linked via reconfiguring switches.

Tree, Tree/Pyramid, Lattice, Sphere, and Whorl Networks

A tree seems an appropriate starting structure for networks. Trees are very simple and regular, and they pack reasonably large numbers of nodes within a given diameter. But trees have faults, including the possibility of complete breaking of all connections if any single component fails, and bottlenecks from uneven loads, especially toward the root.

Trees can easily be improved upon, by judiciously "augmenting" them, that is, adding links to give alternative paths for fault tolerance, to even message loads, and to shorten paths. The augmentations to hex-trees and hyper-trees are good examples of this.

A tree should probably be augmented until each node has the same number of links (the desirable number being 4, 8, or 16). And input, output, and mass memory devices are just more nodes, scattered about where they serve the most purpose. Linking processors via memory seems a good way to add alternative paths and help processors to work together.

Arrays, which have a near-neighbor linking pattern that is obviously appropriate and very powerful for a variety of problems (including image processing, pattern recognition, physical modeling, and number crunching) can be viewed as augmented trees. They have surprisingly good interconnection and fault tolerance properties.

Trees and arrays can be combined in a variety of ways that appear to offer great potential promise. For example, the bud nodes at the end of the tree's branches can be stitched to an array, giving a "tree/pyramid," with the array at the base of the tree. Intermediate nodes can be laterally linked, thus building successively smaller arrays moving toward the root of the tree, giving a "pyramid" or "cone."

The pyramid can be generalized into a lattice-of-pyramids of a lattice-cube, making each array layer as large as the base array. This lattice can have opposite faces linked together, giving a toroidal lattice.

A variety of other overall shapes are possible, including disc, torus, and sphere. The sphere would seem to offer the most compact yet spacious room for assigning a large number of different programs, in a multiprogram environment, to different empty regions of the network.

The individual processor might contain only 2 to 10 devices, and serve as part of a network or array of very simple more or less specialized logic gates. Or it might be a very simple and general 1-bit processor, with 50 to several hundred gates. Or it might be a special-purpose systolic system, with whatever number of gates are needed for the particular process embodied in silicon. Or it might be a more traditional 8-, 16-, or 32-bit CPU, possibly with some reconfiguring capabilities to be used as several 1-bit or 4-bit processors.

A variety of microstructures might be used for combining the individual processors and memories into basic building blocks and modules of such a network. These include directional or omnidirectional, solid or faceted, square or hexagonal structures. It appears that a $3 \times 3 \times 3$ cube or an 18-node three-dimensional hexagonal solid is a reasonable size for the basic module, since a larger module would need too many internal links, whereas such a module can be used to build a wide variety of larger structures.

Any of these structures might be generalized to n dimensions. But three-dimensional structures, when they can be compactly packed with mostly short near-neighbor links, seem sufficient for achieving networks with many thousands or even millions of processors. Three-dimensional networks are more easily built, with much shorter and simpler links. They are appropriate to the one-, two-, and three-dimensional structure of large classes of important real-world problems. And they are more easily understood, so that they will be more efficiently programmed. Thus there appears to be little reason to consider higher dimensions.

To handle input, output, control, and message passing at a distance, especially as the network grows very large, it seems desirable to add an IOC control pyramid. This serves to lace the system to itself, reducing already small distances even further, and allowing for efficient overall control and coordination. Input, output, and mass memory devices are probably best considered to be like any other node, and placed wherever they seem most useful (e.g., at the surface and in the IOC control pyramid).

Exploring and Comparing Promising Alternative Designs

It seems far more important to examine, evaluate, and compare as great as possible a variety of networks than to settle on specific designs. The possibilities are extremely rich, and it is likely that many different network structures will find their niches, and that many of these niches will be far larger than the one found during the past 40 years by the 1-CPU computer. But let me comment on some of the attractive features of several structures that appear especially interesting:

A pyramid (or cone) seems appropriate for problems that regularly converge and reduce information (as in perception) or, conversely, diverge (as when searching a large tree or graph).

A lattice is an attractive general shape for a multiprogramming network. A lattice-of-pyramids would have as good local interconnections and global diameter as the much smaller pyramid.

A sphere seems a bit more general. One might consider building it from hexagonal modules. Or, possibly better, from hexagonal facets, giving an onionlike sphere of successively smaller concentric spheres, the flow of processes oriented toward and/or away from the center.

The following extensions to either pyramid, lattice, or sphere also seem attractive:

An IOC control pyramid would handle input, output, fault tolerance, control and coordination, and message passing.

Hexagonal, possibly faceted, modules would give better shape and directionality of flow.

Moore graphs, or other clusters with desirable properties, might be used as basic building blocks and modules.

Linking of opposite faces would shorten and regularize the interconnection pattern.

Regions of the surface might be specialized, e.g., with simple 1-bit processors, for "sensory" input purposes.

Different types of processors and memories might be scattered about, as appropriate.

Additional Structures: N-Cubes, Stars,
Moore Graphs, Reconfiguring

The N-cube is an interesting basis for large networks, especially when augmented with local clusters of slower processors surrounding each node, or/and with nodes along the edges.

A variety of more or less starlike clusters, some proposed, others still to be worked out, also have promise, especially when they can be tailored to particular flows of processes.

Petersen and Singleton graphs, and other graphs with desirable properties (e.g., good density, symmetry, or many easily used alternative paths) are attractive candidates for basic building blocks in larger networks. A variety of compounding operations are now available to build relatively dense larger graphs from good local cluster graphs of this sort.

Reconfiguring offers, potentially, a great deal of flexibility. It also produces networks that serve the vital experimental purpose of allowing us to observe which particular reconfigurations programmers find most useful, and thus can help appreciably in exploring designs for future networks. But the costs of the extra switches for reconfiguring must be justified. They must either be cheap enough, or use components needed for memory stores or other purposes during their idle time, or be worth the extra costs because they are being used in exploratory research on design.

The Promise, and Problems,
in Future VLSI Technologies

The extravagant advances in technology we have experienced during the past 30 years should continue for at least the next 20 or 30 years. We have good reason to feel that this means 1,000,000 transistor-equivalent devices on each single chip by 1985 or 1990, and 10,000,000 or more within 5 or 10 years after that.

We have reason to fear future bottlenecks, e.g., the present pin-count bottleneck of 64 per chip, which is unlikely to rise above 256 or so and, probably more unsettling, entirely unanticipated technological limits that may emerge in the future. But we have reason to think that such bottlenecks can be overcome (for example, if chips or wafers can be stacked successfully, as in the Etchells—Grinberg—Nudd 1981 3D computer), since we have been so successful in overcoming them up to now. If we can achieve really compact processor designs, then heat dissipation problems may ultimately limit further packing. And I-O may become a fundamental problem because it must take place across a surface whose area is small with respect to the internal volume available for processors. But these limits appear to loom only distantly.

What this means is that we can think of building computers by 1985 or 1990 that have 1,000,000 devices on each of their 1000 or 10,000 (or, if we want to build a complicated supercomputer, even more) chips. The numbers 1000 and 10,000 will probably not grow, since they reflect the basic complexity of wiring large numbers of diverse elements together. Thus the number of devices on each chip will far exceed the number of chips. We will achieve 10^6 to 10^7 (or, possibly, 10^8) devices on each chip, but then get only a 10^3 or 10^4 (or, at great cost, 10^5) additional improvement by wiring chips together.

Therefore most of the complexity of a system will reside on each single chip. The burden of designing and fabricating chips will be enormous (if it continues to be done in the present way). We can expect that automated computer-aided graphics design techniques, of the sort described by Mead and Conway (1980), will appreciably reduce this burden. But designing a 1,000,000 or a 10,000,000 device chip is inevitably a very large task. Probably the best we can hope for is to make chip design (that is, the design of hardware) as easy as is program design and coding (that is, the design of software).

But along with our 30 years of tremendous advances in hardware technology we have seen only the slightest, most laggard advances in software techniques. Roughly, as hardware has improved by six or seven orders of magnitude, programming has improved by one, two, or possibly three.

Probably the most attractive and the most feasible way of making the chip design simpler is to micromodularize.

Large memories are an extreme example of modularization that has been handled very successfully.

The simple, small, 1-bit processor, each with its own regular memory, and with a very simple interconnection pattern tying many of these processors together, may well be another way.

We can also break a complex processor down into a number of simpler modules, much as a program is broken down into a number of much simpler procedures or subroutines, and procedures break down into even simpler instructions. Modularization can make complex processors easier to design and fabricate. And the processor can be designed to reconfigure into, for example, 64 1-bit, 16 4-bit, 8 8-bit, 4 16-bit, 2 32-bit, or 1 64-bit processors.

Graphs whose nodes are very simple processors, or even small handfuls of gates, linked so that information can flow through and be appropriately transformed, can be embodied directly in VLSI chips.

Carefully optimized, highly efficient special-purpose modules of the sort Kung (1980a) is devising for particular systolic algorithms that are relatively frequently used are also intriguing candidates. Bentley and Kung's (1979) tree of modules for database applications is an interesting move toward more general systems of this sort.

An intriguing problem is one of constructing flow graphs (they might be planar or nearly planar, depending upon the chip technology) with a few million nodes in a cluster (to be put on a single chip) and a few thousand clusters, with only 256 or so links between clusters (the number of links allowed by the pin fan-out bottleneck between chips).

We can assume, roughly, that modularization will reduce 200, or even 2000 devices to a single module (a simple 1-bit processor, or an algorithm for fast arithmetic, convolution, edge detection, string matching, etc.). But we still will need designs that combine 500 to 50,000 such modules on a single chip. We know how to design large arrays of simple processors. There is hope these can be packed efficiently, and that micromodular systolic components, and even larger modules of more powerful computers, can be iterated on a chip in a regular, relatively compact manner.

One obvious way of trying to simplify further is to design with modules that are themselves made up of modules, just as procedures call procedures.

Allocating Gates to Networks of Processors and Memory on Silicon

A typical traditional single-CPU computer has roughly $10^6 - 10^8$ gates in its memory, but only $10^4 - 10^5$ gates in its processor − controller.

In sharp contrast, the very large arrays have 10^2-10^3 gates in the memory and $10^2-8\times10^2$ gates in the processor of each of their 10^4 computers. At the extreme, CLIP4 has only 32 bits of memory for each of its 10,000 computers (each processor consisting of roughly 600 gates).

Thus the traditional computer uses roughly 100 to 1000 times as many devices in its high-speed memory as in its CPU. But each of an array's computers uses roughly equal numbers of devices in memory and processor. CLIP has 10,000 computers with 32,000 bits of memory *in toto*. A typical VAX configuration (a medium-sized computer of roughly the same price) has 8,000,000 bits of memory to one processor.

What is the proper ratio of processor to memory (see Shore, 1973; Uhr, 1981c)? Traditionally, the common wisdom has been that memory should be as large as possible, to keep the CPU busy. But it seems more reasonable to think of memory as being large enough to contain all the needed input data and intermediate results. As we develop algorithms and computers that use many different physical processors, each handling portions of these data, we can disperse data to many different appropriately connected much smaller memory stores. Parallel arrays, and especially pipelines whose processors pass intermediate results directly from one to another, need even less memory.

The better the structure of the algorithm as mapped onto the network, the smaller the need for memory to store intermediate results. The greater the percentage of the network working at actively and efficiently transforming information, rather than passively storing information, the more that network will do. Consider the variety of ways in which we can allocate gates to nodes (memories, processors, controllers, and reconfiguring switches); nodes to chips, and chips to networks.

The different possibilities for general-purpose, reconfigurable, specialized, and special-purpose systems seem almost potentially infinite. The variety of potential interconnection patterns (all possible graphs!) is overwhelming. Even though relatively few are attractive, their numbers are also very large. Some seem attractive to me; some seem attractive to others. Quite possibly the most attractive have not yet even been appreciated, or thought of. We still need a basic set of criteria for evaluating and comparing these structures. But we cannot definitively judge such complex structures until we build, use, and evaluate them extensively. That will take several whole new industries, each with a number of manufacturers developing their own sets of user — customers.

Pipe—Network Flow through Combined
Array—Network Clusters

Two major very different types of computer networks that use large numbers of relatively closely coupled processors to work on a single program are being developed today:

The very large SIMD arrays with near-neighbor connections appear to most people to be specialized, possibly overly specialized, for processing arrays of information, whether images or equations. They have especially appropriate local near-neighbor interconnections; but their global interconnections are poor. Yet they are very powerful and very fast (and they are general purpose).

The larger set of different MIMD network structures being explored appear to hold greater promise for a wider variety of algorithms. But the problems of programming and using them efficiently are turning out to be very difficult.

In the successive development of potentially interesting and powerful new architectures in this book, I have worked under the implicit assumption that both types of power are highly desirable, if not necessary. So a major goal has been to combine arrays, with their very useful (I, along with most people actually working with arrays, would say absolutely necessary) near-neighbor connections with networks that give the whole system a more efficient and more flexible interconnection pattern.

Very large arrays are feasible, and can be and have been built, because their near-neighbor connection pattern is simple and cheap (and also the one-bit processor and the SIMD mode, which means only one controller for the entire system, are cheap). They have enormous potential power (for the problems for which they are structured) precisely because of the near-neighbor connections. This is because the pieces of information most relevant to one another are most probably near one another, whether in the array or in the slice of the physical world it models and tries to capture.

But arrays are cumbersome and waste large amounts of time when information must be moved over long distances. And it is difficult to keep all the array's processors busy at useful tasks when they must all execute the same instruction. Nor is it clear how to combine links to distant nodes with the array's near-neighbor links.

Here I suggest that we think about structure in terms of several different types of linkages:

(a) *The processors that actually work together should be clustered into small, compact, appropriately structured groups with all the links they need.* Hyper-trees, de Bruijn network shift registers, and other augmented trees today pack the largest number of computers within a given diameter, resulting in the highest global density. *But good local density and good structuring are more important criteria, and in these they are mediocre.* And other criteria, including structural correspondences to programs' structures, should also be taken into account. Compounds of clusters of Petersen graphs or other dense graphs might have the best properties, in terms of connectivity and local density. But in the development of this book I have favored arrays, which give a continuous, useful, homogeneous clustering (that can easily be decomposed by the program if the programmer desires), and have a structure well suited to a variety of important problems.

(b) *The most distant processors can be linked together, to give short-cut paths that will often drastically reduce the diameter of the structure* (though often the effect on the average distance between nodes will be far less dramatic). We have seen how the buds of trees can be linked in rings or shuffles, how arrays can have their edges wrapped around, and how, in general, a structure can have its opposite faces linked. This both draws together the most distant parts and makes possible a circulating flow of information through the entire system, as though pipe-networking through a loop.

(c) Whenever the near-neighbor and the most distant links are still not sufficient, *periodic internal linkages can be added to lace the structure more closely to itself.* As structures grow larger this will become more and more desirable, if not necessary. This appears to be a very powerful way to introduce links between (formerly) more distant nodes. And it gives us a technique for integrating more or less specialized local structures like near-neighbor arrays or clustered Petersen graphs with one another, within a more general structure.

The array/net described in Chapter 6 is a concrete attempt in this direction. The stacked converging arrays (Chapter 6) and the tree/pyramids (Chapter 16) attempt to place a tree, or a pyramid, over an array in more general and more flexible ways. The subsequent pyramid, lattice, sphere, and whorl structures (Chapters 17 and 18) continue to use all three types of linkages. But they integrate the intermediate lacings more intimately into the structure, and they explore a variety of alternative ways for specializing subregions (e.g., only the base of the pyramid, or the surface of the sphere, might have near-neighbor links).

Networks with millions of computers will almost certainly need good intermediate lacings as well as near-neighbor and most distant connections. They will almost certainly contain very large arrays, if only because arrays are often useful for important types of problems and their simple processors and linking patterns offer the only hope of putting enough processors on a single chip so that 10,000 or so chips (the limit on what can reasonably be built) will contain a million computers.

A Final Word:
The Immediate and More Distant Futures

During the next few years SIMD arrays with 16,000 and more simple 1-bit computers, like Batcher's (1980) MPP, will be built. MIMD networks of more powerful computers will probably have only 50 or 100, or 1000 at the most. For this short-term transition period there are a number of attractive possibilities. For example:

(1) Combine a large array with a pyramid (or some other variation on a tree) that gives it satisfactory global interconnections, control, and asynchronous MIMD processing when needed.

(2) Compound good local clusters (e.g., Moore graphs, arrays) into an overall architecture with good global interconnections.

(3) Augment trees to improve global interconnections and local properties, and emboss good local clusters (whether special purpose, specialized, or general purpose) as desired.

(4) Build three-dimensional MIMD lattices of relatively powerful processors, with appropriate surface nodes linked to SIMD arrays.

Before the turn of the century we shall be in a position to build SIMD arrays with millions of computers and MIMD networks with many thousands, if not millions. The VLSI technologies for designing and fabricating suitable chips with large numbers of simple processors on each; or with one, or a few, powerful processors on each, will be widely available. Whether such systems will actually be built that soon will depend on whether enough people have explored them seriously, using simulations and prototypes, and whether manufacturers and funding agencies appreciate and are able to respond to the profound potential importance of this development.

An attractive possibility for these long-term (1990s) networks is architectures, languages, and operating systems that allow programs to flow, and swirl, through a pipe-network-layered stack of arrays or other structures, whether a lattice, pyramid, sphere, whorl, or some other overall shape. The local structure of interconnections should closely fit the structure of the flow of information through the program. Each processor should be appropriate to its task; in some networks quite powerful, in other networks (or regions in the same network) the simplest of gates. Some might be special-purpose systolic processors with their code built into hardware. Others might be specialized as appropriate, e.g., to 1-bit, or for string matching, or in local arrays or SIMD clusters. Reconfiguring switches might be used for selective restructurings among different regions.

Algorithms and programs should be designed with the network's "anatomy" in mind. The operating system should map (and when necessary remap) the program into the network, using whatever reconfiguring resources are available, to balance the computing load across processors, keep information-flow commensurate with the bandwidth of the links between processors, and maximize the computing bandwidth of the total network. The constant goal should be to execute programs as efficiently and, when there are real-time constraints, as fast as possible.

The hardware of the network, and of its individual components, should be designed keeping in mind the flow of data through the procedures to be executed. This means the processors should be no more powerful than needed; individual memories should be no larger than needed. The higher the percentage of processor gates that are kept busy at useful tasks, the greater the system's total productivity.

It is intriguing to contemplate more "organic" designs of chips, with processors surrounded by memories through which they communicate and pass the results of their transformations. Appreciable savings in chip area might be effected by designing processors according to the natural anatomy of the flow of information through their gates, and by packing areas left empty with highly micromodular memory banks. Arrays can clearly be packed quite densely, and there are several relatively efficient ways of embedding trees in chip grids. The packing capability of different logic designs and different interconnection topologies will become increasingly important as chip area, rather than number of devices, becomes the pertinent measure.

The enormous number of $10^{10}-10^{12}$ devices potentially available in such a multicomputer will open up completely new possibilities for intelligent perceiving—thinking systems, for modeling of large three-dimensional physical phenomena, and for working with large databases of information. Losses of one or two orders of magnitude because of problems in packing or message passing or developing efficient algorithms will be likely, but such losses will be relatively small compared to the enormous total power. On the other hand, such very large systems will not be feasible unless carefully designed modules are appropriately interconnected, and the programs they execute map appropriately into the hardware.

Let us make several rough estimates, and look at the consequences:

A 1-bit computer uses 500 processor gates and 500 memory gates.

A 32-bit computer uses 10,000 processor gates and 1,000,000 memory gates. (Since this probably underestimates the number of memory gates on the 32-bit computer, and memory can be packed more efficiently on chips, I should note that factors of 100 1-bit processors to one 32-bit processor, and 100 to 1000 1-bit computers to one 32-bit computer are reasonable estimates whether gate count or chip area estimates are used.)

Today 4000 to 16,000 1-bit computer arrays, and 50 to 100 computer networks are built or planned. The MPP uses a near-neighbor array and a (very expensive) data remapping capability via buffer memories. The 50-node multicomputers use a few local clusters, a very expensive cross-point switch, or a very high-bandwidth bus or ring.

From 1983 to 1990 VLSI will allow for 250,000 1-bit computers or 1000 large 32-bit computers. Neither a pure two-dimensional array nor a single bus will be feasible. The arrays might be folded into a three-dimensional lattice with problems piped and swirled through them, or augmented with a new set of connections, e.g., a tree or pyramid. The 32-bit computers might be built with reconfiguring switches, as lattices, or as compounds of clusters.

Once the 1,000,000-gate chip can be designed reasonably quickly and economically (because of good computer-aided graphics design systems) and good compact processor designs have been achieved, multicomputers with 10^7–10^9 1-bit processor nodes or 10^4–10^6 32-bit processor nodes, or suitable mixtures of both, will become possible. Further increases of several orders of magnitude will still be feasible. During this period the need for good local and global architectures, and for good languages, algorithms, and mappings, will become painfully apparent. But there is some reason to hope that some of the approaches already being pursued and other approaches that will be discovered during what should be an unusually fruitful immediate future will lead to reasonably powerful and efficient use of these incredible impending riches. Successively compounded highly modular designs appear to be among the most feasible, and the most promising.

Integrating arrays, trees, lattices, and other good types of network topologies into the very large algorithm-structured architectures we will soon have the technology to build is a fascinating challenge, with important practical consequences to the extent we succeed. Today's serial von Neumann bottleneck can (in theory) be widened to whatever underlying degree of parallelness can be used to effect each algorithm. Many hardware and software improvements are needed to succeed at this widening. Ultimately, we can envision architectures that embody the structure of algorithms, and programming languages and operating systems that map these algorithms into multicomputers so that the load is balanced for maximal flow of information through the network structure.

PART V

Appendixes: Background Matters and Related Issues

Applied Graphs, to Describe and Construct Computer Networks

A network, pipeline, or array of computers is naturally described using graphs. Programs are often represented as flow charts, flow graphs, or other graphlike structures. It seems fruitful to conceive of our problem as one of mapping program graphs onto network architecture graphs so that the flow of information through procedures is handled as simply and directly as possible by flowing information messages through physical processors. The branch of mathematics known as graph theory offers useful terminology, and suggestive results and insights.

Defining and Describing Connected Graphs

Some Preliminary Definitions

A "connected graph" is simply a set of "nodes" connected by "links" (each link joined to two nodes, called its "ends" or "parts") such that there is a nodes—links "path" between any pair of nodes. (Terminology is not standard: "vertex" and "edge," "point" and "line" are commonly used synonyms.)

The "degree" of a node is the number of links joined to it.

A "regular" graph is one with all nodes of the same degree.

When there are several paths between a pair of nodes, these form "cycles," since they give paths from a node back to itself. A graph with no cycles is a "tree."

A "complete" graph has a link between every pair of nodes. A "complete bipartite" graph can be partitioned into two subsets of nodes so that each node is joined to every node in the other subset.

A "directed graph" (or "digraph") has one-way links (rather than undirected two-way links). The word "network" is usually used more generally, in this book and elsewhere, to designate computers with several processors and memories, and as a synonym for graph. But in the context of "network flow" problems "networks" are defined as digraphs with nonnegative integer values (designating, e.g., channel capacity or power) assigned to links and two distinguished subsets of nodes, the "source" (where, e.g., raw materials or initial information are input) and the "sink" (where, e.g., finished products or solutions to problems are output).

Using Graph Theory Notation to Describe Computers

It is often convenient to consider a computer as a node, and a link as connecting two computers. We will usually be interested in two-way links, where either computer node can send information to the other. This gives graphs of the sort shown in Fig. 1.

```
        C-C              C-C              C
                         | |             |
                         C-C           C-C-C
  C-C-C-C                                |
                                         C
```

Fig. 1 Examples of graphs that describe simple networks. (Note: C indicates a computer [node]; | and — indicate links.)

Sometimes it is necessary to indicate more detail: If input and output devices are to be specified, they will be different kinds of nodes, and they will often be linked in only one direction. Sometimes the computers' processors are linked together; or their memory stores are linked together; or processors are linked through memory stores. Figure 2 gives some examples.

```
      P-S->O           P-S->O          I->P->O
                        | |             |
                        I->S-P          P-S-P
  P-S-P-S                                |
                                         P
```

Fig. 2 Graphs that detail different kinds of nodes and links. (Note: P = processor; S = store; I = input; O = output; > and < = arrows, indicating directions of message passing.)

We can usually simply use the symbol C for a node (C indicating the computer couple: processor plus memory store). And often we can assume that a bidirectional link joined to one node only is an input — output link (that is, connects to a device that effects both input and output).

Representing Successively Lower Levels of the Network as Graphs

We shall occasionally find it useful to look at processors and stores in more detail, effectively tearing them apart into their components, or specifying more or less specialized processors. For example, a typical computer's CPU might be symbolized by a P (for processor), or with a set of its special processors, e.g., A (for adder), F (for floating-point multiplier), M (for matcher), etc.

Alternatively, it is occasionally convenient to specify a set of processors P(i), as P1, P2,...,Pn. Similarly, different banks of memory stores might be specified S1, S2,...,Sn.

Processors might all be connected to a bus (I shall use |, —, and O as ideograms for buses), or in a spoked-wheel-like subgraph. Or they might be connected in a variety of other ways. Or they might be dispersed into a graph that linked and intertwined them with store nodes.

It is convenient to introduce still another kind of basic component, X, for a reconfiguring switch that allows the path between two processors to be changed (exactly as a switch can be used to route a train over two different paths). This will be necessary to specify "reconfigurable" networks, where the program can send a message to a switch and thus "throw" it to establish a new interconnection pattern. In this way the network is literally physically changed. (See Fig. 3.)

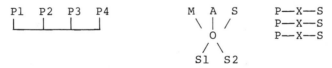

Fig. 3 Examples of graphs giving lower-level details. (Note: [, —, _, |, /, O all indicate links and buses; P = processor, M = matcher, A = adder, S = store, X = switch.)

It is important to note here that, ultimately, a computer is made entirely of switches. The basic stuff of a memory store is a large set of binary switches. When set in one direction the switch signifies a 0;

when set in the other direction the switch signifies a 1. The processor is simply a network of switches that effect the basic logical operations the computer uses for all its transformations.

This means that we can expand the graph of a computer, tearing it apart into successively smaller units, until we reach the basic diagram of its logical operations. The switching networks that are put between computers, to serve to reconfigure them, move us in that direction. But we shall also find this useful when we examine the configurations that will be possible using VLSI (very large scale integration of hundreds of thousands, or millions, of switches on each single chip) technologies, with special appropriately designed chips.

Some Formal Properties That May Be Desirable for Computer Networks

We can describe networks of computers quite straightforwardly in terms of graphs.

This means that results from graph theory, a branch of mathematics that explores the properties of graphs (usually with one kind of node connected by links either bidirectional or directed), should be able to throw light on the problem of designing computer networks. It is illuminating to look at computer networks in this concise way, since it makes salient certain of their interconnection characteristics.

A number of characteristics seem desirable for a computer network (some will be examined below). These include

(a)   As many nodes packed as closely as possible ("density").

(b)   As large a "girth" (the shortest cycle) as possible.

(c)   As many alternative paths as possible between nodes.

(d)   As many symmetries as possible.

(e)   As good "connectivity" and "cohesion" as possible (the number of nodes, and of links, that must be removed before the graph is no longer connected).

(f)   Simple routing algorithms.

(g)   Good techniques for mapping programs onto subnetworks.

Virtually no systematic evaluations of different proposed computer network structures have been made. Nor have these criteria and their importance and value been investigated more than casually. But it may be of some interest to examine density, as an example criterion, and because several rather surprising results have recently been found.

Using Graph Theory to Help Increase Density

Let us look at the key issue of packing as many nodes as close to one another as possible, to get "denser" graphs (i.e., with more nodes packed within a given diameter and degree). Closely related is the minimization of the maximum distance (called the "diameter") between any two nodes.

Here the optimal way to minimize diameter is too costly: connect every node to every other node.

More interesting are solutions that minimize costs, but for them we must assign costs in a common currency to links and to time needed to pass messages (distance).

A good example of a judicious improvement is to establish a "central" node, and to link each other node to that center, giving a spoked-wheel-like graph (often called a "star"). Now of the $N \times N$ pairs of nodes, $N \times N - (N-1)$ will be 2 apart; the other $N-1$ (those involving the center node) will be 1 apart. This small increase in time reduces the number of links from $(N^2 - N)/2$ to $N-1$, which becomes overwhelmingly important as N increases.

But there will still be too many links to the central node (in computer networks, more than four or sixteen links can cost too much).

Let us take a different tack: Minimize the number of links used to connect a given set of nodes, and also the number of links connecting each. The solution is simple, and uninteresting: given N nodes, use $N-1$ links to connect them in a linear string. This gives unacceptably long distances, since the diameter is $N-1$. (Note that this is a one-dimensional collection, like a file or list.) Diameter can be cut in half, by adding one more link that connects the two end nodes.

A simple way to keep the number of links down to any size N is to build an "N-ary" "tree." [A tree is a graph with no cycles, that is, with only one path of links between any two nodes. The central node (called the "root" of the tree) has N or $N+1$ offspring nodes linked to it; each offspring (of offspring) has N offspring. The final ply of offspring, with only one node, are the "buds."] Here we use the same total number of links as in the spoked wheel (which is simply a tree that is all buds except for the root node), but we limit the number of links at each node to N. The price paid is a diameter that increases as N decreases.

A regular tree that is "balanced" (i.e., with all buds equidistant from the root) is a surprisingly compact and efficient way to build a network, in terms of crowding nodes within a small diameter (for example, it has a higher density than an N-cube). But in computer networks links are very cheap compared to nodes, and there is little extra cost in giving all the buds (which in a tree have only one link) as many links as the other nodes. Some of these links can be used for input, output, and mass storage devices.

A tree can often have its diameter reduced substantially by adding "augmenting" links to its buds, turning the tree into a regular graph, as done by Despain and Patterson's (1978) augmented X-trees and Finkel and Solomon's (1977, 1980b) sutures. (The Petersen graph discussed below can be obtained by linking buds to augment a tree.) We can often similarly link distant nodes at the opposite sides of a network or array to give new paths that reduce the diameter and new cycles that improve connectivity.

A desirable property for a network of computers would appear to be a minimum diameter for a given number of nodes, with a reasonable (e.g., 4, 8, or 16) degree, since that gives the shortest paths between communicating processors.

Techniques, often heuristic, for reducing the diameter of graphs have been developed by Elspas (1964), Akers (1965), Trufanov (1967), and others. Friedman (1966) and Storwick (1970) used techniques for combining trees into graphs with relatively high packings. Arden and Lee (1978) developed a technique for generating "multitree-structured graphs" and thereby discovered several relatively dense graphs (all of degree 3). Toueg and Steiglitz (1979) successfully used search techniques to find a number of graphs with still more nodes packed within a given diameter.

An upper "Moore bound" can be put on the number of nodes in a regular graph of degree d and diameter k:

$$N_{(d,k)} = \frac{d\,(d-1)^k - 2}{(d-2)} \qquad (d > 2).$$

These bounds can be achieved only by the "Moore graphs":

The "complete" graphs (those with every node connected to every other node).

The polygon graphs (the regular graphs of degree 2).

Only three other cases—the (10,3,2) (i.e., node=10, degree=3, diameter=2) Petersen and (50,7,2) Singleton graphs shown below, and the possible but undiscovered (3250,57,2). In only a few other very small graphs are the bounds approached at all closely. It seems likely that these bounds are far too high, but there is also good reason to believe that many more graphs with better packing remain to be found— as demonstrated by the steady improvements made over the years by

Elspas, Akers, Friedman, Arden and Lee, Toueg and Steiglitz, and others.

Very recently several new techniques have been discovered that achieve a number of new denser graphs (see Table I):

Imase and Itoh (1981) have proved that de Bruijn networks (1946), which are commonly known as shift registers, give the densest graphs found so far, starting at roughly 15,000 nodes. De Bruijn networks can be drawn as trees that are augmented into regular graphs by linking

Table I

Some of the High-Density (n,d,k) Graphs

d\k	2	3	4	5	6	7	8	9	10
3(Best)	10*P	20*E	34h	56As	78h	122h	176h	311h	525h
&Best:		30TS			72As	120As	164As		
Storwk:	10*P	20*E	28E	36E	60S	66S	90F	138S	216S
MBound:	10	22	46	94	190	382	766	1534	3070
4(Best)	15*E	35A	67h	134h	261h	425h	910er	1360sl	2312s2
&Best:		45A		110ec1	200s3	420ec2		1200s4	2240rs1
Storwk:	15*E	35A	40E	62S	114S	188S	320F	566S	996S
MBound:	17	53	161	485	1457	4373	13121	39365	118097
5(Best)	24*E	48h	126A	262h	505h	1260ec3	2450s5	4690s6	9380s7
&Best:					240ec4	450s8			
Storwk:	24E	36E	126A	120E	232S	442S	850F	1770S	3512S
MBound:	26	106	426	1706	6826	27306	109226	436906	17e6
6(Best)	31E	65h	164h	600ec5	1152s9	2520rs2	6561sr	19683sr	59049sr
&Best:			105Cpl			1728s10	6048sl1	16002ec6	31752 12
Storwk:	31E	55E	105A	462A	447S	867S	1872F	4317S	9465S
MBound:	37	187	937	4687	23437	117187	585937	2929687	14648437
7(Best)	50*HS	88h	252h	992ec7	2880ebl	4680rs3	12250eb2	43200rs4	8640 rs
&Best:				150Cp2	378A	2304rs6	3410sl3	12096rs7	71424sl4
Storwk:	50*HS	80E	150E	378A	1716A	1574S	3626F	9422S	22836S
MBound:	50	302	1814	10886	65318	391910	3351462	141e5	846e5
8(Best)	57E	105E	384eb3	2550ec8	5760eb4	16384sr	65536sr	262e3sr	104e4sr
Storwk:	57E	105E	175E	504E	1716A	1574S	3626F	9422S	22836S
MBound:	82	457	32011	22409	156865	1098057	7686401	538e5	377e6
9(Best)	74E	150Cp3	600eb5	3306ec9	12500eb6	20160eb7	76500rs8	38e4rs9	10e5db
Storwk:	74E	150Cp	240Cp	666E	1904S	5148A	24310A	32706S	94416S
Mbound:	82	658	5266	42130	337042	2696338	216e5	173e7	138e7
10Best:	91S	200Cp4	864eb8	5550ec10	25000eb9	78125sr	39e3sr	19e4sr	98e5sr
Storwk:	91S	200Cp	320E	910E	2780S	6864A	19305A	92378A	170685S
MBound:	101	911	8201	73811	664301	5978711	538e5	484e6	436e7

Best: densest graph to date.　&Best: additional dense graphs.
　　Storwk: Storwick table .　Mbound: Moore bound　*: Maximal
Cp: Cartesian product　1:(4,3)x(2,1)　2:(4,2)x(3,2)　3:(7,2)x(2,1)　4:(7,2)x(2,1)
A: Akers　E: Elspas　F: Friedman　HS: Hoffman-Singleton
P: Petersen　S: Storwick　AL: Arden-Lee　TS: Toueg-Steiglitz
　　sr: shift register (de Bruijn network)
　　h: heuristic search (Leland)
eb: embossed bipartite compound (Li)　1:(5,2)x(2,2)　2:(4,3)x(3,2)
　3:(3,1)x(5,2)　4:(5,2)x(3,2)　5:(4,1)x(5,2)　6:(7,2)x(2,2)
　7:(5,2)x(4,3)　8:(5,1)x(5,2)　9:(7,2)x(3,2)
ec: embossed complete (Uhr)　1:(5,2)　2:(3,3)　3:(4,3)　4:(4,2)
　5:(5,2)　6:(5,4)　7:(6,2)　8:(7,2)　9:(8,2)　10:(9,2)
er: embossed regular compound (Leland-Uhr)　(3,2)x(10,2)
s: split (Leland)　1:(3,3)x(3,4)　2:(3x4)　3:(3,2)x(3,4)　4:(3,3)x(3,4)
　5:(4,3)　6:(4,3)x(4,4)　7:(4,3)x(4,5)　8:(4,2)　9:(5,2)
　10:(5,2)x(5,3)　11:(5,2)x(5,4)　12:(5x4)　13:(6,2)x(6,3)　14:(6,2)x(6,6)
rs: regularized split (Leland)　1:(3,3)x(3,5)　2:(4,3)x(5,2)　3:(5,2)x(6,3)
　4:(5,2)x(6,5)　5:(5,3)x(6,5)　6:(5,2)　7:(5,2)x(5,4)　8:(4,1)x(8,5)　9:(7,2)x(8,5)

their buds to one another and to internal nodes, giving a regular graph whose degree is one greater than the degree of the original tree.

Uhr (1980), Leland and Li (see Leland *et al.*, 1980) have discovered several new "compounding operations" (see Harary, 1969) that use several good small "cluster" graphs, and generate the densest graphs found so far between roughly 500 and 15,000 nodes. These include (among others):

"Emboss" an *n*-node graph into each node of a complete graph of degree *n* (i.e., join each node in the embossed graph to a different link joining the node in the complete graph).

Emboss an *n*-node graph into each node of a complete bipartite graph (one with two sets of nodes, each node linked to all nodes in the other set) of degree *n*.

A general procedure may be emerging: emboss clusters into complete graphs, and complete *n*-partite graphs. Since they compound clusters, the local properties of these graphs can be determined by choosing clusters with desirable local properties.

Leland has coded and used a heuristic search program (see Leland *et al.*, 1980) to generate a number of new graphs, the densest so far, of up to about 525 nodes. The smaller of these new graphs, those that approach the Moore bound (see Table I) appear to have a high degree of symmetry. But it is likely that many of these graphs, which are heuristically augmented trees, do not have as nice local properties as do the compounds. Very recently Bermond *et al.* (1982) have achieved still denser compounds.

Ore (1968), Gewirtz (1969), and others have studied these issues more theoretically. But no general algorithms have yet been developed for constructing maximally connected regular graphs of minimum diameter.

For computer networks, average distance between nodes is probably preferable to diameter as a measure of goodness. A major goal of the programmer and of the operating system should be to use a subnetwork with as small a diameter as possible for each particular program, and thus minimize the possibility that distant nodes would ever need to communicate. Toueg and Steiglitz (1979) successfully used average distance as the criterion to guide their search procedure. Goodman and Sequin (1981) and Despain and Patterson (1978) similarly used average distance to evaluate different hyper-trees and X-trees.

Even better might be a more detailed evaluation, giving the percentage of node pairs at each distance, or graphs that placed some specified cluster size within a minimum diameter. That would help us decide whether there is enough "room" for a particular program.

Evaluating Networks in Terms
of Local versus Global Density

Measures of "local density" would be very useful to evaluate large networks into which programs are mapped onto subgraph clusters. But, so far as I am aware, none have been proposed, much less explored. The following is a first attempt of my own:

For simplicity, consider local density using the diameter of an $(n, d-1, k)$ subgraph (with n nodes, degree $d-1$, diameter k) into which the graph G can be decomposed. [This implicitly assumes that $(d-1, k)$ graphs have been compounded, and that n is a number that will frequently be chosen for the number of computers to assign to one program.]

The following illustrate some of the kinds of interesting comparisons that can now be made:

(1) The $(110,4,5)$ graph compounded by embossing 11 $(10,3,2)$ Petersen graphs into the 11-node complete graph of degree 10 $(11,10,1)$ can obviously be decomposed into those 11 subgraphs, which are known optimal. The $(123,4,5)$ heuristically connected tree, whose global density is somewhat better than $(110,4,5)$, if it can be decomposed at all reasonably (which is unknown, and doubtful), will certainly give poorer subgraph clusters, on the order of $(5,4,2)$.

(2) Looking at larger graphs, where the de Bruijn shift registers achieve the best global density, we find a similar situation. Simply "tessellate" some subgraph known to be dense—e.g, $(10,3,2$—by starting with one subgraph, linking each of its nodes to a new subgraph, linking each of its nodes to a new subgraph, etc., to give a local density of (at least) $(10,4,2)$. In contrast, the de Bruijn network, which, like the heuristically completed tree, is basically locally a tree, will be on the order of $(5,4,2)$, and therefore a rather poorer topology from the point of view of local density.

(3) When larger local graphs are desired, either good compounds or good completed trees can be chosen, and then tessellated, compounded, or "embossed" (i.e., its nodes of degree d replaced by graphs with d nodes, each ported to a different link).

(4) Emboss a graph with the desired local properties into a graph with the desired global properties whose degree equals the number of nodes in the first graph, by joining each link to the original node to a different node in the graph embossed into it.

(5) Emboss complete bipartite graphs into a tree by replacing each node with n copies of an n-node graph and replacing each link in the

original tree by n links from each graph (one from each node). This increases local density and puts many local cluster graphs into close proximity to one another.

To the extent that message passing can be contained within each cluster, so that the need to traverse links between clusters becomes rarer and rarer, there might be fewer and fewer intercluster links, freeing more and more links for other purposes.

This suggests building graphs using subgraphs that are as dense as possible, and/or have whatever mix of properties are deemed most desirable, but linking $1,2,...,n$ of each subgraph's nodes to other subgraphs. Therefore, assuming local density to be the only criterion, we should search for dense clusters that were almost regular graphs of degree d, but missing $1,2,...,n$ joins.

Some Graphs with Maximal Connectivity

A great deal of potentially relevant research has been done on graphs with good properties for communication networks [see Wilkov (1972) for a fine review and McQuillan (1977) for a survey of some highlights]. Much of this research has focused on finding maximally "connected," and "cohesive," graphs (those that remain connected after the largest number of nodes, or links, have been removed), of maximum "girth." (The girth of a graph is the smallest path cycling from a node back to that node.)

An m-regular graph of girth n with the smallest possible number of nodes is called an "(m,n)-cage." Cages are maximally connected regular graphs. "Moore graphs" (Hoffman and Singleton, 1960) are cages that are also of minimum diameter (Wilkov, 1970). They therefore have a very unusual mixture of properties, since they link the largest possible number of nodes within the smallest diameter in a highly regular and symmetric way, with maximal connectivity and cohesion. Unfortunately, very few such graphs exist.

The Petersen graph, a $(3,5)$-cage, is the simplest example (Fig. 4).

The $(7,5)$-cage Hoffman-Singleton graph consists, roughly, of five five-pointed stars and five pentagons connected as shown in Fig. 5 [see Tutte (1966), Harary (1969), and Bondy and Murty (1976), for other examples of cages].

This graph has probably never actually been constructed. The partial construction in Fig. 5 must be completed as follows: Given the ten 5-node graphs as shown and labelled, link vertex i of P_j to vertex $i+jk$ (mod 5) of Q_k.

Since this is a Moore graph it also has an optimal minimum diameter of 2 (with 50 nodes).

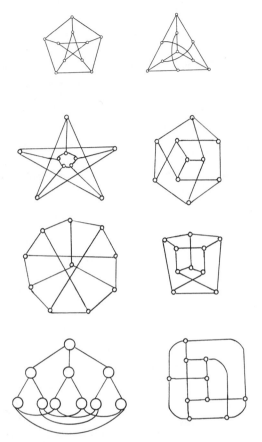

Fig. 4 The Petersen (3,5)-cage (with degree 3, diameter 2). (The same 10-node graph drawn in a variety of ways.) [From Bondy and Murty (1976).]

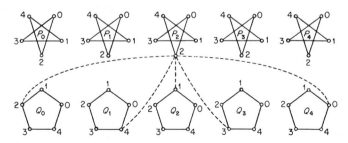

Fig. 5 The Hoffman-Singleton (7,5)-cage. [From Bondy and Murty (1976).]

These graphs might be useful for networks where the operating system, or a program's highest-level procedure, moves to any node. For example, a Petersen network might be constructed with Petersen graphs as nodes, so that each program could work within the individual graph to which it was assigned. This might be continued recursively, giving clusters of clusters (of ...) Petersen graphs. Note that the Petersen graph has a diameter of only 2, so it packs its 10 nodes in a very compact manner. A 2×5 square array, the leftmost and rightmost columns connected ("wrapped around") also has 10 nodes, but a girth of 4 and a diameter of 3.

It might also be of value to emboss a cage into a suitable larger graph, by replacing each node in the larger graph by the cage, to serve as a minimal skeleton to be preserved at all costs as components go bad.

In general, we can think of using some larger structure, like a tree or a tiling, into which to emboss a desirable cluster, whether a Petersen graph, some other graph that approaches the Moore bound, or a graph with a completely different set of desirable formal or structural properties.

Gaining an Intuitive Understanding from, and Being Misled by, Graphs

A graphical representation of a computer network often gives us a better intuitive understanding of that network. And we can trace through the distances between pairs of nodes that are of interest, and examine the graph for other properties of importance, e.g., symmetries, connectivity, or desirable structures. But it can often mislead, since for any but the simplest of graphs there are a variety of ways of drawing the same graph.

Time-consuming computer programs are needed to try to match two graphs, to see if they are identical, or if one can be embossed in the other, or how much the two overlap. The matching of two graphs (that is, the graph isomorphism problem) quickly becomes computationally infeasible (it is not known whether it is NP-complete). Some of the most illuminating results that have been achieved recently about computer networks (see Chapters 3, 7, 14, and 15) have been demonstrations that several seemingly quite different networks are actually identical, or very similar. Interesting examples are the Petersen graph itself, which can be drawn in a number of surprisingly different ways, and the de Bruijn networks, which can be drawn in shift-register fashion or as augmented trees or in the entirely different symmetric-graph fashion used originally by de Bruijn in his 1946 paper. (See Fig. 6.)

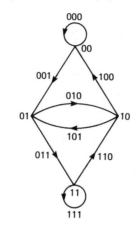

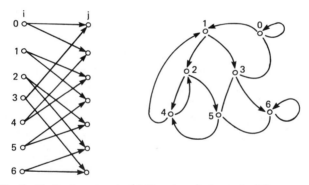

Fig. 6 De Bruijn networks (shift registers), drawn in different ways.

Embedding Data-Flow Graphs in Grids, and in Chip Area and Time

Graph theory can be used fruitfully for a variety of other issues related to multicomputer architectures and programs. But work is just beginning, and I will here simply mention a few possibilities.

The network problem can be viewed, loosely, as the problem of finding the data-flow graph most appropriate to the transformation of information from input data to output results, and finding and using the most appropriate multicomputer network to execute that data-flow graph.

When we represent an architecture as a network (e.g., of processors and memories, or of gates) then the problem is one of matching two graphs—the data-flow graph and the architecture graph (Bokhari, 1981). This is equivalent to the graph isomorphism problem and can therefore be solved only in special, limited, regular cases. Heuristic techniques, which are not guaranteed to be optimal or even to succeed, are often indicated in such situations. Their results must usually be evaluated empirically, by comparing them with previously achieved results and choosing the best. One attractive way of cutting this problem down to size is to work with small local module clusters and successive compounding operations over them.

If we ask programmers to restructure algorithms to better fit the available hardware architecture, we are asking them to try to transform the program graph into one that matches the architecture graph more closely. This is a very hard problem, and we should try to develop computer-aided graphics-based design systems that interact with and help programmers in carrying out this enterprise.

An architecture with reconfiguring capabilities, and an operating system that attempts to map programs into structurally similar regions of the multicomputer, are both confronted with the problems of matching graphs, and of modifying graphs to increase their degree of match.

To maximize efficiency, we might mark working nodes as "active," and try to increase their percentage (Uhr, 1981c). To maximize speed, we might ask how to widen the effective "bandwidth" of the graph, viewed as a channel from the "source" that inputs information to the "sink" into which the results are output.

The grid is a good graph into which to embed architectures (see Chapter 15 for embeddings in the VLSI landscape). As elegant, compact modules are developed, whether for memory, simple logical functions, systolic processors, 1-bit processors, or more complex building blocks, those modules can be used instead. For example, the processor—memory might be designed as a graph of logic gates that embodied the structure of that processor's processes, surrounded by memory gates that spread to fill space where the processor structure did not need gates.

This suggests that the design of a network with 10^4 or more chips, each with 10^6–10^8 transistor-equivalent devices and 256 or so links to other chips, poses a rather interesting set of problems for large finite graphs. Simplifying construction rules convert chip area into the size of the grid, or other graph, that is used as the intermediate structure into which the network is embedded.

The architect who works in VLSI actually uses a set of idealizing and simplifying assumptions that turn the design problem into a graph-like problem. People attempting to apply graph theory to VLSI design idealize further, in the attempt to achieve more general analytic results. The correspondence of both thrusts suggests that we are moving into a new era where very large computers will be designed in far more general ways.

Bits of Logical Design
Embodied in Binary Switches

This appendix examines some basic issues of logical design of computers and their components, and explains how computers do everything using structures of two-valued switches that embody and perform logic.

Underlying all of the computer's instructions, underlying all the procedures a computer will ever execute (whether they process mathematical entities or information), is logic.

This is true because the computer uses sets of logical functions to effect all the transformations made on data. It also uses logic to store, to transfer, and to route data across links and switches.

Underlying the logic is switches.

All the logic is realized with the simple basic building blocks of simple "logic gates," each built from a few binary on-or-off switches.

Underlying switches is solid-state physics.

Only through enormous leaps in our understanding of how to control the movement of electrons in solids have we been able to move rapidly toward ever smaller components packed ever tighter on ever more efficient silicon chips. The development of x-ray microscopy and electron microscopy make possible the precise reduction of the logical design to microscopic size. The development of laser technology makes possible the precise etching of this design, after enormous reduction to electron-microscopic size, on the chip.

Underlying solid-state physics is ecology—clean living, right thinking, patience, and care.

The production of VLSI chips with thousands or millions of devices is one of the most astonishing accomplishments of this complex edifice of high high technology. Each technology entails enormous care, precision, and sophistication. Putting them all together into a total system where the tolerances for errors are microscopic at each step is a perpetual triumph. Everything in the factory must be perfectly controlled and free of contamination. Even the air must be free of the microscopic dust particles that might blemish and ruin a chip. In fact nobody quite knows why some factories simply do not succeed in turning out adequate yields of chips.

Some Basics of Logical Design Using Switches

It turns out that the logical manipulation of information is sufficient for arithmetic and any other kind of mathematics, as well as for the processing and transformation of other types of information.

Logic as the Foundation Underlying Mathematics

Among the many streams of mathematical, scientific, and technological advances that have been necessary flows culminating in the very fast and very powerful modern computer was the development of our understanding of logic as the foundation of mathematics (the manipulation and transformation of numbers, sets of numbers, and other kinds of symbols). A succession of constructivist mathematicians and logicians, from Frege, through Whitehead and Russell to Hilbert, Turing, Post and many others showed how all mathematical thinking could be broken down into basic very simple logical steps.

For logicians and mathematicians, this was a very important development, since it put mathematics on a firm foundation and it clarified the relation between mathematics and logic. But it was not especially useful for actually doing mathematics, since breaking things down into small steps and handling these steps in a precise and regular way is terribly tedious. It not only eliminates most of the fun, it also obscures the intuitive higher-level "understanding" of what one is doing. It turns one into a "machine."

But as soon as we had "computing machines" this development not only gave us confidence that these machines are indeed adequate to their task, it gave us techniques and algorithms for getting them to compute.

A few simple examples are probably the best way to explain how logic can be used to do mathematics.

Two-Valued Logic and Two-State Switches

First, I must mention that today's computers use "binary" (that is, two-valued) logic. This is not necessary. But it is very convenient, because it means that a two-state switch (one that can be thrown into either one of its two possible states—think of them as "on vs. off" or "left vs. right" or "true vs. false" or "1 vs. 0") can physically embody and represent the two-state logical expression.

The two-state switch is not necessary either, for three-state, four-state, ..., switches can be made. But the two-state switch is far cheaper and easier to make and, even more important, it is far more reliable. This follows from the fact that a switch must be able to move along some dimension (e.g., up—down, hot—cold). A switch with only two states can be moved to its extreme low for one state and its extreme high for the other. These are as far apart as possible, and the switch can be driven in a forceful single-minded way; there is no need to try to move it to a correct location, not too near and not too far.

Binary, Decimal, and N-ary Systems
to Represent Numbers and Symbols

The two-state switch and also the simplicity of two-valued logic (as opposed to three-valued, or *N*-valued logics, which also exist) are strong arguments for building computers that do mathematics and symbol manipulation with binary switches. All we need to do is "recode" (that is, convert) whatever other symbols we may want to use into strings of binary digits (often called "bits").

We can use any pair of symbols we choose to represent the two states of a binary digit. Logicians often use ["true" vs. "false"] or ["T" vs. "F"]. Or we might use ["X" vs. "Y"] or ["A" vs. "B"] or ["I" vs. "O"]. If we were only interested in picture processing we might even use ["dark" vs. "light"].

But the most commonly used, and probably the simplest, is ["1" vs. "0"]. The binary digits [0,1] are the first two of the ten digits 0,1,2,3,4,5,6,7,8,9 used in "decimal" (that is, 10-valued) number systems.

We can recode a decimal number into its equivalent binary number by using Table I (or the rule it exemplifies).

Table I

Binary and Decimal Encodings

Decimal	Binary	Symbol	Word
String representing the number or symbol in			
0	0	zero	white
1	1	one	grey
2	10	two	slate
3	11	three	black
4	100	four	brown
5	101	five	yellow
6	110	six	orange
7	111	seven	red
8	1000	eight	purple
9	1001	nine	green
10	1010	ten	blue

A word/symbol like "ten" can be encoded into the equivalent decimal symbol "10" and also into the equivalent binary symbol "1010." We can set up any other more or less arbitrary encoding, simply by writing a code book. (The arabic number system uses code books that can be generated by a simple algorithm.)

Thus (using Table I) "blue" can be recoded to "ten" or "1010," "yellow" to "101," etc. It is in just this simple-minded way that we recode numbers, symbols, words, sentences, formulas, and any other string of symbols into binary strings when we input them to the computer. (Such recoding is routinely done by the computer itself, using simple tables of the sort shown in Table I.)

Binary numbers seem strange to most people, but that is because we have not spent many years learning to use them in basic classes in elementary arithmetic. Actually, they are quite convenient to use. Their only (rather minor) drawback is that a binary number has about three times as many digits as the equivalent number using the decimal system.

Realizing Arithmetic and Handling
Information Using Two-Valued Logic

We are now ready to examine how logic can be used for mathematics.

Storing Data in Memory Banks Made from Ordered Switches

First, a number can be represented by the set of states of an ordered sequence of two-valued switches. Let us define a two-valued switch as being in either of the two states "+" or "−." Now we can recode a decimal integer like 5 into the binary integer 101, and then represent this integer with the ordered set of three binary switches + − +.

(Following the general principle just described, that we can recode from any symbol set to any other symbol set, I will sometimes recode "+" to "1" and "−" to "0" so that we no longer have to bother with the extra step from binary arithmetic to binary switches.)

A computer's memory stores and registers are simply ordered sets of switches.

When "information is stored into a particular word in memory" the sequence of switches that embody that word is thrown into the sequence of states that represents that piece of information.

When "information is fetched from a word in memory" the values of the ordered set of switches that embody that word are read out and used to set the switches in the word (often one of the registers that the processor can then access directly) into which this information is read.

That is, a copy is made of the switch settings of the word in memory, giving the identical switch settings in the register. Because this is often convenient for arithmetic and other processes, this sequence of switch states, that is, a sequence of + and − values (of, e.g., voltage), is often shifted out serially, the lowest-order digit first.

Processing Data

Now we need (among others) a switching net to add two numbers; for example, it will add 2 to 5, giving 7 as the result.

Using binary arithmetic, it adds 10 to 101, giving 111.

Using two-valued switches, it adds + − to + − +, giving + + +.

Some positive quantity (e.g., a positive voltage, or heat) is input to a switch, to put it into the positive + state. This then represents 1. A negative quantity is put into the next-lower-order switch, giving −, or +− when both switches are considered. Three switches are used to represent the 7, all three in their positive states, that is, +++.

How can we build a network of switches that will input +−+, input +−, and output +++?

Consider for a moment how we were taught to do addition: We add the "lowest-order" (that is, the rightmost) digits, try to remember any carries, add the next-rightmost digits plus these carries, and continue, moving from right to left. We can build a switching network that follows exactly that procedure.

It first shifts in the two rightmost digits (− and +).

The rightmost switch in the set of switches representing the binary number 10 is −; the rightmost switch in the set of switches representing the number 101 is +. We therefore need a circuit that, when it inputs + and −, outputs +.

This is got very simply by a logical "EXCLUSIVE−OR" function, one whose output is "true" when either (but not both) of its inputs is "true"—that is, when one of its inputs is "true" and the other one of its inputs is "false."

It is helpful at this point to look at Table II, an example of a "truth table," which shows how logical expressions are evaluated. For example, the second row shows that if A is true and B is false, then not(AorB) is false (since AorB is true).

Table II

Truth Tables for Several Basic Logical Functions

A	B	notA	AandB	AorB	not(AorB)
T	T	F	T	T	F
T	F	F	F	T	F
F	T	T	F	T	F
F	F	T	F	F	T

We also need a logical "AND" that outputs a + value to a (1-bit) "carry register" when both the inputs are +.

Now the next digits are processed, along with a third digit, the one stored in the carry register. This process then continues until all digits have been added together in this way.

The typical 16-bit or 32-bit computer will use a series of 16 or 32 such 1-bit adders, pumping numbers through them. A 1-bit computer like CLIP, DAP, or MPP will cycle numbers through a single 1-bit adder. More elaborate parallel adders that embody parallel algorithms for adding the pair of numbers at each digit in parallel are also used in especially powerful computers, giving a specialized speeded-up arithmetic capability.

Note that to compute a logical operation like AandB or notAorB just the single AND, OR, and NOT gates are needed. This is the basis for the 1-bit logical operations that dominate array computers like CLIP and DAP, and that are built into virtually every traditional computer set of machine-language instructions.

A set of such AND gates can be used to match two characters, or two strings of characters, as in a word, or the address a database management system searches for. This is the basis for the manipulation and transformation of single alphanumeric characters, which are actually 8-bit (1-"byte") strings of bits, strings of characters, and information of any sort.

Note that character or string matching proceeds much as addition does, except that the two bits are ANDed together rather than ORed, and no carries are needed. Either can be effected by cycling bit by bit through a 1-bit processor in "bit-serial" fashion or by pumping a whole chunk (e.g., a 32-bit number or an 8-bit character) through hardwired hardware in a more powerful 8-bit, 16-bit, or 32-bit computer.

This has been a somewhat cumbersome description (yet it is probably still not detailed enough). But it gives the flavor of the intricately detailed, yet basically simple, processing that a computer must do.

We can represent such logical functions with diagrams of logic gates, as shown in Fig. 1.

The actual embodiment of the logic in the particular technology of switches being used will vary. But it is straightforward (although much very sophisticated and expensive equipment may be needed). For example, a two-input gate needs three or four transistors in CMOS technology, but eight to twelve in ECL and TTL. In CMOS technology only one transistor is used for each switch that stores 1 bit of information in the 64,000-bit and 256,000-bit VLSI dynamic RAM memory chips. (These simple switches gradually leak current and are unstable, so they must be "refreshed" "dynamically" every few milliseconds.) A faster technology like TTL needs four or more transistors to store each bit.

It is only because computers can carry out millions, or billions, of such switch-throwing steps each second that they are able to put together much more interesting and more complex functions at still enormously fast speeds.

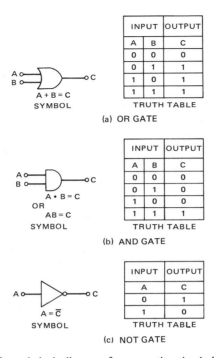

Fig. 1 Example logic diagrams for computing simple functions.

Choosing among Equivalent Logical Expressions or Alternative Technologies

The embodiment of each of the logical functions a particular computer uses will depend upon the choices made at the level of logic design, and also upon the economics of the particular technology used to build the computer. A good deal of human knowledge, experience, and intuition appear to be needed today, to make wise choices at each level, and then to move from the level of logic design to the level of the detailed blueprint of the chip crammed full of devices that realizes that design.

There are a number of ways to realize any particular logical function. We usually think of using "AND," "OR," and "NOT" functions as our primitive functions. But we could also use only the single function "NAND" (that is, "NOT−AND") or "NOR" (that is, "NOT−OR"). Indeed, these are usually preferred when designing computers, since, although more steps, that is, more logic and more devices, are often needed to realize the function, it is simpler to work with a single primitive building block.

Concluding Remarks

Note how logic and switches are used to handle both memory (stores and registers) and also the actual procedures that transform the information stored in these memories. A computer has very large banks of simply organized switches that embody its large memory stores, and separate very large banks of very complexly organized switches that embody its CPU.

A typical medium-sized traditional 1-CPU computer has roughly 10,000 gates in its CPU and 10,000,000 bits (realized using 1 to 12 gates per bit) in its high-speed memory. In sharp contrast, each of the 10,000 1-bit computers in the CLIP4 array has about 600 processor gates and a 32-bit memory (using 8 gates per bit). The differences are striking, especially when we remember that both are general-purpose computers capable of executing exactly the same set of programs.

Memory, processors, and reconfiguring switching networks are all made of the same sort of stuff. They are at bottom very small and simple things. Clearly, they can be torn apart, dispersed, and mixed together.

Component Cost and Chip Packing Estimates for VLSI

A "device," "transistor-equivalent device," "gate," and "element" can be considered roughly equivalent when designating the basic components of a computer. This makes possible some interesting comparisons between different types of computers, and different ways of allocating resources among a computer's major subsystems (e.g., processors, memories, registers, controllers, reconfiguring networks of switches).

But I should emphasize that these are rough equivalences. Typically, a gate is built from 2 to 6 or more transistors. For example, Shima (1978) gives counts for the Z8000 microprocessor chip of 5833 gates and 17,500 transistors. A two-input logic gate usually needs 2 to 12 transistors, not just 1. It is possible to store 1 bit of memory with only 1 transistor (as on the 64K-bit and 256K-bit dynamic RAMs); but frequently 2 to 12 are used. Different technologies allow for, and encourage, a variety of different designs. Some designs are faster, or more stable, or more modular, or dissipate less energy—therefore a whole set of trade-offs must be considered when particular circuits are chosen.

The following are necessarily rough estimates for not yet developed technologies during the 1980s and 1990s. I am assuming that the device count on a single chip will continue to grow, roughly doubling every 12 to 18 months. This has been the case for over 20 years, and enough is already known about how to improve upon several alternative existing technologies so that most people expect this rapid rate of increase to continue for at least the next 10 or 20 years (see, e.g., Sutherland *et al.*, 1976).

We can think of and expect four major packing densities:

(a) LSI chips with roughly 50,000 devices (1977−1984);
(b) VLSI chips with roughly 100,000 devices (1979−1987);
(c) VVLSI chips with roughly 1,000,000 devices (1985−1995);
(d) VVVLSI chips with roughly 10,000,000 devices (199?−20?1).

A typical small computer might be built from 1, 100, or 1000 chips, a medium-size computer from 2000−7000, a large computer from 10,000−100,000 or even more. The determination of the marketplace cost of a computer is rather ethereal, as with the cost of perfume. But when the number of chips fabricated, computers sold, competition, and marketplace demand are held constant, a computer built in large quantities might cost 2 to 10 times the cost of the chip, and chips with reasonably high yields that sell in large quantities might cost from $1 to $20.

A Quick Look at the Modular
Way Computers Are Built

Once a chip is mass produced (by the millions) its price is driven down to a few dollars ($1 to $2 for small high-yield chips; $10 to $100 or more for large low-yield chips that push the state of the art).

Each chip is mounted in a plastic or ceramic package whose pins plug into a socket on a board on which the circuits between chips are printed. The boards slide into slots in a chassis, and are connected by cable to one another, to input and output devices, and to the power supplies. Thus the computer is a big black box that packages little black box chassis packages of board packages of chips.

The chip is the smallest package in physical size (roughly 4 to 8 mm square); but it is *by far the largest package in terms of the thousands or millions of components it contains.* A board (roughly 20 by 30 cm) typically holds 10 to 40 chips, and a chassis holds 10 to 30 boards. Most computers today use only a single chassis, but a large system might have 5 to 20.

A computer is "large" if it has a large number of chips—roughly, 10,000 or so. A single 10,000,000-device chip will be far more powerful than a medium-sized 1000-chip 1981 computer. The single chip has a complexity (in terms of device count) of 10^4 to 10^7 or 10^8, while the linking of a number of chips into a total computer increases this complexity by only 10^2 to 10^5 or 10^6.

A computer is built of chips of enormous power. But bad bottlenecks and weak links occur in the connections between each chip, profligate with devices, and its external world. For each chip has only 12 to 64 pins, each pin acting as a link over which a stream of 1-bit information can be sent (several of these pins must be linked to the power supply). A 32-bit computer needs 32 pins just for (parallel) input−output to memory, or/and message passing to other computers.

In contrast to the steady doubling of devices on each chip that we can almost certainly expect every 12 to 18 months, relatively little increase in the number of pins can be counted on. Probably 128 pins per chip, and possibly 256 pins per chip, will be achieved in 10 or 20 years. There are other techniques, including using the whole wafer, or using lasers to drill holes into which pins can be set, or stacking and interconnecting wafers, that may be developed to the point where they give adequate yields to be commercially viable. But these are much less certain than is the steady increase in packing density.

Some of the Determinants of Packing Densities and Device Counts

The exact costs of the basic components with which networks are built will vary greatly with the particular technology used. Increased packing depends upon refining technologies for lowering power consumption and heat dissipation, and for making masks of the circuits on the chip, e.g., by using x-ray or electron-beam lithography that allows for greater reductions to smaller devices. As we move into higher and higher packing densities, the area used on a chip becomes a more valid measure than the traditional count of devices. Chip area is itself a function of the number of devices, since each device needs space. Also heavily involved are heat dissipation and the ability to configure devices close together with simple and short connecting links.

The faster the technology the harder the circuit is driven, the greater the energy dissipated as heat, the finer the allowable tolerances and, in general, the lower the packing densities.

The yield of chips without defects goes down drastically as chip size increases. Murphy (1964) shows that yield y is related to defect density d and chip area a by the formula $y = ke^{-da}$ (k is a proportionality constant). Typical defect densities are around 10 to 20 cm^2. For example, increasing a $D = 20$ 5 × 5 mm chip 10%, to 5.5 × 5.5 will lower yield from 4% to 2.8%.

A simple, regular design will usually lead to a greater packing density. Today we typically see two to ten times as many devices on a memory chip as on a processor chip of the same area. Memories iterate very small modules many thousands of times. And their basic components, the binary switches that are used to store binary digits, are built into very simple strings, the words, which themselves form very simple blocks.

The simpler the processor, the simpler is the module, with greater possibilities for tighter packing (usually).

The simpler the pattern of links connecting components, and the nearer the neighbors being connected, the greater the possibilities for tighter packing (usually).

So it would appear that, at least after more experience has been gained in designing such chips, many simple computers on a single chip should achieve a packing density a good bit better than that achieved for a single powerful processor (but less than that achieved for memory chips).

Cost Estimates for Specific Types of Devices

Ultimately, transistors and the links between them are the basic building blocks in a modern computer. They chiefly serve as switches, or gates, whether to store information or to effect logical transformations. So I shall use the terms "switch," "gate," "transistor," and "device" interchangeably.

With the above proviso that chip area is an alternative, possibly preferable, measure of costs (see Thompson, 1979, 1980), I will (as most people do today) make estimates chiefly in terms of "transistor—equivalent device" (or "switch") counts, with occasional comments about packing issues. Essentially, the number of devices packed in a given area will depend upon how compactly the graph of the architecture being laid out can be mapped onto a grid or some other simplifying topology that can be printed onto the chip by using the chosen technology. For example, trees of reasonable degree and most planar graphs can (if some crossovers are allowed) be embedded in linear area (Valiant, 1981). Brent and Kung (1980), Storer (1980), and a growing number of other researchers are examining issues of embedding graphs that represent and execute algorithms on chiplike grids. Uehara and Van Cleemput (1981) and others are developing layout techniques for particular technologies.

The components are discussed below:

Processors

A logic "gate" (e.g., to compute a logical AND, OR, NAND (Not-AND), or NOR (Not-OR) typically needs 2 to 12 devices.

The very simplest general-purpose 1-bit processors need from 50 to 800 devices (see Table I). In several cases a substantial percentage of the hardware is used to enhance the processor beyond the bare bones. For example, CLIP uses roughly 100 devices to effect the parallel logic over the eight near neighbors, and the MPP uses roughly 250 devices for the very fast 32-bit registers each of its 16K processors has been given to speed up arithmetic.

Table I

The Range of Processor Sizes and Components

General-purpose 1-bit processors	
System	Device total
STARAN	50
DAP	100
CLIP4	600
MPP	800

General-purpose N-bit processors for traditional serial computers	
Type	Device total
8-bit micro	1,000 – 8,000
16-bit mini	2,000 – 18,000
32-bit medium	5,000 – 20,000
32-bit large	20,000 – 50,000
Supercomputer	50,000 – 200,000

An n-bit processor needs roughly n times the number of switches. But a powerful 32-bit processor will usually be given a lot of extra hardware, e.g., for floating-point arithmetic, parallel arithmetic, and parallel input—output. A "super computer" (e.g., the Cray-1) will be given still more, still faster hardware, including pipelines of processors for vector operations.

Memory Stores

These need from 1 to 12 switches per bit. Memory has typically been built from "flipflops" (which flip or flop into either of their two states) that use 2 to 6 transistors. But today's "dynamic" RAM memories (which leak slightly and therefore lose their charges, which must periodically be dynamically "refreshed") use only 1 transistor—equivalent device per bit. The faster memories dissipate more energy and need more devices and more breathing-space.

Following Patterson *et al.* (1979), I will use the estimates of 1 switch for relatively slow dynamic RAMs and 4 switches for static RAMS (that are roughly four times as fast) (see Jecmen *et al.*, 1979; I. Lee *et al.*, 1979; J. M. Lee *et al.*, 1979; Pashley *et al.*, 1979).

Registers and Cache Registers

A register is simply a small memory store. But it is usually made of faster technology (since it must be accessed more often and is small). So we should estimate 4 to 12 devices per bit.

Ports and Switches

A multiport processor, memory, or register needs (a) a switch to choose the port, and (b) an address register to store the code that controls this choice.

Both switch and register need $\log P$ bits for P ports. Each needs 1 to 4 devices per bit. Device counts for plausible numbers of ports are given in Table II.

Table II

Rough Device Counts for Multiport Memories or Processors

Number of ports—switches	Speed and technology	
	Slow/cheap	Fast/expensive
1—2	4	16
3—4	8	32
5—8	16	64
9—16	32	128

Reconfiguring Networks

A reconfiguring network is just a switching network. It may be used to connect nodes of any type, e.g., processors to processors, or processors to registers, or processors to memory. (Note that a network with several links to each processor or memory store can be thought of as a reconfiguring network where a processor or store replaces and can, when needed, serve as a switch.) A typical reconfiguring network needs $n \log n$ switches between n nodes. It also needs $n \log n$ registers for addressing, one for each switch. Therefore it grows rapidly in cost as the number of nodes being reconfigured grows, as shown in Table III.

Table III

Estimated Costs of Reconfiguring Networks of Various Sizes

Nodes in reconfiguring devices (using slow/cheap technology)	
Nodes in network	Total switches (Each address register)
4	8
16	64
32	160
64	384
128	896
256	2,048
512	4,608
1,024	10,240

The Controller

I have included the controller only in the estimates of n-bit computers, since the simple 1-bit processors are used in SIMD mode, with one controller for the entire system. Their controllers are sometimes relatively simple, because of the low-level 1-bit operations performed by machine-language code and the way they are synchronized to effect the same operation with the same addresses for data. But in fact the MPP controller is a very large and sophisticated piece of hardware, since it handles very fast parallel input—output, reconfiguring of data via buffer memories, and arithmetic, among other processes, using very sophisticated chips driven by a very fast 40-MHz clock.

There must be a link between the controller and every processor; in arrays this is usually handled with an array of wires to the program registers, much like the array linking to registers for input—output. It is especially difficult to estimate the size of a controller, which can vary from only 1000 or so devices to over 100,000, depending upon the variety of functions it handles and its sophistication.

Table IV summarizes these various component costs, showing a range of plausible device counts.

Table IV

Estimates of Component Costs[a]

Component	Number of Switches (Gates) Needed (Range)	
	Min/Slow	Max/Fast
1-bit processor	50	500
N-bit processor	$50N$	$500N$
1-bit memory	1	10
N-bit registers	$4N$	$8N$
N-way ports	$2 \log N$	$8 \log N$
N-way switches	$\log N$	$\log N$
Reconfiguring nets	$N \log N$	$N \log N$
SIMD controller	1,000	80,000
N-bit proc. controller	5,000	100,000

[a] Note: Powerful controllers may use 80,000—200,000 devices. SIMD systems use one controller for all processors.

Allocating Resources to Different Types and Mixes of Processors

We can move in either of two extreme directions, toward

(1) Two ever bigger buckets, for (a) a superpowerful processor and (b) as gigantic as possible a memory;

(2) As many processors as possible, each (a) as small and weak as possible, with (b) a memory no larger than needed, and (c) whatever switches are needed to pass information and, possibly, reconfigure.

This gives supercomputer chips, and chips with minuscule special processors and intermediate memories. There appear to be several other promising combinations: another 8 to 32 chips crammed with additional memory for a supercomputer processor chip; a few minicomputers; or many simple 1-bit computers; or a mixture of processors and memories of different sizes.

Superprocessors, a Few Miniprocessors, Many 1-Bit,
  or Still More Specialized Processors

The following are plausible goals, once 10,000,000-device chips are achieved.

Super. A single chip will probably never become too large to house only one single computer if we want to give that computer a very large on-board memory. Thus a 10,000,000-device chip will barely hold a 1,000,000-byte memory (plus a large 200,000-device CPU). So an additional 4 to 32 memory chips might be used for each CPU.

Mini. Another attractive configuration is one with a small number of 4, 8, or 16 computers on each chip. Each would be the size of today's "micro" and "mini" computers. Each would have good access to its own (relatively small) on-chip memory, and reasonably good access to larger memory stores on other chips. But more than 16 computers would be severely limited because dividing the small number of pins on a chip among them all would leave each with too small an input−output channel.

Many. Computers might be used with 1000 to 2000 devices each (e.g., 1-bit 300-gate processors with 1000 bits of memory), possibly with an occasional controller amortized over 16 or 64 processors, to give a mixed MIMD−SIMD system. A 32×32 or 64×64 array of such computers might be put on each chip.

Very Many. Toward the small extreme, since a general 1-bit processor can be built from 50 devices and a useful memory from another 10 or 20, we can conceive of putting as many as 500,000 50-device computers on a single 10,000,000-device chip (to be more reasonable, make that 200,000 or 100,000).

Minuscule. Even more extreme, this simple 1-bit processor can be torn apart, so that a smaller set of 4 to 20 switches might be used to compute each particular logical function. Interspersed between these switches would be just enough switches to serve as intermediate memory registers.

Mixture. Sometimes a specialized chip that mixes a number of different types of processors, of different sizes and powers, or/and with

Table V

Possible Packings of Future VLSI Chips with Computer Resources

Per chip	Node	Processors	Memory	Configuration	Nodes/chip
		Number of devices per component			
10,000	25	15	10	0	400
100,000	25	15	10	0	4,000
1,000,000	25	15	10	0	40,000
10,000,000	25	15	10	0	400,000
10,000	50	40	10	0	200
100,000	50	40	10	0	2,000
1,000,000	50	40	10	0	20,000
10,000,000	50	40	10	0	200,000
10,000	200	100	80	20	50
100,000	200	100	80	20	500
1,000,000	200	100	80	20	5,000
10,000,000	200	100	80	20	50,000
10,000	500	300	150	50	20
100,000	500	300	150	50	200
1,000,000	500	300	150	50	2,000
10,000,000	500	300	150	50	20,000
10,000	2,000	1,000	600	400	5
100,000	2,000	1,000	600	400	50
1,000,000	2,000	1,000	600	400	500
10,000,000	2,000	1,000	600	400	5,000
10,000	5,000	3,000	1,500	500	2
100,000	5,000	3,000	1,500	500	20
1,000,000	5,000	3,000	1,500	500	200
10,000,000	5,000	3,000	1,500	500	2,000
10,000	20,000	5,000	10,000	5,000	0.5
100,000	20,000	5,000	10,000	5,000	5
1,000,000	20,000	5,000	10,000	5,000	50
10,000,000	20,000	5,000	10,000	5,000	500
10,000	50,000	9,900	40,000	100	0.2
100,000	50,000	9,900	40,000	100	2
1,000,000	50,000	9,900	40,000	100	20
10,000,000	50,000	9,900	40,000	100	200

different specialties, might be used. Such a system might be harder to
design and to pack compactly than the others. But a rough idea of how
many of each type of component might be used can be got by taking

different sized processors and memories from Table V (which gives some idea of the range of possibilities) until the estimated number that the chip being contemplated could hold is reached.

When a processor is not on the same chip as its memory, the constraints become severe. One or two processors can be given enough parallel links to memory. But, because of the stringent limits on the number of pins to a chip (40 or at most 64 today; probably no more that 128 or 256 in 10 or 20 years, when the 10,000,000+device chip reaches maturity), any more will be severely slowed down by the necessary serial access to memory.

This means that traditional small computers, needing at least 64,000 bytes of memory, can at most be packed only 4 to 16 to a chip!

But there may be little point to that. One might as well pack a single powerful processor on the chip, giving it plenty of parallel links to off-chip memory, which will itself take several more chips. This is an attractive vision, and one that many people hold—only one design for the largest possible 1-chip processor, with a family of more powerful computers, depending upon the number of memory chips added (and also the number of such computers put into a total network).

Only when we take a very different approach, using relatively weak processors, each with an extremely small (from the traditional perspective) memory, do we reach a second attractive alternative strategy for allocating the real estate on a chip.

The pin fan-out limits not only mean that processor-to-memory links quickly become a bottleneck, but also that there will not be enough links to and from the chip to handle input and output (including message passing) to that chip fast enough to keep its processors busy. Only when the chip becomes a self-contained module, able to work for a relatively long time (with respect to the input—output times needed), can pin limits be overcome.

This means, for perception programs, that each chip should be able to process an array large enough so that it will almost always resolve and be large enough to surround each meaningful object of interest and most of its relevant context. For other problems it means that the set of processors on a single chip should on the average spend most of its time working on a program, rather than waiting for input and output.

Enough memory must be accessible to the processor or processors on a chip to keep the need for I-O from slowing down the system. This means enough memory on the chip and/or few enough processors on the chip so enough pins can be dedicated to each to access enough memory off the chip. See Table VI for useful example packings for different types of chips.

Table VI

Example Packings for Useful Chips of Different Types

		Number of devices per component			
Per chip	Node	Processors	Memory	Configuration	Nodes/Chip
Minuscule specialized processors					
10,000	25	15	10	0	400
100,000	25	15	10	0	4,000
1,000,000	25	15	10	0	40,000
10,000,000	25	15	10	0	400,000
1-Bit general processors					
10,000	50	40	10	0	200
100,000	50	40	10	0	2,000
1,000,000	50	40	10	0	20,000
10,000,000	50	40	10	0	200,000
100,000	200	100	80	20	500
1,000,000	200	100	80	20	5,000
10,000,000	200	100	80	20	50,000
10,000,000	500	300	150	50	20,000
1,000,000	2,000	1,000	600	400	500
10,000,000	2,000	1,000	600	400	5,000
4 to 16 Minicomputers					
50,000	5,000	2,000	3,000	0	10
100,000	20,000	3,000	16,500	500	5
1,000,000	100,000	5,000	95,000	0	10
10,000,000	1,000,000	10,000	990,000	0	10
One Supercomputer					
1,000,000	1,000,000	50,000	950,000	0	1
10,000,000	10,000,000	100,000	9,900,000	0	1

Summary Discussion: Toward a More Powerful Distribution of Resources

The fine details of the logic, the number of devices that will be needed to realize that logic, and the chip area these devices will need, will vary with the particular kind of chip and technology used. But the basic component devices remain roughly constant. This offers a good perspective from which to examine networks.

What percentage of a network's resources should be allocated to processors, and what percentage to high-speed memory?

Conventional 1-CPU serial computers have traditionally allocated 99%, 99.99%, or even more of their basic devices to memory. The overriding consideration has been to keep the CPU busy. The larger the memory the more likely that all needed data will be available and ready when needed. The memory is made large enough to store millions of bits (using one to eight or more gates per bit); the processor typically has from 5000 to 100,000 gates.

The very large 1-bit arrays use around 100 to 500 gates per processor, and as little as 32 bits of memory per processor. Pipelines, where each processor immediately transforms information output to it by its preceding processor, can use even more gates in the processor and even fewer gates in each processor's memory.

We thus see in the extremes serial computers with substantially less than 1% of their resources in processors, and parallel arrays and pipelines with 90% or more of their resources in processors. Processors actively transform data; they do work. Memory passively holds data. The higher the percentage of resources we can put into processors and use effectively, the more efficient will be the resulting system (see Uhr, 1981c).

It seems likely that the data-flow requirements of the programmed algorithm determine the amount of memory needed. If an algorithm is properly dispersed over an array or network of computers, and the flow of information among processors is simple and efficient (as in a well-structured program that is properly mapped onto an appropriately structured network), and intermediate output data are used as soon as possible and abandoned quickly (as is especially true in pipelines), the total storage requirements for the total system may well be less than, or at most the same as, the total needed by a 1-CPU computer.

The use of processors with just the right amount of power for their tasks offers another major possible source of improvements. When 1-bit logic is needed (as in Boolean logic, and when looking for complex compounds of features), the 1-bit processors are extremely efficient. For working with small numbers (e.g., representing gray scale or intensity or weights), 4-bit and 8-bit processors are often sufficient. The average length of information instructions must gobble up will often be appreciably less than 32 bits (especially when the mix includes 1-bit features, 3-bit intensities, 4-bit weights, and 8-bit symbols). In such cases 8-bit, 4-bit, or 1-bit processors can be attractive alternatives, even though they may occasionally have to be used in word-serial mode.

For particular algorithms and transformations, special-purpose processors of the sort Kung (1980a,b), Foster and Kung (1980), and Leiserson (1980) have been designing become very attractive. If we can design efficient systems that combine a variety of such appropriately special-purpose processors with general-purpose computers we will open up still another major source of power.

Increasing the usable percentage of resources in processors will increase efficiency, and it would appear from the considerations discussed above that major increases are possible. They will depend upon solving the whole set of interlocking problems involved in the joint design of algorithm, program, data flow, and network architecture. But each step toward improvements will increase efficiency and allow for a higher percentage of devices in processors that do significant work. Even 10%, 20%, or 40% would be very significant improvements.

We can ask what, roughly, are good percentages for allocating resources to memory, processors, and reconfiguring switches; whether to allocate resources in big or little, or littler, packages, and whether to allocate resources to reconfiguring switching networks that allow for convenient reallocation of resources by the programmer when desired.

Very small processor modules plus a limited reconfiguring capability offer intriguing (and virtually unexplored) possibilities for reconfiguring around faults, both to work around portions of the chip as they go bad and to increase the yield of acceptable chips during the manufacturing process.

We can seriously consider a vast range of possibilities, but only a few appear to be attractive. These include

(1) Processors can be very small, with far fewer than 50 gates, interspersed between memories just big enough (e.g., storing 4 to 128 bits) to handle the intermediate data in a flow of information pipelining through these processors. These might be built into carefully crafted special-purpose systolic processors, or into micromodular logic arrays.

(2) As many as possible of the flexible and general-purpose but simple 1-bit processors can be packed on each chip.

(3) Processors (and their associated memory stores) can be on the order of microcomputers, minicomputers, or medium-sized computers, possibly still packed 8 or 16 to a chip.

(4) Processors can be large and powerful, packed one to a chip, with a large amount of additional memory off the chip.

(5) Reconfiguring switches can be interspersed, to change the topology of individual processors (e.g., from 1 32-bit to 48-bit to 32 1-bit processors) or/and networks (e.g., from array to tree), and also to shunt around faults.

(6) In addition, the whole system can have a mixture of such processors, with particular regions specialized.

Design
of Basic Components
and Modules
for Very Large Networks

This appendix defines and uses the basic components of computers to construct some of the building blocks and network modules described in this book.

Finite-state automata (Moore, 1956, 1964), Petri nets (Petri, 1962, 1979; Peterson, 1977; Brauer, 1980), parallel program and flow graph schemata (Karp and Miller, 1969), parallel RACs (Lev *et al.*, 1981), and a number of other formulations (see Baer, 1973) have been used to describe a computer and the flow of information through its transforming processes. The development that follows (along with other discussions throughout this book) is in their spirit, though closest to Bell and Newell's (1971) PMS (processor, memory, switch) notation.

My purpose is to establish a common framework for viewing the structure of

a single computer,

successively more detailed components,

individual logic gates,

the whole network of computers,

programs as processes through which information flows.

Basic Components and Representations
of a Computer Array or Network

A "multicomputer network" is composed of
  nodes;
  links connecting them;
  exchange switches that allow the links to branch.

Nodes (Networks, Computers, Processors,
  Stores, Switches, I-O Devices)

A "node" can be any of the following types of components:

  network (N, Net);
  computer (C, Comp);
  processor (P, Proc);
  (main) memory store (M, S, Mem);
  register store (R, Rgstr);
  cache store (CS, Ch, Cache);
  exchange switch (X, Switch);
  input transducing device (I, In);
  output transducing device (O, Out).
  (The controller will be assumed, and ignored, in this discussion.)

A network can thus contain networks.

A "computer network" must contain at least one processor and at least one memory store. (Several processors can share a single controller, and the several different functions of a controller can be broken apart and dispersed. But these alternative possibilities will usually be ignored.)

A "computer" (short for "1-CPU computer") typically contains one and only one processor, plus one main memory, some kind of input and output capability, plus other components, including registers and secondary memories. Here there can be variations, as in serial computers with special added processors (e.g., CDC 6600), several CPUs (e.g., Univac 1100), or pipelines (e.g., Cray-1).

Occasionally we will want to examine and specify how these components are constructed; e.g., a large memory from small memories, or a processor from special-purpose processors. Ultimately, we can reach the level of gates and binary switches.

Links, Ports, and Composition Rules

A "link" (L, indicated in figures by a short line is two way unless an arrowhead indicates that it is one way, e.g.,

A link connects a pair of nodes. When direction is specified it is to indicate the flow of information. It can be assumed, unless specified by arrowheads, that links are bidirectional.

Wherever a link connects to a node, there is a port. Thus each node always has one port for each link to it, and each link always touches two ports, of the two nodes it connects. Since the link completely implies the ports, ports need not be shown. But it should be remembered (and is sometimes made explicit, when appropriate), that a switch is needed wherever a node has more than one port: N ports require a $\log_2 N$ switch.

Hierarchies of Components (in Terms of Size and Speed)

Any node can be given a tag to indicate successively larger (usually slower) instances. In particular tags can be used for the following:

Memories can be tagged M, Ma, Mb, ..., to indicate successively slower, larger mass memories, including disks and tapes.

Processors can be tagged Pa, Pb, ..., to indicate special-purpose processors (within a single processor, or separate).

Registers can be tagged R, Ra, Rb, ..., to indicate successively larger (often slower) registers, including the cache (a small very fast memory best thought of as a large set of registers).

An exchange switch can be numbered X, X1, X2, ..., to indicate its size in bits. The components it is linked to can be indicated within brackets; e.g., P−X[P,P,M] links two processors and one memory to one processor through one switch.

Memories, caches, and registers can be numbered, to show their word size and (if a second number is given) the number of words.

Building Blocks, Modules, and Facets

Building blocks, modules, and facets are simply particular (small) networks that seem worth describing, collecting, and using as components for larger networks.

Modules and building blocks might be of the following types:

(1) faceted,
(2) compact; and of six general shapes:

   (a) cubes and flattened cubes,
   (b) pyramids and cones,
   (c) related to hexagonals,
   (d) spheres,
   (e) Moore graphs and other dense, symmetric graphs,
   (f) some other specially chosen cluster.

I will use the terms "module" and "building block" or "block" interchangeably. But usually "module" will suggest a larger to some extent self-contained component, and "block" will suggest a more basic, close to primitive, object, useful for building bigger structures.

"Facet" implies that the network is relatively flat (that is, almost two dimensional, but not necessarily planar) with some orientation that reflects the orientation of some array (e.g., the surface of a sphere, or the base of a pyramid).

Specifying Flow, Channel Capacity, and Power

The nodes and links in these networks can be given attributes, either informally for descriptive purposes or precisely for purposes of simulation or analysis. Thus the word size and number of words can be specified for a memory. A second key attribute is speed (access time to fetch a word from memory; time for a processor to execute an instruction; gate delay).

The most important overall attributes are the channel capacity, or bandwidth (of a switch, I-O device, wire, bus, or other link), and the power of a processor. Often it is not possible to assign precise numbers (e.g., a processor's power is a complex function of its speed, bandwidth, and the mix of instructions and algorithms pumped through it).

Network flow graphs (Ford and Fulkerson, 1962; Lawler, 1976), are directed graphs whose links are assigned nonnegative integers (channel capacity), with two usually disjoint sets of nodes, the "source" and the "sink." Networks of the sort examined in this chapter and used in figures throughout this book can have a channel capacity assigned to each link, and be treated as network flow problems.

Constructing Building Blocks from Basic Components

We will start with the very simplest of building blocks: A "computer" is a building block that combines one processor and one memory (a "couple") or several memories.

```
C-1M        C-2M      C-3M       C-4M

M-P        M-P-M     M-P-M        M
                       |          |
 P                     M       M-P-M
 |                                |
 M                                M
```

Fig. 1 Examples of "computer" building blocks. [Note: Two equivalent drawings are shown for a one-memory (C-1M) computer.]

A "complete computer" must also have input and output devices (the controller will be ignored).

```
I-P-M    M-P-O    M-P-M-I
  |        |         |
  O        I        M-O
```

Fig. 2 Examples of "complete" computers.

The term "computer" when used in the context of a network will almost always mean a computer without input and output devices, which will be handled separately. But when talking about a single-CPU stand-alone computer, I will usually be referring to a complete computer.

Computers (and complete computers) also include a variety of other more complex, and much more complex, systems. Many memories can be added, including the very small very fast memories called registers. Two major types of additions are made:

(a) A hierarchy of successively slower and larger and more massive memories.

(b) One or more registers between the processor and the memory, including a cache memory.

Figure 3 gives several examples of hierarchized computers.

```
1R,1M,1Ma          2R,1M              2R,3M,2Ma,1Mb

P—R—M              P—R                M—P—R—M
   |               |  |               |  |  |  |
  Ma               R—M               Ma R—M—Ma—Mb
```

Fig. 3 Computers with a hierarchy of mass memories and registers. (Note: Ma,Mb,Ra,Rb,..., refer to slower memories, registers.)

Some Alternative Possible Types
of Linkages between Components

Networks are usually shown as though simply connected, e.g.,

C-C (that is computer-to-computer, or "CC").

But the following alternatives will usually be possible, and often preferable:

P-M-P (processor-to-processor via memory, or "PMP");
P-R-M (processor-to-memory via registers, or "PRM");
P-R-Ch-M (processor-register-cache-memory, or "PRChM");
P-X[P,P] (processor-to-processors via switch, or "PX[PP]");
P-X[M,M,M] (processor-to-memories via switch, or "PX[MMM]");
P-X[P,P,M,M] (several components linked to the switch).

The flow of information between input and output devices and other components will usually be through registers and switches, but then linked to memories, not processors.

Figure 4 illustrates some (but not all) of the possible ways in which processors can be linked to processors, whether directly, or through memories, registers, or switches.

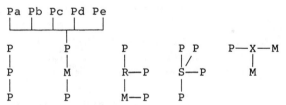

Fig. 4 Examples of processor-memory-processor links. (Note: ⊥ indicate the bus for special-purpose processors Pa,...)

Direct processor-to-processor links can badly tie up the network, since any message passing will interrupt the relatively valuable processor from its normal work.

Memories can be used as cheap switch buffers for linking processors. This introduces potential contention problems and delays, but they can be avoided in several ways (the level of detail of the present constructions does not indicate these variations). Each memory can have two ports or actually be two memories, and then two processors can be kept from both using the same memory at the same time (either by the controller, the operating system, or by the programmer).

Input, output, and mass memory devices can similarly be linked directly to processors, or to memories, registers, or switches. Since they almost always are involved in relatively large transfers of data, direct links to processors are too disruptive. Transfers to buffering memories are the usual flow.

A Reconfigurable Processor—Store Group

A very attractive microstructure for a single VLSI chip is to pack as powerful as possible a processor, along with as large as possible a memory store, but to build the processor so that it can be successively reconfigured into successively larger numbers of simpler processors, each with its own (smaller) portion of memory.

For example, assuming a 1,200,000-device chip, we might pack a 16×16 array of 1-bit processors, each using up to 4500 devices, e.g., 500 for processor and 4000 for memory. Now reconfiguring could turn this into 8 32-bit processors, each made from 16,000 devices, each with a 128,000-bit memory.

We might even facilitate a whole succession of reconfigurings, successively coalescing the 1-bit processors into larger and larger wholes: 256 1-bit processors, 64 4-bit, 32 8-bit, 16 16-bit, 8 32-bit, 4 64-bit.

Even with a 100,000-device chip we can consider a chip reconfigurable between 16 1-bit processors, or 4 4-bit, 2 8-bit, or 1 16-bit processors.

Such a chip would allow the programmer to specialize the different parts of the hardware network to the program's requirements. A simple reconfiguring network between processors and memory, like the Flip network on the ASPRO—STARAN chip, might further be added to help these processors access data, from what could now be treated as a single common memory.

When 1-bit processors are used to execute 32-bit arithmetic they are (at least) 32 times slower. But when 32-bit processors are used to do 10-bit arithmetic they waste 1/3 of their hardware; when they do 1-bit logic they waste 31/32 of their hardware.

Alternatively, a network with specialized regions, working in a parallel parallel mode, can try to allocate portions of programs to appropriate specialized subnetworks. (But making the necessary decisions and shipping programs and data take time.)

Two-Dimensional Building Blocks

We are now in a position to look at some of the basic building blocks of several arrays and networks of interest.

Figure 5 shows several two-dimensional square-array building blocks.

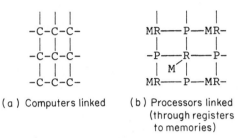

(a) Computers linked   (b) Processors linked
(through registers
to memories)

Fig. 5 Building blocks for typical two-dimensional square arrays. (Note: Figure 5(b) is not complete—it should show a link from each memory to its own register, and from each register to the four neighbor processors.)

Figure 6 gives details of one embodiment of 5(b) that has each processor linked to one register (its "own" register), and each memory linked to one register (its "own" register), making a processor – register – memory group. Then the registers are linked to their four square near neighbors.

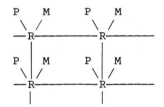

Fig. 6 Processor – register – memory groups linked through registers.

It is too cumbersome to specify all the details of transfer from memory to register to register to processor, and back. So one typically shows only processors and memories, or only computers, occasionally mentioning that data are transferred to registers (or to caches), and shifted among registers, and/or over switches.

Some Attractive Two-Dimensional "Facet" Building Blocks

Figure 7 shows the near-neighbor array connections to

(a) the four square neighbors one away,
(b) the eight square neighbors one away, and
(c) the six hexagonal neighbors one away.

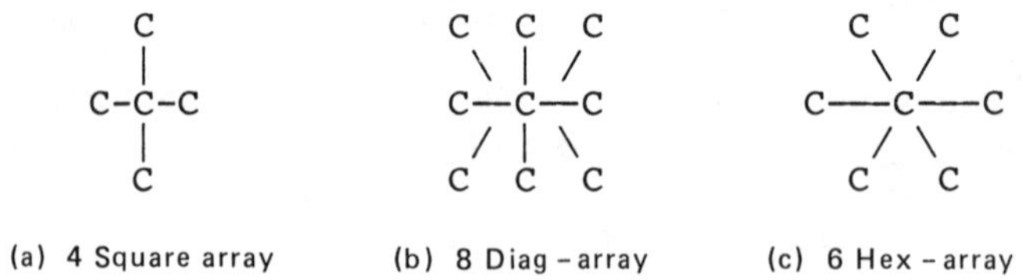

```
       C                    C   C   C                  C      C
       |                     \  |  /                     \   /
     C-C-C                  C —C— C               C —— C —— C
       |                     /  |  \                     /  \
       C                    C   C   C                  C      C
```

(a) 4 Square array   (b) 8 Diag – array   (c) 6 Hex – array

Fig. 7   Square, diagonal, and hexagonal connections.

These are all building blocks within a 3×3 hexagonal subarray (or "window"). Each can be continued out into an entire array, or into some larger subarray. Or different arrays can be tessellated (tiled) out.

Larger windows can be used for the basic building blocks. Thus a 3×5 would have 24 links from its center processor to each of the 24 neighbors; a 7×7 would have 48, and so on. These probably need too many ports and links to be practicable. But subsets can be selected, linking to cells near the horizontal and the vertical axes as in Fig. 9 below.

Larger modules can be built by tiling these smaller modules together. Several variants are possible, with zero overlap, or with successively greater overlap. Zero overlap means that each cell is linked to only one center. Complete overlap means that every cell is a center cell. Overlaps in between are obtained as follows:

The window is $R \times C$ in size.
The center cells can be spaced D apart.
When center cells are one apart we get complete overlap.
When center cells are R and C apart we get zero overlap.

The integers between 1 and R, and between 1 and C give successively greater overlap.

Hexagonal Facets, Hexagonal Solids;
Moore Graphs, Dense Clusters

Most of this book examines square arrays and windows. But hexagonal windows that form "facets" in larger structures that are composed from and more or less reflect the structure of hexagons seem simpler, more regular, and more compact. But they are harder to visualize or talk about. Much work is needed to develop more good algorithms for hexagonal processes; but see Golay (1969), Preston (1971) and Mersereau (1979).

Hexagons are the most efficient way to pack spherical objects (the cones in living eyes pack into hexagons).

```
P P P P P P P P P P P P P P P P P
 P P P P P P P P P P P P P P P P P
P P P P P P P P P P P P P P P P P
 P P P P P P P P P P P P P P P P P
P P P P P P P P P P P P P P P P P
 P P P P P P P P P P P P P P P P P
P P P P P P P P P P P P P P P P P
 P P P P P P P P P P P P P P P P P
P P P P P P P P P P P P P P P P P
```

Fig. 8 An example of a hexagonal array.

In two dimensions each cell in a hexagonal "facet" consists of one center cell and six nearest neighbors. In three dimensions, each cell has eighteen nearest neighbors. Circular disks will pack only one way in two dimensions—into hexagons. Spheres will pack into hexagonal solids, but two different ways are possible: each sphere nestles in the hollow between four others, but adjacent hollows can be opposite or diagonal.

Hexagonal tiling can be used in an array of any shape. But it seems to go well with circular shapes (increasingly large hexagons with additional tiling give increasingly less jagged circles). So it seems especially suited for overall cone and sphere structures.

The Moore graphs (see Appendix A and Bondy and Murty, 1976) have an especially interesting set of properties, since they are optimally dense, have maximum girth, and are highly connected and perfectly symmetric. Unfortunately, only two nontrivial ones are known, the (10,3,2) (10 nodes, degree 3, diameter 2) Petersen graph (1891) and the (50,7,2) Singleton graph (Hoffman and Singleton, 1960).

The 10-node Petersen graph and the 50-node Singleton graph are attractive candidates for local clusters tiled or compounded into larger networks. (A 10-node tree of degree 3 has a diameter of 4.) [See Chapter 14, Appendix A, Leland *et al.* (1980), and Bermond *et al.* (1982) for several possible compounding operations.]

Petersen graphs might be tiled into an array as follows: Draw each graph as a pentagon connected to an inner star (as in Appendix A, Fig. 4). Connect each adjacent pair of graphs using two links, one to an inner and one to an outer node. (A great variety of other tilings are possible.) Each 10-node Petersen graph could then be used as a 3×3 subarray, with the 10th node serving a variety of ancillary purposes.

A hexagonal array of graphs can be built from hexagonal modifications of Petersen graphs: Use 12 nodes, formed in an outer hexagon and an inner six-pointed star, but otherwise like the Petersen graph. (Note that this is not a Petersen graph. Its diameter is 3, but for each node only 2 other nodes are farther than 2 away.) Now lay these graphs out as an hexagonal array, connecting each adjacent pair by linking 2 nodes, as above.

Still other candidates for modules are listed in Appendix A, Table I. And we can expect to uncover more good new candidates for clusters. Those found to date have used density as the overriding criterion. As we get a better grasp on the characteristics of good computer networks we shall be able to use a variety of other criteria to search more wisely.

Since the array structure seems especially suited to a wide variety of tasks, including image processing and real-world modeling, the development of architectures in this volume focuses on arrays. Petersen graphs do not have the simpler (but more costly) near-neighbor connectivity of arrays. But Petersen graphs and others with desirable properties could be substituted for arrays, and for single computer nodes, in a variety of designs. And they might well be indicated at the "deeper" levels of a network.

N-Dimensional Compact Modules

Let us look now at some three-dimensional modules. (They can easily be generalized to N dimensions. But since networks are built in, and are often used to model, the real three-dimensional world, and there appear to be attractive designs for three-dimensional networks of 1,000,000 or more computers, I will not discuss them here.)

$3 \times 3 \times 3$ Cubes

The smallest module with a center is $R = C = D = 3$. This gives a $3 \times 3 \times 3$ cube with 26 cells surrounding the center cell. It is feasible, but expensive, to link the center to all 26 neighbors. More plausible are square links (to 6 neighbors), diagonal links (to the 8 corner diagonals), or square plus diagonals; these alternatives are shown in Fig. 9. Note that there are a number of other possibilities, some of which might occasionally be attractive.

```
           L                    L  L         L  L
  L     LCL     L                     C
           L                    L  L         L  L
      (a)  Square                  (b)  Diagonal
```

Fig. 9 Attractive linkages to subsets in a $3 \times 3 \times 3$ cube. (Note: The array is sliced in thirds; L = linked node, C = center node.)

$2 \times 2 + 1$ and $3 \times 3 + 1$ Pyramids, $2 \times 2 + 1 + 2 \times 2$ Lattices, and Hexagonal Solids

Other shapes are possible. In particular, a pyramid seems most attractive. Here there seem to be two good candidates:

(a) $2 \times 2 + 1$: one parent links to four offspring, and

(b) $3 \times 3 + 1$: one parent links to nine offspring.

In either case, several variants are possible:

(1) links are one way (usually from offspring to parents);

(2) links are two-way;

(3) links are added between offspring;

(4) (in the $3 \times 3 + 1$ case) diagonal links are added between offspring. (Links between offspring are almost always two-way.)

Note that the $3 \times 3 + 1$ can be a part of a $3 \times 3 \times 3$. Therefore it can be used to build a pyramid, or a lattice built from $3 \times 3 + 3$ modules can be treated by the program as though containing pyramids built from $3 \times 3 + 1$ modules.

But the $2 \times 2 + 1$ must converge from the four cells in a 2×2 array to a single cell. This is handled straightforwardly in a pyramid. In a lattice the following three ways seem attractive:

(a) Alternate layers, $2 \times 2 + 1 + 2 \times 2 + 1 + \cdots$; that is, sandwich a layer with one processor connected to each 2×2 in each adjacent layer.

(b) Use hexagonal solids, which naturally have one processor nestled between and connected to six others in the next layer.

(c) Keep all layers the same, e.g., $2 \times 2 - 2 \times 2 - 2 \times 2 - \cdots$, linking between layers with any desired degree of overlap.

Solids Larger than $3 \times 3 \times 3$ Cubes

Cubes (and other shapes) larger than $3 \times 3 \times 3$ can be built. To keep ports down to reasonable numbers, this is best done by linking the small $3 \times 3 \times 3$ cubes together, or by simply using nearest-neighbor links (e.g., the six square links or the fourteen square plus diagonal links) to build larger blocks. If special resources, e.g., mass memory or "IOC modules" (i.e., pyramids or other appropriately structured modules used for input, output and/or control purposes; see Chapter 17) are to be interspersed occasionally, a module of such building blocks can be constructed. For example, a $4 \times 4 \times 4$ block of 14-connected processor—memory groups can be built, and then a $4 \times 4 \times 4$ module of these blocks, with a node of the network's IOC pyramid at its center.

Otherwise only selected cells in the solid can be linked. But it makes little sense to link nodes on the surface of the window (and there will always be only six square and eight diagonal surface nodes) unless the nodes on a path to those nodes are also linked.

More Powerful $3 \times 3 \times 3$ Modules and IOC Modules

Consider a "$3 \times 3 \times 3$ module," which extends a $3 \times 3 \times 3$ cube, as follows:

Roughly, it contains "inward" processors, each looking at a 3×3 subarray of information, possibly sharing this with the up to 5 other processors that may be looking at the other of the six faces of the cube, and outputting its results into a store that is itself a cell in a face of a neighboring structure. The following is a more detailed description:

Such a "$3 \times 3 \times 3$"module will have six "faces." Each face has a 3×3 two-dimensional array of nine memory stores. All these stores link inward to one processor, which can be thought of as "inward, toward the center of the module."

Thus each of the six faces of the module links inward to one processor.

Each of the memory stores has nine links outward from the module, interfacing to processors in a 3×3 array of other modules.

There are several possible variants:

(a) "Packed" modules have a different inward processor for each face (that is, six inward processors for the six faces in a three-dimensional cube).

(b) "Sparse" modules have a different inward processor for each dimension (that is, three inward processors for the three dimensions in a three-dimensional cube).

(c) "Minimal" modules have only one inward processor for all faces.

In addition, there can either be

(1) An additional IOC processor (giving an IOC module) or
(2) No additional processors.

Therefore, in a $3 \times 3 \times 3$ module there can be either one, two, three, four, six, or seven inward processors.

Each face processor links not only to the nine memory stores in its face, but also to one memory store of its own and to an IOC store, a memory store that it shares with the IOC processor (and therefore with the module's other processors).

Therefore each processor has eleven links to stores—the nine in its face, its own, and the one it shares with the other processors.

The processor's "own" store is itself also a member of a 3×3 face of an adjacent module (one that is next in the direction established by the flow from that processor's face to it).

The IOC processor (if it exists) has a link to the IOC store, and it also has links to the one, three, or six face processors.

The IOC store has links to the IOC processor (if it exists), to the one, three, or six face processors, and also to an external node.

We can distinguish between the "face structure" and the "internal structure" of the module:

The face structure consists of the face array of memory stores plus the links from these stores to external nodes, and to the processor nodes internal to the module.

The internal structure consists of the set of processors associated with the faces, each with its own store, and the IOC processor and the IOC store.

The IOC processor is linked to the IOC store and (optionally) to each of the face processor's stores.

Each face processor is linked to the IOC store, which is also linked to some external node.

A similar set of modules can be constructed for hexagonal solids.

The IOC module can be used to link its nodes, and also the nodes in its neighbor modules, to distant parts of the network. In a large system it is likely that such modules should not exist everywhere, but only scattered about. That is most simply accomplished by using an $R \times C \times H$ block of $3 \times 3 \times 3$ modules whose center module is an IOC module.

Building Larger Structures from Basic Modules

These modules can now be used to build larger structures.

Trees

A $3 \times 3 + 1$ or a $2 \times 2 + 1$ can be used to build a 9-tree or a 4-tree, simply by making each offspring bud the parent of a new module (of the same type). This assumes there is no geometrical metric, as imposed by the array for which these modules were developed as subarrays, or windows.

Indeed, the $3 \times 3 + 1$ is simply a diameter-2 tree with 10 nodes of degree 9, the $2 \times 2 + 1$ a tree with 5 nodes of degree 4. So we can more generally define and generate trees using $n \times m + 1$ modules.

Pyramids and Cones

The $2 \times 2 + 1$ and $3 \times 3 + 1$ (and more generally an $N \times M + 1$ can be used to generate pyramids, exactly as trees were generated above, except that some of the modules have links between siblings. To get a tree/pyramid, link siblings only at the last layer/ply, that is, at the buds of the tree. To get a pyramid, link siblings within each layer.

When such a system is constructed from $3 \times 3 + 1$ modules with diagonal links, its geometry is rather closer to a cone, in that the distance from apex to each of the four "corners" is the same as the distance to the middle of each "face." To get a truer cone, one could judiciously pare and smooth the surfaces.

Lattices, Cubes, Spheres, and Other Overall Architectures

Any of the $3 \times 3 \times 3$ modules can be used to generate most of the other structures examined in this book. One can simply keep adding these modules to give the shape desired. Or, when the shape is relatively regular, one can do this more systematically, as follows:

To generate a lattice, first generate an $R \times C \times 3$ array from $3 \times 3 \times 3$ modules. Then thicken this solid by attaching identical arrays (attaching modules just as they are attached within the array) until the desired height is achieved. When $R = C = H$ this generates a cube.

An egg, sphere, or other shape can then be got by paring away the surface.

Alternatively, a sphere can be generated from a pyramid, by using a suitably large array as its base and then judiciously linking the pyramid's surface nodes. Or pyramids can be combined into a sphere: put many pyramids next to each other, merging all apexes.

Using Two-Dimensional Facets to Build Larger Structures

Sets of 2×2, 3×3, 7hex, 49hex, or other larger facets can be combined into larger structures.

The simplest, and it appears the most attractive, way is to treat the facet as a set of offspring, linking each cell to a common parent. This establishes a flat pyramid. The line through the apex and the center of the base establishes a "facet direction."

The facet is part of a larger array, for example a (hollow) sphere. Make its apex/parent a part of a (smaller) array, for example a (concentric hollow) sphere, and continue this as far as desired, or until the root has been reached.

A two-way faceted structure can be built by connecting in the opposite direction. Indeed, many of the "solid" pyramid and lattice structures already described can be used as though directional-faceted systems.

Faceting seems most useful for spheres. Here the converging flow is nicely handled by the facet's convergence, while the small facet surfaces lend themselves to approximating the continual curvature of the sphere.

Examples of Languages
and Code
for Arrays and Networks

Here follow several examples of programs and program fragments coded in some of the parallel languages, to give a feeling for their flavor.

The procedures for arrays appear to be shorter and simpler in their structure than are conventional programs for traditional serial computers. The languages for networks are a far more difficult problem, since they must be able to handle asynchronous concurrent processes.

The program fragments in PascalPL0 (the language developed to compile to CLIP3 or CLIP4 assembly code) also give a taste of CLIP code, since both source PascalPL0 code and object CLIP3 code are shown in the listing.

The program in PascalPL1 is complete. It was successfully executed under the Berkeley Pascal system on the VAX.

Examples of Data Flow and Functional Languages

A Simple Construct in the VAL Data-Flow Language

The following example* of the VAL data-flow language (Ackerman and Dennis, 1979) builds an array.

```
FORALL I IN [1,10]    %{comment: range definition}
  Mean,SD : REAL        %{ get the mean}
    := Stats(A[I],B[I],C[I]);  %{and SD for Ith entry}
  Plus2    : REAL
    := Mean + 2 * SD; %{compute result}
CONSTRUCT   Plus2    %{specify result to be put in array}
ENDALL
```

A Simple Example of a Functional Language

Flynn and Hennessy give the following example:[†]
The problem: given x of type Data (defined below), construct a function that returns true if x contains the integer value y. The following function Search will solve this problem if it is invoked with Search (x,y).

```
CONST  Size = 100;
TYPE   Index = 1 ... Size;
       DataRecord = RECORD
                    CASE IsLeaf:boolean OF
                    true:(value:integer)
                    false:(child:^data)
                    END;
       Data = ARRAY [Index] OF DataRecord;
VAR    SearchTree:Data;
       Target:integer;
```

*From J. McGraw, Data flow computing: Software development. *IEEE Trans. Comput.* **29**, 1095−1103. Copyright © 1980 IEEE.

[†]From M. J. Flynn and J. L. Hennessy, Parallelism and representation problems in distributed systems. *IEEE Trans. Comput.* **29**, 1080−1086. Copyright © 1980 IEEE.

```
FUNCTION  Search(x:Data;target:integer):boolean IS
          {OR the result}
INSERT OR IN
    {construct a boolean array of comparison results}
    FOR i:Index BE  {Name the value of the ith entry}
    LET t:DataRecord; t = x[i]
    IN  {compare if t is a leaf, else invoke recursively}
    IF t.IsLeaf THEN t.Value = Target
    ELSE Search (t.Children ^, Target).
```

Examples of Languages for SIMD Arrays

PICAP's PPL (Picture Processing Language)

The following shows PPL code for PICAP, from Gudmundsson (1979, p. 17).

```
PROGRAM THRESH
//EDT THRESH
**LIS 1,99
*0001 PROGRAM THRESH
*0002 MATR LAPL
*0003 0  -1  0
*0004 -1  4  -1  0
*0005 0  -1  0
*0006 TEMP T1,T2,MASK(M,SH)
*0007 X  X  X     X  X  X     X  X  X
*0008 X  Q1 X 0  X  X  X 15 X  X  15 3
*0009 X  X  X     X  X  X     X  X  X
*0010 CHAIN THR:T1*T2
*0011 PICT ZERO
*0012 INT MED,SUM
*0013 BOOL FLAG
*0014 PROC ABS
*0015 BEGIN
*0016 DRO=TVIN    R3=DRO     #INPUT FROM CAMERA
*0017 PHIST                  #PRINT HISTOGRAM OF ORIGINAL
*0018 ZERO=0    RO=ZERO      #CLEAR RO
*0019 R2=LAPL(R3)   ABS(2)   DRO=R2
*0020 DO                     #MANUAL THRESH. OF LAPLACE PICTURE
```

```
*0021     PRINT('THRESHOLD') READ(I)
*0022     IF I=0 THEN LEAVE FI
*0023     Q1=<I
*0024     R1=THR(R2)      DRO=R1
*0025 OD
*0026 R2=MASK(R1)               #EXTRACT FROM ORIGINAL
*0027 DRO=R2  PHIST
*0028 MED=(4096-HIST(0))/2 #START LOCATING VALLEY IN HISTOGRAM
*0029 SUM=0   I=0
*0030 WHILE SUM<MED DO
*0031     I=I+1   SUM=SUM=HIST(I)
*0032 OD
*0033 IF HIST(I-1)>=HIST(I)   THEN
*0034     WHILE HIST(I+1)<=HIST(I) DO
*0035         I=I+1
*0036         IF I=15 THEN FLAG=FALSE LEAVE FI
*0037     OD
*0038 ELSE
*0039     WHILE HIST(I-1)<HIST(I) DO
*0040         I=I-1
*0041         IF I=1 THEN FLAG=FALSE LEAVE FI
*0042     OD
*0043 FI
*0044 IF FLAG THEN                         #IF VALLEY FOUND
*0045     PRINT(I:5X,'VALLEY:',1)
*0046     Q1=<I+1
*0047     RO=THR(R3)      DRO=RO          #THRESHOLD ORIGINAL
*0048 FI
*0049 END
**APP 32
*0032 OD
0033 FLA=TRUE
**** FLA      NOT DECLARED ****
0033 FLAG=TRUE
0034 $
**OUT
//RUN
HISTOGRAM:
  0.  1.  2.  3.  4.  5.  6.  7.  8.  9.  10.  11.  12.
  0   1   2   4  17  16  52  48  59  42   57   49   77
  13. 14. 15.
  129 291 3252
```

An Example of Pixal Code (Levialdi et al., 1981)

The following program does gray-scale relaxation. It uses four masks that increasingly weight the difference in gray level between the center pixel and its neighbors. It continues for *k* iterations (*k* is experimentally determined, or chosen by interactively applying the program to a picture and monitoring the results of each iteration on a graphics terminal).

```
PROGRAM  gray-relaxation
BEGIN
   GRAYARRAY image  [1:100,1:100];
   REALARRAY prob   [1:100,1:100];
   REALMASK  m1      [1:3,1:3] OF (" ... ");
   REALMASK  m2      [1:3,1:3] OF (" ... ");
   REALMASK  m3      [1:3,1:3] OF (" ... ");
   REALMASK  m4      [1:3,1:3] OF (" ... ");
   GRAY min, max, den;
   INTEGER i,j,k;
   COMMENT : the values of m1...m4 must be computed;

   BEGIN
     min := max := image [1,1];
     FOR i := 1 TO 100  DO
     FOR j := 1 TO 100  DO
     BEGIN
       IF image[i,j] < min  THEN  min := image[i,j];
       IF image[i,j] > max  THEN  max := image[i,j]
     END
   den := max - min;
   PAR
     prob := (image-min)/den
   PAREND;
   FOR i := 1 TO k  DO
   PAR
     prob := amax1(sum(overweigh(m1,prob))),
                  sum(overweigh(m2,prob))),
                  sum(overweigh(m3,prob))),
                  sum(overweigh(m4,prob)))
   PAREND
   END
END
```

Douglass' (1981) MAC, for SIMD and Asynchronous
MIMD Networks

The following process "createregions" scans an image and segments it into regions of approximately uniform color and texture.

```
PROCESS createregions :
  numofregions : 0..maxregions
          {numofregions counts the number
          of region processes created}
  regions : ARRAY[1..maxregions] OF ^ PROCESS region;
  {definitions of PROCs and FUNCs and INIT section go here}
END;   {of process createregions}
```

The following algorithm for stereo image region matching assumes that both images have been segmented into region descriptions, and that "createregions" has associated a process with each region in one of the images (image A). First, each image process in image A creates a list of regions from image B as candidates for a match. Next, process "region" initiates a matching process, for each candidate, that returns a value expressing the goodness of fit. Finally, "selecthighest" chooses the candidate with the highest matching score. Process "createregions" has activated a graph of processes of the form:

```
regions:  ARRAY[1..numofregions] OF ^ PROCESS region
```

where "region" is defined in the PROCESSTYPE section as:

```
PROCESS region:
  matchingregion : INTEGER;
       {index of corresponding region in image B}
  candidate : ARRAY[1..maxcan] OF candidatedescriptions;
   {an array holding a list of regions
    that are candidates for match}
  numcan : 0..maxcan;
           {the actual number of candidates found}
  matcher : ARRAY[1..maxcan] OF ^ PROCESS match;
              {process "match" is defined below}
```

```
PROC createcandidates;
   {selects 0 to maxcan regions
    from image B as candidates for match
    with this region and stores them in candidate}
PROC selecthighest;
{sets matchingregion to candidate with highest match value}

PROC domatch;
   VAR i : INTEGER;
   BEGIN
     FOR i := 1 IN [1..numcan] :
     BEGIN  ACTIVATE(matcher[i]);
       FORK(matcher[i]^.
            compare(self,candidate[i]#candidate[i].metric));
     END;
   END;
INIT
   {initialization of process "region"}
END;   {of process region}

PROCESS match;
        {compares two regions using the PROC "compare"}
   PROC compare(reg1,reg2:regionindices#matchscore:INTEGER);
        {"compare" scores the match
         between regions reg1 and reg2,
         and returns a score in matscore}
   INIT
END {of process "match"}
```

Example Fragments of PascalPL0 Code
and the CLIP3 Object Code

The following figures are two fragments from a much longer Pas-
calPL0 program.

Figure 1 shows the start of the program, which lists the PascalPL0
source code, as comments (the ";" indicates a comment follows).

Figure 2 shows the start of the actual CLIP3 code. Each PascalPL0
statement is repeated (preceded by " ;****IN= "), followed by the
CLIP3 statements it generates.

```
;ICTL MODES= CLIP3, SQUARE,
;ICTL INAMES= 0
;   2; PASCALPL PROGRAM FOR 15 OBJECT-NAMES
;   3;   OBJECTS ARE =♦  ABCOUCTTPFSTPDF
;ICTL DNAMES= 0
;T AND F MUST BE USED FOR 0,1 IN BLN EXPR
;DON'T PUT INTEGER NAME OF B IN MASK EXCEPT AFTER '+'
;4       START ♦INSCOPE♦ ;
;5          0 = -0(1-9)*0    ; ; CONTOUR
;6          1 = 0(*15←)      ; ; 4 EDGES
;7          2 = 0(*26←) ;
;8          3 = 0(*37←) ;
;9          4 = 0(*48←) ;
;10           ♦ERASE♦1,2,3,4, ;
;11          <1=♦     1111.A..A.A2A21 ;
;12     L11    <2=♦       12...222222.2A1 ;
;13     L12    <3=♦       ..111.11AAA2A21 ;
;  14;   4 SQUARE ANGLES
;15     L13    <4=♦       21...2.2B2222A1 ;
;16     SQANGLS 5 = + 4(♦12,2)*2(♦54,2) ;
;17          =♦     1.BAB112C22.2B ;
;18     L27    6 = + 2(♦34,2)*4(♦76,2) ;
;19          =♦     1.BA....C22.2B ;
;20     L39    7 = + 2(♦12,2)*4(♦56,2) ;
;21          =♦     1....2..C22.2B ;
;22     L51   10 = + 2(♦78,2)*4(♦32,2) ;
;  23;   SQUARE-EDGED COMPOUNDS
;24          =♦     1...B212C22.2B ;
;25     L64   11 = + 5(♦1,3)*7(♦7,3)     ; ; SQUARE
;26          =♦     ..B.A..1B23.B.. ;
;27     L80   11 = + 6(♦5,3)*10(♦1,3)    ; ; SQUARE
;28          =♦     ..B.A..1B23.B.. ;
;29     L96   11 = + 3(♦1,3)*1(♦3,3)     ; ; ← (INVERTED V)
;30          =♦     2.A.B.21B..3.3. ;
;31     L112  12 = + 1(♦7,3)*3(♦5,3)     ; ; V
;32          =♦     A.A.1..1B....3. ;
;33     L128  12 = + 12(6,4)*11(♦2,4)    ; ; DIAMOND
;  34;   3 LONG EDGES
;35          =♦     ..B1..11BB9.B5. ;
;36     L148  12 = 1(1,9)*1(0) ;
;37          =♦     2.AA.......1.1A ;
;38     L160  13 = 2(2,8)*2(0) ;
;39          =♦     12AA.23.332B28A ;
;40     L172  14 = 3(3,8)*3(0) ;
;41          =♦     2.AA...1...1.1A ;
;42     L184  11 = + 11(♦2,6)*12(3♦4,4)*14(♦31,4)*4(6,4) ;
;43          = 1(5)      ; ; IMPLIES A-LETTER STRONGLY
;44     L231  11 = + 13(8,3)*5(♦43,3)   ; ;FLAG-PART
;45          =♦     11..B1...3.B.B. ;
;46     L246  11 = + 2(8,2)*2(4,2)    ; ; PARALLEL VERTS
;47          =♦     1.A.13.2A11.A.. ;
;49     L256  11 = + 1(84,2)+3(84,2)*13(84,1)  ; ;POLE+BRANCHES
;  49;     CURVES - LOCAL + SHIFTED
;50          =♦     1.B...5.B...... ;
;51     L270  11 = 0(*1-4)*-0 ;
;52          12 = + 1(♦3,2)*4(♦23,2) ;
;53          11 = 11+12 ;
;54          =♦     .2.2.1.1B ;
```

Fig. 1 The beginning of an example PascalPL0 program.

```
         ;CLIP PROGRAM STARTS HERE.
         ;IN= SHOWS INPUT SOURCE CODE AND COMMENTS
         ;SUBROUTINES NEEDED AND ADDED ARE -
         ;    UPWT,DNWT,MAX.
         ;****IN= ;
         ;****IN= ;; PASCALPL PROGRAM FOR 15 OBJECT-NAMES
         ;****IN= ;; OBJECTS ARE =* ABCOUCTTPFSTPDF
         ;****IN= START *INSCOPE* ;
0200     START: LD  0D, 0D, 0, S        ; 1
0201        PR A      ; 2
0202           HI 50       ; 3
0203           EG 0        ; 4
0204           BR 0, 4, START       ; 5
0205     LD C, C, 15      ; 6
0206     PR A      ; 7
         ;****IN= 0 = -0(1-8)*0   ;; CONTOUR
0207          LD   0, C, 0    ; 8
0210          PR 0(1-8)-A, P*A,  S  ; 9
         ;****IN= 1 = 0(*15<)    ;; 4 EDGES
0211          LD   0, C, 1    ; 10
0212          PR 0(15)-A, -P*A,  S  ; 11
         ;****IN= 2 = 0(*26<) ;
0213          LD   0, C, 2    ; 12
0214          PR 0(26)-A, -P*A,  S  ; 13
         ;****IN= 3 = 0(*37<) ;
0215          LD   0, C, 3    ; 14
0216          PR 0(37)-A, -P*A,  S  ; 15
         ;****IN= 4 = 0(*48<) ;
0217          LD   0, C, 4    ; 16
0220          PR 0(48)-A, -P*A,  S  ; 17
         ;****IN= *ERASE*1,2,3,4, ;
0221             LD  C, C, 16         ; 18
0222          PR 1      ; 19
0223          LD 16, C, 16       ; 20
0224          PR 0(2)A,P*A, S     ; 21
0225          LD 16, C, 16       ; 23
0226          PR 0(4)A,P*A, S     ; 24
0227          LD 16, C, 16       ; 26
0230          PR 0(6)A,P*A, S     ; 27
0231          LD 16, C, 16       ; 29
0232          PR 0(8)A,P*A, S     ; 30
0233     LD 1, 16, 1      ; 32
0234          PR A*P    ; 33
0235     LD 2, 16, 2      ; 34
0236          PR A*P    ; 35
0237     LD 3, 16, 3      ; 36
0240          PR A*P    ; 37
0241     LD 4, 16, 4      ; 38
0242          PR A*P    ; 39
0243     LD 0, 16, 0      ; 40
0244          PR A*P    ; 41
         ;****IN= <1=* 1111.A..A.A2A21 ;
0245             LD  1, C, 1         ; 42
0246        PR A      ; 43
0247        PR * -A      ; 44
0250        BR 1, P, L11          ; 45
0251        LR   1 (1)         ; 46
0252        LR   1 (2)         ; 47
0253        BS   UPWT      ; 48
0254        LR   2 (1)         ; 49
0255        LR   1 (2)         ; 50
0256        BS   UPWT      ; 51
0257        LR   3 (1)         ; 52
0260        LR   1 (2)         ; 53
0261        BS   UPWT      ; 54
0262        LR   4 (1)         ; 55
0263        LR   1 (2)         ; 56
```

Fig. 2 Continuation of the example PascalPL0 program.

Examples of Statements and a
Simple Program Coded in PascalPL1

A Simple Four-Layer Pyramid Program for Pattern Recognition

The following program uses a number of simple feature detectors and compound characterizers to successively transform an input image through a four-layer converging pyramid of successively smaller arrays. A full-blown program would need many more such transforms.

Statements 1 − 9 declare variables, including the feature and characterizer arrays (hor1, boat3, etc.) using ordinary Pascal code. (Array dimensions are declared for 16 × 16 arrays, since this was a toy demonstration program.) Note how 13 − 17 reinitialize the feature arrays to contain all zeros.

Statement 24 looks at the center and its four square neighbors, for each cell in the retina, and computes the weighted difference, to threshold it.

Statements 32 − 35 look at the appropriate neighbors for each of the first-layer features, adding to the weight for that feature in the cell corresponding to each location where it is found. For example, 32 looks for a 1 (which means "above threshold" from the previous thresholding operation) in the center cell and in its East and West neighbors, and adds 1 to the weight of hor1 (horizontal in layer 1).

Statements 41 − 49, 55 − 58, and 64 − 67 apply sets of similar feature detectors and compounding characterizers for each layer. (These were chosen without much thought, for demonstration; they would need to have their weights tuned, after trying them out on a range of example input scenes.)

Statements 73 − 76 output four arrays, showing where each of the highest-level objects was found. Similar "write(...)" statements could be used to output features and objects implied at previous layers. A simple procedure (in ordinary Pascal) could be added to choose the object with the highest weight.

```
1   program recognize(input,output);
2   const ninputs = 10;
        xmin = 0; xmax = 15; ymin = 0; ymax = 15;
3   type xindex = xmin..xmax; yindex = ymin..ymax;
4     arraytype = array [xindex,yindex] of integer;
5   var trials : 0..ninputs;
        arrayx : xmin..xmax; arrayy : ymin..ymax;
6     inret,retina,hor1,vert1,diagne1,diagnw1:arraytype;
7     langle2,vangle2,cross2,xangle2:arraytype;
8     tepee2,deck2,flagpole2,
          house2,house3,boat3,jumble3 : arraytype;
9     cross3,church4,house4,boat4,jumble4 : arraytype;

11  ||procedure sense;
12  begin
13  ||dim *merge from [0..15,0..15];
14  |set   retina,hor1,vert1,diagne1,diagnw1 := 0;
15  |set langle2,vangle2,cross2,
          angle2,tepee2,deck2,flagpole2 := 0;
16  |set house2,house3,boat3,jumble3,cross3 := 0;
17  |set church4,house4,boat4,jumble4 := 0;
18    ||read(inret);
19  ||end;

21  ||procedure retinathresh;
22  begin
23  ||dim *merge from [0..15,0..15];
24    ||if inret[+(0:0*5,-1:0*-1,1:0*-1,0:1*-1,0:-1*-1)] > 0
          |then retina+1;
25    write('THRESHOLDED retina is   ');
26    |write(retina);
27  ||end;

29  ||procedure retinatransform;
30  begin
31  ||dim *merge from [0..15,0..15] to [2,2];
32    ||if retina[0:0,0:-1,0:1] > 0 ||then hor1+1;
33    ||if retina[0:0,-1:0,1:0] > 0 |then vert1+1;
34    ||if retina[0:0,-1:-1,1:1] > 0 ||then diagnw1+1;
35    ||if retina[0:0,-1:1,1:-1] > 0 ||then diagne1+1;
36  ||end;
```

```
38   ||procedure layer1transform;
39   begin
40   ||dim *merge from [0..7,0..7] to [2,2];
41     ||if hor1 * vert1 > 0  ||then  cross2+27;
42     ||if diagne1 * diagnw1 > 0
             ||then xangle2+35, vangle2+3;
43     ||if vert1[+(-1:0,0:0,1:0)] > 2
             ||then flagpole2+39;
44     ||if hor1[+(0:-1,0:0,0:1)] > 2
             ||then deck2+45;
45     ||if hor1[0:0] * vert1[-1:-1] > 0
             ||then langle2+17;
46     ||if diagnw1[0:-1] * diagne1[0:1] > 0
             |then vangle2+19, xangle2+3;
47     ||if diagnw1[0:1] * diagne1[0:-1] > 0
             ||then tepee2+19;
48     ||if hor1[*(-1:0,1:0)] * vert1[*(0:-1,0:1)] > 0
             |then house2+77;
49     ||if hor1[*(-1:0,0:0)] * vert1[*(0:-1,0:0)] > 0
             |then house2+8;
50   ||end;

52   |procedure layer2transform;
53   begin
54   ||dim *merge from [0..3,0..3] to [2,2];
55     ||if cross2 * xangle2 > 0 |then jumble3+99;
56     ||if flagpole2 * deck2[1:0] > 0 |then boat3+133;
57     ||if house2 > 0 |then house3+777;
58     ||if cross2 > 0 |then cross3+875;
59   |end;

61   |procedure layer3transform;
62   begin
63   ||dim *merge from [0..1,0..1] to [2,2];
64     ||if cross3[-1:0] * house3 > 0 |then church4+999;
65     ||if house3 > 0 |then house4+777;
66     ||if boat3 > 0 |then boat4+133;
67     ||if jumble3 > 0 |then jumble4+235;
68   ||end;
```

```
70  |procedure apex;
71  begin
72  |dim *merge from [0..0,0..0] to [1,1];
73     | |write(house4);
74     | |write(church4);
75     | |write(boat4);
76     | |write(jumble4);
77  |end;


81  begin {program}
82    for trials := 1 to ninputs do
83      begin
84        sense;
85        retinathresh;
86        retinatransform;
87        layer1transform;
88        layer2transform;
89        layer3transform;
90        apex;
93      end;
94  end.
```

*An Example of PascalPL1 Output for a Simple Conditional
Statement*

The following shows the PascalPL1 conditional statement (labeled
INPUTz), and then the Pascal code output by PascalPL, for

```
||if featurei[+(0:1,0:-1)]
      * featurej[+(1:0,-1:0)]  >  11
  ||then  labeli+19, labelj*2, labelk-33
  ||else  labell*2, labelm+27;

(*INPUTz1=   ||if featurei[+(0:1,0:-1)]
               * featurej[+(1:0,-1:0)]  >  11  *)
      {IF STATEMENT BEING HANDLED HERE.}
   for arrayx :=   0 to   2 do
     begin
       for arrayy :=   0 to   2 do
         begin
           arrayxloc := arrayx + 0;
           arrayyloc := arrayy + 1;
```

```
                        {(**KEEPS IT IN BOUNDS!**)}
        beyondborder := false;
          if arrayxloc < 0 then beyondborder := true;
          if arrayxloc > 2 then beyondborder := true;
          if arrayyloc < 0 then beyondborder := true;
          if arrayyloc > 2 then beyondborder := true;
        if beyondborder = true   then
           TEMPARRAY1[arrayx,arrayy] :=   0
          else
           begin
             tempstore :=
                 featurei[arrayxloc,arrayyloc] *  1;
             TEMPARRAY1[arrayx,arrayy] := tempstore;
           end;
     arrayxloc := arrayx +  0;
     arrayyloc := arrayy + -1;
                        {(**KEEPS IT IN BOUNDS!**)}
        beyondborder := false;
          if arrayxloc < 0 then beyondborder := true;
          if arrayxloc > 2 then beyondborder := true;
          if arrayyloc < 0 then beyondborder := true;
          if arrayyloc > 2 then beyondborder := true;
        if beyondborder = true   then
           TEMPARRAY2[arrayx,arrayy] :=
              0   +  TEMPARRAY1[arrayx,arrayy]
          else
           begin
             tempstore :=
                 featurei[arrayxloc,arrayyloc] *  1;
             TEMPARRAY2[arrayx,arrayy] :=
                 TEMPARRAY1[arrayx,arrayy]
                     +   tempstore;
             if border < 2   then
               if TEMPARRAY2[arrayx,arrayy] > 1   then
                 TEMPARRAY2[arrayx,arrayy] := 0;
           end;
      TEMPARRAY0[arrayx,arrayy] :=
           TEMPARRAY2[arrayx,arrayy];
     arrayxloc := arrayx +  1;
     arrayyloc := arrayy +  0;
```

```
                    {(**KEEPS IT IN BOUNDS!**)}
        beyondborder := false;
          if arrayxloc < 0 then beyondborder := true;
          if arrayxloc > 2 then beyondborder := true;
          if arrayyloc < 0 then beyondborder := true;
          if arrayyloc > 2 then beyondborder := true;
        if beyondborder = true   then
          TEMPARRAY1[arrayx,arrayy] :=   0
         else
           begin
             tempstore :=
                 featurej[arrayxloc,arrayyloc] *  1;
             TEMPARRAY1[arrayx,arrayy] :=  tempstore;
           end;
     arrayxloc := arrayx + -1;
     arrayyloc := arrayy +  0;
                    {(**KEEPS IT IN BOUNDS!**)}
        beyondborder := false;
          if arrayxloc < 0 then beyondborder := true;
          if arrayxloc > 2 then beyondborder := true;
          if arrayyloc < 0 then beyondborder := true;
          if arrayyloc > 2 then beyondborder := true;
        if beyondborder = true   then
          TEMPARRAY2[arrayx,arrayy] :=
                0    +  TEMPARRAY1[arrayx,arrayy]
         else
           begin
             tempstore :=
                 featurej[arrayxloc,arrayyloc] *  1;
             TEMPARRAY2[arrayx,arrayy] :=
                     TEMPARRAY1[arrayx,arrayy]
                         +   tempstore;
           if border < 2   then
             if TEMPARRAY2[arrayx,arrayy] > 1   then
               TEMPARRAY2[arrayx,arrayy] := 0;
           end;
     TEMPARRAY0[arrayx,arrayy] :=
           TEMPARRAY0[arrayx,arrayy]
              *  TEMPARRAY2[arrayx,arrayy];
     end;
 end;
```

```
(*INPUTz2=        ||then  labeli+19, labelj*2, labelk-33  *)
    for arrayx :=   0 to   2 do
      begin
        for arrayy :=   0 to   2 do
          begin
                      (**modify if found greaterthan value**)
            if ( TEMPARRAY0[arrayx,arrayy] > 11 )  then
              labeli[arrayx div xshrink,arrayy div yshrink] :=
                labeli[arrayx div xshrink,arrayy div yshrink]
                    + 19;
                      (**modify if found greaterthan value**)
            if  ( TEMPARRAY0[arrayx,arrayy] > 11 )  then
              labelj[arrayx div xshrink,arrayy div yshrink] :=
                labelj[arrayx div xshrink,arrayy div yshrink]
                    * 2;
```

```
(*INPUTz3=              ||else  labell*2, labelm+27; *)
                      (**modify if found greaterthan value**)
            if  ( TEMPARRAY0[arrayx,arrayy] > 11 )  then
              labelk[arrayx div xshrink,arrayy div yshrink] :=
                labelk[arrayx div xshrink,arrayy div yshrink]
                    - 33;
            else
                          (**else  do the following:**)
              labell[arrayx div xshrink,arrayy div yshrink] :=
                labell[arrayx div xshrink,arrayy div yshrink]
                    * 2;
                          (**else  do the following:**)
              labelm[arrayx div xshrink,arrayy div yshrink] :=
                labelm[arrayx div xshrink,arrayy div yshrink]
                    + 27;
          end;
      end;
```

APPENDIX F

Organizations in Crystal, Cell, Animal, and Human Groups

An organization of many computers is a very new beast, and a very strange one. It is not at all clear whether techniques that have been evolved, whether by natural evolution or human ingenuity or trial and error, for organizations of more than one individual can throw light on our problem of how to organize groups of more than one computer.

But it should be instructive to examine the similarities and the differences between these different kinds of groups, and to explore ways in which we might extrapolate our experience and understanding of already existing organizations to help us in building new architectures of many highly organized computers.

Virtually nothing of this sort has been done, and this chapter will only scratch the surface.

I shall examine some of the kinds of organizations that exist in nature-evolved and man-made systems. I shall explore whether there are enough similarities between these and systems of computers to justify trying to use some of our knowledge about the former in designing the latter. And I shall try to suggest some general principles and some specific organizations that seem to be indicated.

Natural Organizations

Nature has evolved a fabulous variety of organizations animal, vegetable, and mineral; molecular, intracellular, among cells, among organs, among individuals, and between groups. Many of these have information-processing components, or can be viewed in that way. For example, when exposed to a certain substance a cell may become permeable to certain ions, allowing them to pass through the cell membrane to the cell's interior. One might say that the cell first receives and then acts upon information that tells it to absorb this (possibly useful) substance.

But from the wealth of such examples, let us simply focus on the primary one that nature has evolved for processing information — the nervous system (with brief glances at the ants and the bees, and at crystals).

The Structure of Crystals and Other Organized Inorganic Matter

The structures of crystals should be mentioned here, since they may well tell us a great deal about how to structure and pack very large numbers of processors in regular ways. More generally, solid-state physics now underlies the development of VLSI computer chips, and the interactions of particles and flows of forces will become increasingly pertinent as network components grow smaller. For example, the structural features of superconductive crystals profoundly influence the flow of current (messages) through them.

The Organization of Neurons

Here we see a structure that combines a great variety of different organizing techniques (far from understood or fully appreciated by us). The brain, and especially the sensory — perceptual systems, is highly organized, with many precise and specific, often hierarchical and multilayered, subsystems. But there is also an enormous amount of variability in local details. And a large but unknown amount of redundancy and error-correcting capability is built in, probably to handle this variability and also to handle the inevitable aging and dying of individual neurons in the brain (neurons, of all the body cells, are the one kind that do not regenerate).

The brain, then, gives us an interesting example of a highly organized system with many sub(sub...)systems, yet with a good deal of (mostly local) variability and disorganization, along with powerful techniques for handling (both ignoring and making use of) this variability.

One has the impression (though our understanding of the brain is not deep enough to say with any degree of certainty) that the individual neurons process information and, occasionally, each may make local decisions, but, essentially, acting as slaves or drones. And the total system has evolved to expect rather fallible results from each individual neuron, so that some kind of error correction, possibly polling and averaging over a handful, or even hundreds or thousands of neurons, appears to be routinely used.

No single ultimate master controller has been identified in the brain, whether anatomically, physiologically, in psychological experiments, or by introspection. Nor is there any obvious simple single hierarchy moving upward through managers and bosses that suggests there might be some final single superboss, or some small committee of boss neurons. Therefore people like Warren McCulloch have suggested the brain is an "heterarchical" rather than an "hierarchical" system, with information converging toward, and the flow of transformations of this information being controlled by, many rather than by a single locus.

It seems quite likely that there is no simple strict hierarchy, as in a pyramid or tree of processors with a single top-level processor at the apex or root in control. But there are certainly structures that appear to transform and converge information in a regular cascading fashion.

Organizations of Workers, Such as Ants and Bees

Today's computer programs, even the most "artificially" intelligent, are still far less intelligent than animals far lower in the phylogenetic scale than *Homo sapiens*. The first program that adequately models an insect like a fly or a bumblebee will be a major landmark. From that perspective, it seems foolhardy indeed to take human intelligence and human organizations as models for today's computer networks, rather than the organizations of the far simpler insects.

In insects we appear to find slavelike, dronelike subservience and obedience to a commonly held goal. There may be limited discussions of how best to go about doing things. Certainly there are elaborate dancelike ways of transmitting information about dangers, or foods. But there seems to be little if any group decision making, and little need to pass information that each individual will then evaluate, leading to that individual's decision, leading finally to a group decision.

Human-Made Organizations:
To Produce Goods, to Make Decisions

Human beings have developed a variety of organizational patterns for groups of individuals. There would appear to be a potential wealth of hints for organizing large numbers of computers, both from studying the social and political systems that we actually use, and from examining the results of social-psychological experiments on different group structures performing different kinds of tasks.

But there is one major caveat that should be kept in mind, one that may well suggest that this body of information is not relevant (at least for today). For human beings are far more flexible, adaptable, and in general intelligent than a computer (at least until we develop truly intelligent computer programs). Each human being in the group may perform some specific role, possibly a very small and simple role, in the job of the total group. But that person is at the same time exercising her/his entire intelligence, and very often it is just this independent intelligence that plays a crucial role in keeping the group on track.

Groups Organized to Produce Some Goods

When we human beings organize ourselves to get some job done, for example to make cars or to run a bank, we usually seem to handle things in one or a combination of the following ways:

(a) We work as a group, each member taking some part of the task she/he feels capable of completing in a reasonable amount of time with the tools at her/his disposal. This entails a great deal of intelligence on the part of every individual, in judging the magnitude of the tasks and their own capabilities.

(b) We set up an "assembly line" that assigns an appropriate, rather small and of equal size, subtask to each individual. Then we pipeline materials through this sequence, so that each person repeatedly performs the same set of operations, once on each product turned out. (Note that the assembly line was the "great invention" of modern mass production. But now there is a trend to return to the more involving and more interesting less formal group approach.)

(c) We impose a hierarchical organization of managers over the workers. The workers in a group or on the assembly line have "foremen" who assign tasks, watch, monitor, and admonish. These foremen in turn are bossed by low-level managers who, in turn, are bossed by several successively higher-level managers. At some point a manager may be called a "vice-president" and, finally, a "president."

The army is an especially clear-cut and simple example of such an hierarchical organization, moving from squads of private soldiers bossed by corporals up through sergeants and lieutenants to several levels of generals.

The almost classic picture of organization is one where each manager bosses some small number (seven plus or minus about three) of underlings, until the bottom level of workers is reached.

This suggests that each level is strictly slave to its boss, and knows about and acts upon information and/or material that comes either from its immediate boss or its immediate underlings. But in reality this will rarely if ever be the case. Rather, in addition to this "formal" hierarchy and chain of command, human organizations almost always have a number of crucially important informal channels, through friends and acquaintances.

To summarize: We often divide jobs into smaller tasks, attempting to make them equal in size and to assign them to individuals who have the tools and the skills to handle them well. We also tend to put workers under a hierarchy of managers, with each boss responsible for about seven underlings. We often pretend that this is the complete table of organization, but almost always there is an informal network connecting the members of the group that plays a vital role in making that group work.

Decision-Making Groups

Many of our groups serve purposes other than the production of goods. We try to come to common agreements, as when a political party chooses a candidate and the entire populace elects a leader. We join together for some common purpose, whether to worship or to change some law or to decide on a single course of action.

In such cases there may often be some component of producing goods or results. But the chief problem is one of getting people to agree, to change their minds and attitudes, to achieve a workable consensus.

It is here where we have developed a variety of forms of democratic, socialistic, and anarchic organizations, as well as the quite autocratic organizations that are typical of the job/production situation. From the point of view of the freedom, integrity, and psychic well-being of the individual human being, democratic freedoms seem the most important value of a society like ours. But it is not at all clear that they are appropriate models for organizations of computers, for at least two major reasons:

First, an organization like a democracy or an anarchy is based on the premise that each individual is a highly intelligent entity, and in fact the most capable of deciding upon its own course of action.

Second, it seems clear that this is not the best organization for getting the most work done. When we have to put out a fire or harvest a crop we revert to one of the production-oriented kinds of organization. We stop trying to know our own minds and change each other's minds; we start working. The more slavelike, the more dronelike the better. When we are highly motivated, and in complete agreement, we can work hard, in a "single-minded" way, without needing to spend time coming to agreements and motivating one another.

The Problems with a Democracy or an Anarchy of Computers

It is intriguing to think of a network of processors as a democracy (even an informed and enlightened democracy) that discusses and decides what might best be done next. Certainly that is a congenial model for someone who believes (as I do, along with most people who live in and value democracies) that decisions and actions that affect a group should be fair and agreed to by that group. (I am assuming that the values and goals of the individual and the group are similar enough, and individuals perceive the furthering of group goals beneficial enough to their own goals, so that individuals will pursue the group's goals.)

Another closely related alternative is what I shall call a "creative anarchy" in which each member of the group both commits her/himself to working for the good of the group and for the group's goals, and also studies the present situation, decides what she/he can most productively do for the group, and simply does it.

These are the ways I personally would like to see groups and societies function.

But these are the (ideal and utopian) ways for human beings to act when they make decisions that affect all of them, and, to a greater or lesser extent, depending upon the issue, when they implement these decisions and try to solve problems toward that implementation. And one reason we have not come close to such democratic or anarchic societies is because their members must be rather unusual people—intelligent, well informed, wise, fair, unselfish.

Even given such groups of such people, when it comes to getting work done, and especially getting work done on tasks that have been thoroughly analyzed and planned out, participatory democracy and creative anarchy are probably *not* the ways to get these jobs done as quickly and as efficiently as possible. On the contrary, now we need slaves and drones; we need individuals who will do what they are supposed to do, without question or hesitation.

I am making this argument in such detail because the idea of a democracy, or an anarchy, of computer processors is, at least superficially, a very attractive one, and one with which I would like to be in sympathy (for the reasons I have given).

But the computers in today's networks, or in any networks that we can envision (until we have solved all the problems of artificial intelligence, so that we can make each computer at least as intelligent as a human being) are not intelligent, informed, wise, etc. On the contrary, what first glimmerings of any of these traits they might have are only given to them by tedious programming.

Nor is the major reason for participatory democracy—to get enthusiastic agreement so that each individual is treated fairly, and will in turn participate fully and vigorously, that is (from the point of view of work), try hard to do a good job—relevant to a network of computers. There is no need to motivate a computer, or to keep it happy, or to make it feel it is being treated fairly.

All one accomplishes if one asks a computer to pass and evaluate messages, mull things over, and try to arrive at a decision for itself, or a consensus with other computers, is to use up a lot of that computer's machine cycles and precious time.

Nobody has ever suggested that we follow the independent-actor decision-making procedures Hewitt (1973, 1977; Kornfeld and Hewitt, 1981) and others are investigating for networks of processors when we use a traditional 1-CPU computer to execute our programs. But we could clearly do so, since most programs consist of a large number of small procedures. We could treat each procedure like a separate processor, and ask the program's set of procedures to discuss and decide what to do.

But that seems patently silly and frivolous (or maybe I am suggesting a valid new extension of this research, and I am too blind and dogmatic to realize its value). The programmer has spent a lot of time, effort, and (even) intelligence in figuring out exactly what to do and how to do it; all that the program could do is slow things down and mess things up. If we could take the burden off the programmer, and have the program do more of this dreary task, that would be great. But much work has been done on this problem, with little progress. And nobody has suggested that the way to do it is to code procedures to handle at run time some of the issues that the programmer did not or could not think through when coding and debugging the program.

We have a far better grasp of how to write programs and execute them on a 1-CPU computer than we do on how to use a large network of computers to execute a mix of arbitrary programs efficiently. Nobody has considered turning the procedures in a traditional program into a group that tries to decide what to do using the techniques of a town hall or other participatory democracy. But the problems of using a network efficiently are greater still, and much less well understood. The fact that we do not understand them suggests that we need help. But asking a computer system of very simple and unintelligent processes, organized in a very loose, unproductive, and to my mind inappropriate way, smacks of handwaving and a reliance on self-organizing techniques without any idea of why or how something of value might emerge.

I guess I am arguing so strongly against the democratic alternative because I do not especially like the dictatorial, slavelike society of computers I am about to propose.

But we are being charitable to today's computers when we call them slaves or drones. I personally would argue that some of the most promising artificial intelligence programs have enough first glimmerings of intelligence to be considered alongside some of the lower examples that nature has evolved. But the very simple programs that people are considering giving processors in networks, so that they can pass messages and arrive at their own decisions, are not of this sort at all—they have no glimmerings of intelligence; they are very simple and completely dumb.

Computers in a network should be given responsibilities of this sort as we develop their capabilities to the point where they are able to take these responsibilities. Just as with children and with an uneducated citizenry (whom we do not allow to participate fully in the democratic decision-making process), so we should not expect or ask computer programs to participate until they are capable of doing so, until there is some hope they will make a constructive contribution.

And we should (for these purposes but not for intelligent computers in general) try to keep computers cool and unemotional, so that they will (as they presently do) work at their hardest all the time, whether or not they are doing what they might prefer. It is here that, ultimately, we may be able to get better and fairer decisions from computers, since they need not have self-centered egos, feelings, and needs. They will be able to concentrate on substantive problems of how best to get work done. And they will be ready to accept improvements and solutions, simply because they are the most reasonable.

I do not mean to argue against the research of Hewitt and others who are exploring group decision making for networks; there are very interesting issues of organization of groups from the perspective of computer networks. But this research is better thought of as a branch of the artificial intelligence of social systems, rather than as an attack on the problem of developing usable networks of computers that will actually execute larger programs faster.

Groups, and the Operating Systems That Run Them

Democratic, anarchic, and similar organizations of groups of individuals designed to give each individual as much freedom and as much weight as possible in the decision-making process do not seem at all appropriate as models for computer networks. For they assume that each individual is intelligent, knowledgeable, and wise. Computers are without any intelligence unless we give them programs that make them intelligent. But we have not yet developed our understanding of intelligence to the point where we can do that. And even if we had, it seems frivolous to make computers in a network waste their time executing such programs.

Much of human decision making centers on emotional and interpersonal issues. In order to work effectively, much less in a highly motivated, steady way, human beings must feel that they agree with the goals and techniques of the work. But computers simply carry out the instructions in their program; in a certain sense they are always 100% motivated, and in perfect agreement.

When we human beings have a clear-cut task, and a well-described set of procedures for carrying out that task, and we are strongly motivated and in agreement that this should be done, we work in a single-minded, dedicated manner. But this is the typical situation for computers, as they execute the programs given them.

Executing a program on a large network of many computers, and executing it as efficiently as possible, is a far more complex job than executing it on a traditional 1-CPU computer. We do not yet know how to program such complex procedures efficiently. We do not even have good programming languages for networks, good techniques for devising parallel algorithms, or a good understanding of our problem. But it is foolhardy to conclude, "therefore we should have the network of computers take over this burden." If we have not yet found solutions, and we have not managed to write programs for traditional computers to find solutions, we certainly are not going to see the network of computers come up with any solutions by some group decision-making procedure evolved on the fly, one that wastes precious time during the period in which the poor program is trying to get executed.

I have been discussing networks of more or less independent actors. But these arguments apply to any kind of network or operating system that is not carefully structured to execute programs efficiently. An operating system that ignores the structure and the flow of information through programs may efficiently assign processes to all processors. But it will succeed in creating a babble of unnecessary message-passing delays that can too easily swamp the whole system (yet may look like a great flurry of important activity).

I have ignored the mechanisms that implement and maintain the group structure—the managers, politicians, bureaucrats. Operating systems can be designed in any conceivable bureaucratic mold, since with computers we can (and must) start from scratch and have potentially infinite possibilities. Some operating systems are unobtrusive, gentle, even a bit "user friendly" (albeit still rather rigid and dumb). Others give the feeling of mad bureaucrat—robot amok. The purpose of computers and their operating systems is to compute, that is, to help people develop and execute programs to solve problems. The operating system's *raison d'être* is to help users run programs as quickly and as efficiently as possible.

Messages
(Information Passed
between Processors)

Carl Hewitt (1973, 1977; Kornfeld and Hewitt, 1981) and others have pointed out that processors in a network often must send "messages" (that is, information) to one another.

Message passing can become a major preoccupation of processors, and a major burden on a network. Indeed it appears that the need to send, receive, decode, and decide what to do about messages is the major source of degradation of most of the networks that have been built to date. Handling message passing is probably our biggest headache; it may be the key problem in getting networks, especially MIMD networks of relatively independent processors, to work with a reasonable degree of efficiency.

In this appendix I shall examine

(1) the different kinds of information that might be sent as messages, and

(2) different ways of handling the problems message sending generates.

What Is a Message?

The word "message" was chosen because people were focusing on, and wanted to emphasize, the need of several processors working together in the same network to "talk to one another" and coordinate what they are doing.

Message Messages

For example, one processor might find that it needs a resource it does not own, e.g., a printer, or a large database. It might need to "request" a position on the queue for the printer. It might not "know" where the database is, on what disk or set of disks it is presently stored.

Or processors might need to allocate parts of a large program among themselves:

In an autocratic network programs might be input to a "master" that then assigned parts to different subservient processors, sending them messages to that effect. Often the master would not know precisely how busy each processor was (except for the simple situation where a processor becomes completely idle and sends a message to the master to that effect). So the master would have to send messages asking whether the processor is free, or would be likely to be able to help soon.

In a more democratic network the processors would need to send many more messages back and forth, to find out which were busy, which wanted to participate, which had the appropriate set of resources, etc.

No matter what the network's "social" structure, messages are needed to request the use of a resource (e.g., the operating system requests a processor, or a processor requests a tape reader), to give the status of a resource (e.g., "I'm busy," or "tape drive 7 is free, you can have it"), to inform other nodes in the system about changes in status (e.g., "I'm done printing," or "processor q doesn't respond—it may have gone down"), or to inform about more intimate matters (e.g., "I'm disk-swapping," or "I'm disk-swapping at too high a rate; reduce the size of my job," or "I'm waiting most of the time").

I have given a range of examples to indicate some of the kinds of messages that networks might be asked to handle. These are, I think, traditional messages, ones that we might think of as message messages.

There are several other very important kinds of information that must be passed between processors and other nodes in a network that I want to mention now. From the point of view of the computers in the network, they are no different from message messages. All of them impose the same kind of drain on the message sender's and the message receiver's time and resources.

Data as Messages

A program is input to a computer system with the (usually implicit) message "Execute!"—that is, RUN THIS PROGRAM! That's a nice short and simple message, rather like the message a light switch sends the light or radio to turn it on.

But along with that message will (almost always) be a set of input data on which that program will act. Often the program will continue to input data, as needed by the program, or because users have sent it messages that more data are now present, to be gobbled up and processed.

But all these data are simply more information that must be passed by input devices to the system, and then passed between the system's processors, memory stores, and other resources, under the control of messages or a fixed procedure. Some of these "data" (e.g., formatting information in output statements) look suspiciously like "message messages." Other "data" (e.g., information to be formatted) look like data.

We might try to make a distinction that excluded all the different kinds of data from the domain "message." But it would ultimately be a vague and arbitrary decision. And it would be quite unnecessary; we have no reason to bother. For from the point of view of the computer system, all messages are alike. They all entail the passing of information, which then entails the processing, absorbing, and proper using of this information.

The Program as a Message

The program itself must also be input to the system, along with the data and the message "EXECUTE!." We do not normally think too much about that, because one-processor computers input the program to that computer's memory, and there it sits.

But multicomputer networks may have to ship programs and pieces of programs back and forth a great deal, especially if their operating systems reallocate procedures among processors.

Since a program is often a very large hunk of information, passing programs as messages can become a very large burden. Passing procedures may seem better, since procedures are usually quite short (in terms of lines of code). But when we remember that a procedure uses other procedures and may depend upon global declarations, the necessary attendant information is often quite large. Either this must all be passed along with the program, or there will be a constant need to pause, pass messages requesting the needed information, wait, and pass and decode this information.

Contending with Messages:
More Coordination, Hardware, or Silence

Message passing can become a very large burden on a network. Since all programs, and all data, are just messages, and there can be a variety of other kinds of message messages as well, messages can come in very large hunks, and from every direction.

The way the network is set up (with respect to both its physical structure and the operating system's algorithms) will have a tremendous influence on the message-passing burden. This influence can be exercised in two opposing directions:

(1) The network can be given extra resources dedicated to message handling, so that the program-executing parts of the system can work relatively unhampered.

(2) The system can be structured in such a way as to reduce the burden of message handling.

Special Message-Handling Hardware

The ARPANET distributed network (McQuillan and Walden, 1977) is a good example of a system that dedicates major resources to messages. Each installation in the network is given a special relatively large and powerful computer, the IMP, just to handle messages. The network might best be thought of as a network of IMPs, of message handling computers, to each of which is linked a workhorse, a traditional large computer whose chief job is to handle the local mix of programs given it. (Occasionally this workhorse is linked to a local distributed network.) .

The ARPANET has a very large burden of messages, since one of its major purposes is to let people ship papers and programs, as well as shorter pieces of electronic mail, around the country. Messages come frequently; they are often very large; their routing over the large and complicated ARPANET is itself an intricate problem.

Wittie's (1978) MICRONET is a second interesting example of a system with major hardware additions dedicated to message passing.

For Wittie has turned each one of his processors into a couple, wedding it to a second "communications" processor whose sole job in life is socializing and handling the mail (or, less facetiously, message handling).

The present MICRONET uses specially designed and built communication processors, paired with PDP-11/23s that serve as workhorses. Wittie's plan and hope is to to design and build a chip, using VLSI technologies, that would contain this pair of processors. His estimate (personal communication) is that the communications processor will be the larger and more powerful of the two. So he is allocating over half of his (processor) resources to message handling! (Note that at the same time he is freeing the lesser half to devote just about all of its time to work.)

Related examples of interest are systems that handle input and output (which are two of the major message-passing burdens) with special hardware that keeps them from idling the working processors. Thus the MPP (which has very large input and output burdens, since it must input 1000×1000 images and then output their enhanced versions) has I-O lines and registers associated with each processor so that processors can be working at the same time as, in parallel with, input and output (Batcher, 1980). Similarly, Siegel (1979; Siegel *et al.* 1981) and his colleagues are designing PASM with two-port memories so that input−output can use one port while the processor uses the other.

In all these cases, relatively expensive resources are being allocated to message handling. In the ARPANET, probably most of the system; in MICRONET the better half; in MPP and PASM smaller but still hefty amounts.

Reducing the Need for Message Handling

Some systems need far more messages than others. A democratically or anarchically organized network can easily be swamped with message messages. Systems that try to reallocate programs and procedures will often need to do a lot of shipping of large hunks of code (unless that factor is taken into account in the reallocation algorithms).

But the whole system can be designed with the reduction of message handling as a major goal. This would appear to be a desirable goal, simply because message handling is a burden and a drain on computing resources.

The system can be made autocratic. Here the SIMD arrays are a good example. These systems impose extreme straitjackets on how to do things, the benefit being well-structured, efficient programs (for those parallel algorithms for which they are appropriate). In terms of message passing they eliminate all overhead, since all processors send and receive messages over the same set of near-neighbor links, and there is no possibility of contention.

The message-passing protocol can be simplified and speeded up. Much of the protocol can be eliminated when the programmer chooses or is willing to structure message flow. Much of the rest can be wired into hardware.

The programmer can be made aware of the dangers of messages, and the need to help structure the program so that it can be mapped efficiently onto the architecture. The operating system can try to effect an appropriate mapping, and even ask the programmer to help in a common interactive enterprise.

New, possibly more stringent, structural criteria may need to be developed, to break programs down into small separate procedures that are sufficiently independent that they can work efficiently on different processors, with little need to pass messages. (A major criterion of efficiency is that message passing, which gets no work done, be minimized.)

If each processor can just do its own work, and can expect and get from near-neighbor processors the data needed to keep it busy, message passing is drastically decreased. If each processor can continue to execute the same instruction (as is the case in pipelines), then even the fetching and decoding of the next program instruction (which traditional serial computers must always do) is eliminated—giving a substantial decrease in message handling, since that is all this is.

Summary and Conclusion

One solution to the propensity for message passing to swamp multicomputer networks is to do nothing, effectively limiting the network to handling programs where different processes execute independently, with little need to interact. But that simply evades one of our basic research goals, to develop networks that handle a wide variety of information flows. And it keeps an unknown (probably very high) percentage of problems from being effectively programmed for networks.

In general, the appropriate structuring of the hardware architecture, the operating system, programming languages, and individual programs so that all fit together as nicely as possible is probably the key to reducing message passing. The links between processors and memory can then be given sufficient bandwidth to handle what remains (mostly data, and occasionally small hunks of code) with minimal delays. Almost all of the resources are then used for work, rather than for messages.

To the extent that this is not possible, additional hardware resources, whether parallel I-O, communication processors, or higher-bandwidth links, may be needed and indicated, especially if real-time constraints must be met.

APPENDIX H

What Is " 'Real' 'Time' "? —
A Metaphysical Aside

It is only appropriate that a discussion of something "real" should be our major foray into metaphysics.

There are two primary reasons why we need very large networks, including arrays.

Some problems are simply *too large* for any conventional 1-CPU computer to handle (e.g., perception, brain modeling, weather).

Some problems must be executed under severe time constraints, in "*real time*" (e.g., robot control, missile guidance).

But what do we *really* mean by "real time"?

"Time" is one of our most intriguing mysteries. And the philosophical issues surrounding the "real" go very deep. But when we work in the context of computers, often the meanings of concepts that are extremely conjectural and vague become crystal clear.

Examples of Word Magic
and Word Distortion, to Clarify

For example, one frequently reads that a computer program "understands" a command input to it. For example, an artificial intelligence program that answers questions about the weather, or a second program that answers questions about how to cook an egg, "understands" the question asked and also enough about the weather or about cooking to reply: "On July 17, 1980, it rained 0.57 inches in Toronto," or "Fry the egg in butter."

Such programs "understand" in what I am pretty sure is almost exactly the same sense in which a program "understands" when the instruction says "add" or "read." Or the way a light "understands" and turns on when its switch is thrown.

In all these cases, the system makes the entirely appropriate (sometimes one can even say "perfect") response. Within the context of the system's (very simple and limited) world, and the repertoire of possible internal transformations and external response behaviors it can make, it not only "understands" but it also "knows" everything pertinent well enough to "solve its problems," "decide what to do," and "act appropriately."

I am throwing around all these most-high-sounding words both because I think what I have just written is true, and to make ironically clear how vague these words and concepts are, and how much they depend upon the context in which they are used. (The unfortunate thing is that these and similar words are too often used to suggest that the very limited "understanding," "thinking," "perceiving," etc., that a question-answering program that applies a mechanical parse to the question and looks up the answer in a table and—the crucial thing—is demonstrated in such a way as to conceal its limitations and befuddle are key steps toward making computers intelligent. People who make such claims do not, of course, go on to say that a light switch is intelligent, but they should be asked to explain the difference.)

The Limited Meaning within the Computer Context Is the Whole Meaning

I use these examples to point out two things:

(a) Many of our most important concepts are exceedingly vague, and, therefore, the words and statements we use to examine them are inevitably loose.

(b) When they are used to talk about computers and the programs that computers execute, we can examine the precise domain of computer plus program quite thoroughly. We know exactly what is going on. We can understand such concepts completely, since they refer exactly to this limited, completely understood (in theory) domain.

We do not have to solve all the mysteries of "time." Rather, the computer itself has a very precise clock that establishes its basic "cycle time," "keeps time," and coordinates its basic operations. This is the computer's "time"; the computer "knows" it well. Rather rigidly, but well: for example, the computer finishes exactly the same program, the same sequence of operations, in exactly the same time, every time.

But sometimes some need is imposed on the computer system to execute a program and output some information not in its own good time, but rather in somebody or something else's time.

Occasionally this means the program should store its results internally and wait until its clock tells it to output the results, e.g., when it accumulates information about how long each user has been logged on the system during the past 24 hours, and outputs a summary at a fixed time each day. Here there are no problems; the procedure is quite straightforward.

"Real Time" as the Press of "External" "Time"

But sometimes the program needs and takes more time to arrive at and output the results than somebody, or some thing, has to wait. This, I would assert, is all that we mean by "real time": an other's time.

If the "thing" is a television—mechanical arm system looking at a conveyor belt in an assembly line, where objects (say Coke bottles) are packed together in cases and move past the television camera at a rate of eight rows of six bottles each second, "real time" means the "perceptual" system must direct and position the arm to properly cap each bottle, taking about twenty milliseconds per bottle.

A robot crossing the street must figure out that the light has turned green before it turns red again. Indeed it must also get its mechanical legs moving, evade any manholes and potholes and negotiate the curbs, all in the "real time" of traffic lights and bicycles.

A missile-seeking missile must compute the missile's trajectory, recompute whenever the sought missile uses evasive tactics to change trajectories, and do this fast enough to direct itself on a trajectory that shoots that missile down.

There are many such examples, each with its own "real-time" constraints. Most have relatively similar "concepts" of "real time." For most are asking that things be finished in the "ordinary" time needed by most animals and objects that people and other animals have evolved to interact with (to catch and eat, to evade). Some (e.g., ever faster missiles) act in much faster "real times," and, therefore, press our systems beyond any conceivable limits. Some of course move so slowly that they pose no real-time constraints.

On the Possibilities of Changing and Manipulating Real Time

I should mention one more issue: The cases of Coke bottles are an interesting example of a common phenomenon: we could slow down the moving belt. We can sometimes manipulate, reengineer, the environment external to the computer system that poses its problems and establishes its constraints and demands.

Often, as in the case of Coke, this will mean less work done, less money, more competition from unionized American workers and faster Japanese robots, and so on.

Often such a slowdown has many ramifications: if we allowed 60 minutes instead of 60 seconds for the traffic light to change, the city around the robot would stall.

Usually it is impossible simply to slow things down: if we want robots that catch mice and balance themselves as they walk or play with our dogs, the robots must work in the real time of animals and gravity—that is, the "real" "real time" of mother nature's natural laws.

Summary, and an Impossible Possibility

The computer has its own "real time," established by the clock that tells it how fast to run. Whenever a computer is given a program that it must execute faster than this can comfortably be done, either that computer must stop doing enough of the other things (e.g., double-checking its results, time sharing with other programs) it normally does, or we must find another, faster computer.

Very large and highly parallel arrays and networks become crucially important here, since only by increasing the number of computers working in parallel can we continue (potentially with virtually no limits) to speed up programs, and thus meet more and more really stringent time constraints.

It seems appropriate to end by mentioning a way of getting a program to (in theory) compute any arbitrarily complex function in real time, as fast as possible. Simply use table lookup. That is, the program can be given a table that enumerates each possible input condition along with the correct output. For very small functions this is often used as a quick, not unreasonably more expensive, alternative. This is the combinatorially explosive program par excellence. For chess it would enumerate all possible moves; for language it would enumerate all possible sentences. Thus (exponential) increases in space can (in theory but not in reality) be used to achieve almost any possible speedup in time, when time is really of the essence.

Suggestions for Further Reading

Most of the arrays and networks have been described only in papers. These are referred to throughout this book, chiefly in Chapters $2-7$ and 14. But see especially:

Earlier Systems. Unger (1958); SOLOMON: Slotnick *et al.* (1962); ILLIAC III: McCormick (1963); ILLIAC IV: Barnes *et al.* (1968); PEPE: Evensen and Troy (1973); STARAN: Batcher (1974).

Arrays. CLIP4: Duff (1978); DAP: Flanders *et al.* (1977b); MPP: Batcher (1980,1982).

Pipelines. PICAP: Kruse (1978, 1980); Cytocomputer: Sternberg (1978, 1980); systolic arrays: Foster and Kung (1980), Kung and Song (1982).

Networks. C.mmp: Wulf and Bell (1972), Wulf *et al.* (1981); Cm*: Swan *et al.* (1977), Arden and Lee (1978); X-Tree: Despain and Patterson (1978), Goodman and Sequin (1981); Sullivan *et al.* (1977); MICRONET: Wittie (1976, 1978), Wittie and von Tilborg (1980), Manara and Stringa (1981).

Reconfigurable Networks. PM4: Briggs *et al.* (1979); PASM: Siegel *et al.* (1979).

There are surprisingly few books or survey papers on parallel multi-computer network architectures.

Lorin (1972) is a good (but early) examination of parallel systems.

Thurber (1976) is the most extensive survey, but it does not examine the newer systems.

Kuck's (1978) text on architecture has an unusually heavy emphasis on parallel algorithms, and interesting examinations of ILLIAC IV.

Stone (1975; 2nd ed., 1980b) and Baer (1980b) examine architecture design, with some focus on networks.

Fu (1978) surveys several key arrays.

Foster (1980) describes a few key systems in some detail.

Tanenbaum (1981) focuses on loosely distributed networks.

Mead and Conway (1980) is an outstanding introduction to VLSI technology and design. It takes a comprehensive look at the near future, when one will be able to design a whole chip (and therefore a whole network) much as one today writes a program.

See Clark (1980) for a clear description of VLSI technology, Thompson (1980) for VLSI models and bounds, and Valiant (1981) for embedding techniques.

Kung (1980a) examines a wide range of parallel algorithms and the types of structures for which they are most appropriate.

For original papers, see the proceedings of the
Annual International Conference on Computer Architecture;
Annual International Conference on Parallel Processing.

Probably the best continuing source for papers on architectures, languages, and algorithms is the
IEEE Transactions on Computers.

Computer Surveys has devoted several issues to parallel network architectures [in particular, see Baer (1973), Thurber and Wald (1975), and Kuck (1977)], and to other topics relevant to this book.

Electronics has frequent notices of new computers, and articles describing and comparing computers and technologies and projecting future trends.

For Image Processing and Pattern Recognition

Tanimoto and Klinger (1980) have edited a book that describes several of the parallel-serial "pyramid" and "cone" perceptual systems being developed for image processing and pattern recognition.

Uhr (1973) presents and examines a variety of parallel-serial programs that might appropriately be executed by arrays and networks.

Hanson and Riseman (1978) is probably the most comprehensive collection of papers on perceptual systems. It includes a number of papers about programs that do, or might, map nicely onto arrays or networks.

Rosenfeld and Kak (1982) is a recent revision of a good, solid text that emphasizes the lower levels of picture processing.

Suen and De Mori (1981) cover auditory as well as visual image processing and pattern recognition.

The chief sources of original papers are the

Proceedings International (Joint) Conferences on Pattern Recognition;
Proceedings International Joint Conferences on Artificial Intelligence;

and the journals

Computer Graphics and Image Processing;
Pattern Recognition;
IEEE Transactions on Pattern Analysis and Machine Intelligence.

Several good recent collections of papers can be found in Kuck, *et al.* (1977), Feilmeier (1977), Buzbee and Morrison (1978), Onoe, *et al.* (1981), Duff and Levialdi (1981), and Preston and Uhr (1982). Their papers range over architectures, programming systems, and algorithms and user programs. The first three emphasize numerical problems; the last three focus on image processing.

For background readings, see, e.g., Kraft and Toy (1979) on hardware, Bondy and Murty (1976) on graph theory, Lawler (1976) on network flow, Minsky (1967) on automata.

References

Ackerman, W. B. (1979). Data flow languages. *Proc. AFIPS Natl. Comput. Conf.* **48**, 1087−1095.

Ackerman, W. B., and Dennis, J. B. (1979). VAL: A value-oriented algorithmic language: Preliminary reference manual. *Lab. Comput. Sci. Tech. Rep.* **218**, MIT.

Agrawal, D. P. (1980). Graph theoretic analysis and design of multistage interconnection networks (unpublished paper, Wayne State Univ.).

Agrawal, D. P. (1981). A pipelined pseudoparallel system architecture for motion analysis. *Proc. 8th Ann. Symp. Comput. Archit.*, pp. 21−35.

Akers, S. B., Jr. (1965). On the construction of (d,k) graphs. *IEEE Trans. Electron. Comput.* **14**, 488.

Akers, S. B., Jr. (1971). A rectangular logic array. *Proc. IEEE Symp. Switching Automata Theory,* Vol. 12, pp. 79−90.

Akl, S. G., Barnard, D. T., and Doran, R. J. (1980). Simulation and analysis in deriving time and storage requirements for a parallel alpha-beta algorithm. *Proc. Int. Conf. Parallel Process.*, pp. 231−234.

Amdahl, G. M. (1967). Validity of the single processor approach to achieving large scale computing capabilities. *Proc. AFIPS Spring Joint Comput. Conf.* **30**, 40−54.

Anderson, G. A., and Jensen, E. D. (1975). Computer interconnection structures: Taxonomy, characteristics, and examples, *Comput. Surveys* **7**, 197−213.

Andrews, G. R. (1981). Parallel programs: Proofs, principles and practice. *Comm. ACM* **24**, 140−146.

Anonymous (1979). Untitled unpublished paper on ASPRO, Goodyear-Aerospace, Akron, Ohio.

Arden, B. W., and Lee, H. (1978). A multi-tree-structured network. *Proc. Fall Compcon '78* pp. 201−210.

Arvind, Gostelow, K. P., and Plouffe, W. (1978). An asynchronous programming language and computing machine. *Inform. Comput. Sci. Dept. Tech. Rep.* **114**.

Backus, J. (1978). Can programming be liberated from the von Neumann style: A functional style and its algebra of programs. *Comm. ACM* **21**, 613−641.

Baer, J.-L. (1973). A survey of some theoretical aspects of multiprocessing. *Comp. Surveys* **5**, 31−80.

Baer, J.-L. (1980a). Techniques to exploit parallelism. *Comput. Sci. Dept. Tech. Rep.* 80-90-01, Univ. of Washington.

Baer, J.-L. (1980b). "Computer Systems Architecture." Comput. Sci. Press, Potomac, Maryland.

Barnes, G. H., Brown, R. M., Kato, M., Kuck, D. J., Slotnick, D. L., and Stokes, R. A. (1968). The ILLIAC IV Computer. *IEEE Trans. Comput.* **17**, *746−757.*

Baskett, F., Chandy, K., Muntz, R., and Palacios, F. (1975). Open, closed and mixed networks of queues with different classes of customers. *J. Assoc. Comput. Mach.* **22**, 248−260.

Batcher, K. E. (1973). STARAN/RADCAP hardware architecture. *Proc. Sagamore Comput. Conf. Parallel Process.*, pp. 147−152.

Batcher, K. E. (1974). STARAN parallel processor system hardware. *Proc. AFIPS Natl. Comput. Conf.* **43**, 405−410.

Batcher, K. E. (1976). The flip network in STARAN. *Proc. Int. Conf. Parallel Process., pp. 65−71.*

Batcher, K. E. (1980). Design of a massively parallel processor. *IEEE Trans. Comput.* **29**, 836−840.

Batcher, K. E. (1982). Bit-serial parallel processing systems, *IEEE Trans. Computers.* **31**, 377−384.

Bell, C. G., and Newell, A. (1971). "Computer Structures: Readings and Examples." McGraw-Hill, New York.

Benes, V. E. (1965). "Mathematical Theory of Connecting Networks and Telephone Traffic." Academic Press, New York.

Bentley, J. L., and Kung, H. T. (1979). A tree machine for searching problems. *Proc. Int. Conf. Parallel Process.*, pp. 257−266.

Bermond, J-C., Delorme, C. and Quisquater, J.-J. (1982). Grands graphes non dirigés de degré et diameter fixes, *Technical Report,* Philips Research Laboratory, Brussels, Belgium.

Bisiani, R. (1980). The harpy machine: A data structure-oriented architecture. *5th Workshop Comput. Archit. Non-Numer. Process.*, pp. 128−136.

Bogdanowicz, J. F. (1977). Preliminary design of a partitionable multimicroprogramm-able microprocessor system for image processing. Tech. Rep. EE 77−42. School of Electrical Engineering, Purdue University, Lafayette, Indiana.

Bokhari, S. H. (1981). On the mapping problem. *IEEE Trans. Comput.* **30**, 207−214.

Bondy, J. A. and Murty, U. S. R. (1976). "Graph Theory with Applications." Elsevier, New York.

Boral, H. (1981). On the use of data-flow techniques in database machines. *Comput. Sci. Dept. Tech. Rep.* **432**, Univ. of Wisconsin.

Brantley, W. C., Jr., Leive, G. W. and Sieworek, D. P. (1977). Decomposition of data flow graphs on multiprocessors. *Proc. AFIPS Natl. Comput. Conf.* **46**, 379−388.

Brauer, W., ed. (1980). "Net Theory and Applications." Springer-Verlag, Berlin and New York.

Brent, R. P., and Kung, H. T. (1980). The chip complexity of binary arithmetic. *Proc. 12th Ann. Symp. Theory Comput.*, pp. 190−200.

Briggs, F., Fu, K. S., Hwang, K., and Patel, J. (1979). PM4−a reconfigurable multimi-croprocessor system for pattern recognition and image processing. *Proc. AFIPS Natl. Comput. Conf.* **48**, 255−265.

Brinch-Hansen, P. (1973a). Concurrent programming concepts. *Comput. Surveys* **5**, 223−245.

Brinch-Hansen, P. (1973b). "Operating System Principles." Prentice-Hall, Englewood-Cliffs, New Jersey.

Brinch-Hansen, P. (1975). The programming language concurrent Pascal. *IEEE Trans. Software Engrg.* **1**, 199−207.

Brinch-Hansen, P. (1978). Multiprocessor architectures for concurrent programs. *ACM SIG Comput. Archit. News* **7**, 4−23.

Burks, A., ed. (1966). "Theory of Self-Reproducing Automata." Univ. of Illinois Press, Urbana.

Buzbee, B. L., and Morrison, J. J., compilers (1978). "Proceedings of the 1978 LASL Workshop on Vector and Parallel Processors." Los Alamos National Laboratory, Los Alamos, New Mexico.

Chazelle, B., and Monier, L. (1981). A model of computation for VLSI with related complexity results. *Comput. Sci. Dept. Tech. Rep.* 81−107, Carnegie-Mellon Univ., Pittsburgh.

Chu, W. W., Holloway, L. J., Lan, M-T., and Efe, K. (1980). Task allocation in distributed data processing. *Computer* 13, 57−69.

Church, A. (1936). The calculi of lambda-conversion. *Ann. Math.* 6.

Clark, W. A. (1980). From electron mobility to logical structure: A view of intergrated circuits. *Comput. Rev.* 12, 325−356.

Clos, C. (1953). A study of non-blocking switching networks. *Bell System Tech. J.* 32, 406−424.

Codd, E. F. (1968). "Cellular Automata." Academic Press, New York.

Coffman, E. G., Jr., ed. (1976). "Computer and Job-Shop Scheduling Theory." Wiley, New York.

Connell, H. E. T. (1976). A multi-minicomputer network for optical moving target indication. *COMPCON Fall* 76.

Conway, M. E. (1963). A multiprocessor system design. *Proc. AFIPS Fall Joint Comput. Conf.* 24, 139−146.

Conway, R. W., Maxwell, W. L., and Miller, L. W. (1967). "Theory of Scheduling." Addison-Wesley, Reading, Massachusetts.

Cook, R. P. (1980). The StarMod distributed programming system, *Proc. COMPCON 80,* Sept.

Cordella, L. P., Duff, M. J. B., and Levialdi, S. (1978). An analysis of computational cost in image processing: A case study. *IEEE Trans. Comput.* 27, 904−910.

Cornell, J. A. (1972). Parallel processing of ballistic missile defense radar data with PEPE. *COMPCON '72* pp. 69−72.

Cryer, C. W. (1979). Successive over-relaxation methods for solving linear complementar-problems arising from free boundary problems. *In* "Proc. Sem. Free Boundary Problems." (E. Magenes, ed.). Pavia, Italy.

Cyre, W. R., Frank, A. A., Redmont, M. J., and Rideout, V. C. (1977). WISPAC: A parallel array computer for large-scale system simulation. *Simulation* 30, 165−172.

de Bruijn. D. G. (1946). A combinatorial problem, *Proc. Ned. Akad. Wet. Sect. Sci.* 7, 758−764.

Denning, P. J., and Buzen, J. P. (1978). The operational analysis of queueing network models. *Comput. Surveys* 10, 225−261.

Dennis, J. B., and Weng, K. K. (1977). Applications of data flow computation to the weather problem. *In* "High Speed Computer and Algorithm Organization" (D. J. Kuck, D. H. Lawrie, and A. H. Sameh, eds.), 145−157. Academic Press, New York.

Dennis, J. B., Boughton, G. A., and Leung, C. K. (1980). Building blocks for data flow prototypes. *Proc. 7th Ann. Symp. Comput. Archit.,* pp. 1−8.

Dertouzos, M. L. (1971). Time bounds on space computations. *IEEE 1971 Symp. Switching Automata Theory,* 12, 182−187.

Despain, A. M., and Patterson, D. A. (1978). X-tree: A tree structured multi-processor computer architecture. *Proc. 5th Ann. Symp. Comput. Archit.,* pp. 144−151.

DeWitt, D. J. (1979). DIRECT—a multiprocessor organization for supporting relational database management systems. *IEEE Trans. Comput.* 28, 395−405.

Dionne, R., and Florian, M. (1979). Exact and approximate algorithms for optimal network design. *Networks* **9**, 283−307.

Doran, J. E. (1968a). Experiments with a pleasure seeking robot. *In* "Machine Intelligence" (D. Michie, ed.), Vol. 3, pp. 195−216. Edinburgh Univ. Press, Edinburgh.

Doran, J. E. (1968b). A simulated robot/environment system: Progress and problems (unpublished manuscript, Dept of Machine Intelligence, Edinburgh Univ.).

Douglass, R. J. (1978). A computer vision model for recognition, description, and depth perception in outdoor scenes. Ph.D. Dissertation, Computer Science Department, University of Wisconsin, Madison (unpublished).

Douglass, R. J. (1980). The requirements of a language for asynchronous parallel image processing. *Proc. Int. Conf. Parallel Process.*, 1980.

Douglass, R. J. (1981). MAC: A programming language for asynchronous image processing. *In* "Languages and Architectures for Image Processing" (M. J. B. Duff and S. Levialdi, eds.), pp. 41−51. Academic Press, New York.

Duff, M. J. B. (1976). CLIP4: A large scale integrated circuit array parallel processor. *Proc. IJCPR-3* **4**, 728−733.

Duff, M. J. B. (1977). Personal communication.

Duff, M. J. B. (1978). Review of the CLIP image processing system. *Proc. AFIPS Natl. Comput. Conf.* **47**, 1055−1060.

Duff, M. J. B. (1979). "Final Report on Workshop on Higher-level Languages for Image Processing." University College, London.

Duff, M. J. B. (1982). Parallel algorithms and their influence on the specification of application problems. *In* "Multicomputers and Image Processing" (K. Preston, Jr., and L. Uhr, eds.), pp. 261−274. Academic Press, New York.

Duff, M. J. B., and Levialdi, S., eds. (1981). "Languages and Architectures for Image Processing." Academic Press, New York.

Dyer, C. R. (1979). Augmented cellular automata for image processing. Ph.D. Dissertation, University of Maryland, College Park (unpublished).

Dyer, C. R. (1981). A quadtree machine for parallel image processing. *Inform. Engrg. Dept. Tech. Rep.* KSL 51, University of Illinois, Chicago.

Dyer, C. R. (1982). Pyramid algorithms and Machines. *In* "Multicomputers and Image Processing" (K. Preston, Jr., and L. Uhr, eds.), pp. 409−420. Academic Press, New York.

Dyer, C. R., and Rosenfeld, A. (1981). Parallel image processing by memory-augmented cellular automata. *IEEE Trans. Pattern Anal. Mach. Intell.* **3**, 29−41.

Elspas, B. (1964). Topological constraints on interconnection-limited logic. *Switching Circuit Theory Logical Des.* **164**, 133−147.

Enslow. P. H. (1977). Multiprocessor organization—a survey, *Comput. Surveys* **9**, 103−129.

Estrin, G. (1960). Organization of computer systems—the fixed plus variable structure computer. *Proc. AFIPS Winter Joint Comput. Conf.* **15**, pp. 33−40.

Etchells, R. D., Grinberg, J., and Nudd, G. R. (1981). The development of a novel three-dimensional microelectronic processor with ultra-high performance. *Proc. Soc. Photo-Opt. Instrum. Engrg.*

Evensen, A. J., and Troy, J. L. (1973). Introduction to the architecture of a 288-element PEPE. *Proc. Sagamore Conf. Parallel Process.*, pp. 162−169.

Falk, H. (1968). Reaching for a gigaflop. *IEEE Spectrum* **13**, 65−69.

Farber, D. J., and Larson, K. C. (1972). The system architecture of the distributed computer system—the communications system, *Symp. Comput. Networks,* Polytechnic Institute of Brooklyn, April.

Feilmeier, M., ed. (1977). "Parallel Computers—Parallel Mathematics." Int. Assoc. Comput. Simulation, New York.

Finkel, R. A., and Solomon, M. H. (1977). Processor interconnection strategies. *Comput. Sci. Dept. Tech. Rep.* 301. Univ. of Wisconsin.

Finkel, R. A., and Solomon, M. H. (1980a) The Arachne kernel. *Comput. Sci. Dept. Tech. Rep.* 380. Univ. of Wisconsin.

Finkel, R. A., and Solomon, M. H. (1980b). The Lens interconnection strategy. *Comput. Sci. Dept. Tech. Rep.* 387. Univ. of Wisconsin.

Fishburn, J. P. (1979). An optimization of alpha-beta search. *Comput. Sci. Dept. Tech. Rep.* 375. Univ. of Wisconsin.

Fishburn, J. P. (1981). Analysis of speedup in distributed algorithms. Ph.D. Dissertation, Computer Sci. Dept., University of Wisconsin, Madison (unpublished). *(Also Comput. Sci. Dept. Tech. Rep.* 431.)

Flanders, P. M. (1979). FORTRAN extensions for a highly parallel processor. "Supercomputers," Infotech State of the Art Rep. Infotech Int. Ltd., United Kingdom.

Flanders, P. M., Hunt, D. J., Parkinson, D., and Reddaway, S. F. (1977a). Experience gained in programming the Pilot DAP, a parallel processor with 1024 processing elements. *In* "Parallel Computers—Parallel Mathematics" (M. Feilmer, ed.), pp. 269−273 Int. Assoc. Comput. Simulation, New York.

Flanders, P. M., Hunt, D. J., Reddaway, S. F., and Parkinson, D. (1977b). Efficient high speed computing with the Distributed Array Processor. *In* "High Speed Computer and Algorithm Organization" (D. J. Kuck, D. H. Lawrie, and A. H. Sameh, eds.), pp. 113−128. Academic Press, New York.

Flynn, M. J. (1972). Some computer organizations and their effectiveness. *IEEE Trans. Comput.* **21**, 948−960.

Flynn, M. J., and Hennessy, J. L. (1980). Parallelism and representation problems in distributed systems. *IEEE Trans. Comput.* **29**, 1080−1086.

Ford, L. R., Jr., and Fulkerson, D. R. (1962). "Flows in Networks." Princeton Univ. Press, Princeton, New Jersey.

Forgy, C. L. (1977). A production system monitor for parallel computers. *Tech. Rep., Comput. Sci. Dept,* Carnegie-Mellon University, Pittsburgh, Pennsylvania.

Foster, C. C. (1980). "Computer Architecture," 2nd ed. Van Nostrand, New York.

Foster, M. J., and Kung, H. T. (1980). Design of special-purpose VLSI chips: Example and opinions. *Proc. 7th Ann. Symp. Comput. Archit.,* pp. 300−307.

Friedman, D. P., and Wise, D. S. (1978). Aspects of applicative programming for parallel processing. *IEEE Trans. Comput.* **27**, 289−296.

Friedman, H. D. (1965). A design for (d,k) graphs. *IEEE Trans. Electron. Comput.* **EC-14**, 488.

Fu, K. S. (1978). Special computer architectures for pattern recognition and image processing—an overview. *Proc. AFIPS Natl. Comput. Conf.* **47**, 1003−1013.

Fuller, R. H., and Bird, R. M. (1965). An associative parallel processor with applications to picture processing. *Proc. AFIPS Fall Joint Comput. Conf.* **27**, 105−116.

Fuller, S. H. (1976). Price/performance comparisons of c.mmp and the pdp-10. *Proc. 3rd Symp. Comput. Archit.,* pp. 195−202.

Fung, L. W. (1977). A massively parallel processing computer. *In* "High Speed Computing and Algorithm Organization" (D. J. Kuck, D. H. Lawrie, and A. H. Sameh, eds.), pp. 203−204. Academic Press, New York.

Galil, Z., and Paul, W. J. (1981). An efficient general purpose parallel computer. *Proc. 13th Ann. Symp. Theory Comput.,* pp. 247−262.

Gannon, D. (1981). On mapping non-uniform P.D.E. structures and algorithms onto uniform array architectures, *Proc. 1980 Int. Conf. Parallel Process.* 100−105.

Gary, J. M. Analysis of applications programs and software requirements for high speed computers. *In* "High Speed Computing and Algorithm Organization" (D. J. Kuck, D. H. Lawrie, and A. H. Sameh, eds.), pp. 329−354. Academic Press, New York.

Gemmar, P. (1982). Image correlation: Processing requirements and implementation structures on a flexible image processing system (FLIP). *In* "Multicomputer Algorithms and Image Processing" (K. Preston, Jr., and L. Uhr, eds.). pp. 87−98. Academic Press, New York.

Gerritsen, F. A., and Monhemius, R. D. (1981). Evaluation of the Delft Image Processor DIP-1. *In* "Languages and Architectures for Image Processing" (M. J. B. Duff and S. Levialdi, eds.), pp. 89−203. Academic Press, New York.

Gewirtz, A. (1969). Graphs with maximal even girth. *Canad. J. Math.* **21**, 915−934.

Gilbert, B. K., and Harris, L. D. (1980). Advances in processing architecture, display and device technology for biomedical image processing. *IEEE Trans. Nucl. Sci.* **27**, 1197−1206.

Gilbert, B. K., Kenue, S. K., Robb, R. A., Chu, A., Lent, A. H., and Swartzlander, E. E., Jr. (1981). Rapid execution of fan beam image reconstruction algorithms using efficient computational techniques and special purpose processors. *IEEE Trans. Biomed. Engrg.* **28**, 98−115.

Giloi, W. K., and Berg, H. (1978). Data structure architectures—a major operational principle, *Proc. 5th Ann. Symp. Comput. Archit.,* 175−181.

Golay, M. J. E. (1969). Hexagonal parallel pattern transformations. *IEEE Trans. Comput.* **18**, 733−740.

Gonzales, M. J., Jr. (1977). Deterministic processor scheduling. *Comput. Surveys* **9**, 173−204.

Goodman, J. R., and Despain, A. M. (1980). A study of the interconnection of multiple processors in a data base environment. *Proc. Int. Conf. Parallel Process.* pp. 269−278.

Goodman, J. R., and Sequin, C. H. (1981). Hypertree. a multiprocessor interconnection topology. *Comput. Sci. Dept. Tech. Rep.* 427, Univ. of Wisconsin.

Goodwin, R. J. (1973). A design for distributed-control multi-processor computer systems. *U.S. NTIS, AD Rep.* AD-772 883.

Granlund, G. H. (1981). GOP: A fast and flexible processor for image analysis. *In* "Languages and Architectures for Image Processing" (M. J. B. Duff and S. Levialdi, eds.), pp. 179−188. Academic Press, New York.

Gregory, J., and McReynolds, M. (1963). The SOLOMON computer. *IEEE Trans. Comput.* **12**, 774−780.

Griswold, R. E., Poage, J. F., and Polonsky, I. P. (1968). "The SNOBOL4 Programming Language" Prentice-Hall, Englewood Cliffs, New Jersey.

Gudmundsson, B. (1979). An interactive high-level language system for picture processing. *Pap.Conf.* (reprinted in "Picap II," Repts. Dept. Electrical Engineering, Linkoping University, Linkoping, Sweden, 1980).

Gylys, V. B., and Edwards, J. A. (1976). Optimal partitioning of workload for distributed systems. *COMPCON Fall 76* pp. 353−357.

Halstead, R. H., and Ward, S. A. (1980). The MUNET: A scalable decentralized architecture for parallel computation. *Proc. 7th Ann. Symp. Comput. Archit.,* pp. 139−145.

Halton, J. H., and Terada, R. (1978). An almost surely optimal algorithm for the Euclidian traveling salesman problem. *Comput. Sci. Dept. Tech. Rep.* 335, Univ. of Wisconsin.

Halton, J. H., and Terada, R. (1982). A fast algorithm for the Euclidean traveling salesman problem, *SIAM J. Comput.* **11**, 28−46.

Hanaki, S., and Temma, T. (1982). Template-controlled image processor (TIP) project. *In* "Multicomputer Algorithms and Image Processing" (K. Preston, Jr., and L. Uhr, eds.). pp. 343−352. Academic Press, New York.

Handler, W. (1975). A unified associative and von-Neumann processor—EGPP and EGPP array. *Lect. Notes Comput. Sci.* **24**, 97−99.

Handler, W. (1977). Aspects of parallelism in computer architecture. *In* "Parallel Computers—Parallel Mathematics" (M. Feilmeier, ed.). Int. Assoc. Math. Comput. Simulation, New York.

Handler, W., Schreiber, H. and Sigmund, V. (1979). Computation structures reflected in general purpose and special purpose multi-microprocessor systems. *Proc. Int. Conf. Parallel Process.*, pp. 95−102.

Hanson, A. R., and Riseman, E. M., eds. (1978). "Computer Vision Systems" Academic Press, New York.

Harary, F. (1969). "Graph Theory" Addison-Wesley, Reading, Massachusetts.

Hawkins, J. K., and Munsey, C. H. (1963). A parallel computer organization and mechanizations. *IEEE Trans. Comput.* **12**, 251−262.

Hewitt, C. (1973). A universal modular ACTOR formalism for artificial intelligence. *Proc. Int. Joint Conf. Artificial Intelligence, 3rd*, pp. 235−245.

Hewitt, C. (1977). Viewing control structures as patterns of passing messages. *Artificial Intelligence J.* **8**, 323−364.

Hobbs, L. C., and Theis, D. J. (1970). Survey of parallel processor approaches and techniques. *In* "Parallel Processor Systems, Technologies, and Applications" (L. C. Hobbs *et al.*, eds.), pp. 3−20. Spartan Books, New York.

Hoffman, A. J., and Singleton, R. R. (1960). On Moore graphs with diameter 2 and 3. *IBM J. Res. Develop.* **4**, 497−504.

Holland, J. H. (1959). A universal computer capable of executing an arbitrary number of sub-programs simultaneously. *Proc. AFIPS Fall Joint Comput. Conf.* **13**, 108−113.

Huen, W., Greene, P., Hochsprung, R., and El-Dessouki,O. (1977). A network computer for distributed processing. *Proc. AFIPS Natl. Comput. Conf.* **46**.

Hwang, F. K., and Lin, S. (1079). On the construction of balanced switching networks. *Networks* **9**, 283−307.

Imase, M., and Itoh, M. (1981). Design to minimize diameter on building-block network. *IEEE Trans. Comput.* **30**, 439−442.

Ison R. (1977). A programming language for a polyprocessor, *Tech. Rep. Dept. Appl. Math. Comput. Sci.*, University of Virginia, Charlottesville.

Iverson, K. E. (1962). "A Programming Language." Wiley, New York.

Jafari, H., Lewis, T. G., and Spragins, J. D. (1980). Simulation of a class of ring-structured networks. *IEEE Trans. Comput.* **29**, 385−392.

Jecmen, R. M., Ebel, A. A., Smith, R. J., Kynett, V., Hui, C.-H., and Pashley, R. D. (1979). A 25ns 4k Static RAM. *Dig. Tech. Pap., ISSCC* **79**, 100−101.

Jelinek, J. (1979). An algebraic theory for parallel processor design. *Comput. J.* **22**, 323−345.

Jensen, K., and Wirth, N. (1973). "PASCAL User Manual and Report," Springer-Verlag, Berlin and New York.

Jones, A. K., and Schore, P. (1980). Experience using multi-processor systems—status report. *Comput. Surveys* **12**, 121−165.

Jordan, H., Scalabrin, M., and Calvert, W. (1979). A comparison of three types of multiprocessor algorithms. *Proc. Int. Conf. Parallel Process.*, pp. 231−238.

Karp, R. M., and Miller, R. E. (1969). Parallel program schemata. *J. Comput. System Sci.* **3**, 167−195.

Karp, R. M., and Miranker, W. L. (1968). Parallel minimax search for a maximum. *J. Combin. Theory* **4**, 19−35.

Kartashev S. P., and Karteshev, S. I. (1979). Adaptable pipeline systems with dynamic architecture. *Proc. Int. Conf. Parallel Process.*, pp. 222−230.

Kartashev S. P. and Karteshev, S. I. (1981). Reconfiguration of dynamic architecture into multicomputer networks, *Proc. Int. Conf. Parallel Process.* pp. 133−140.

Kautz, W. H. (1971). An augmented content-addressed memory array for implementation with large-scale integration. *J. Assoc. Comput. Mach.* **18**, 19−33.

Kleene, S. C. (1936). General recursive functions of natural numbers. *Math. Ann.* **12**, 340−353.

Kleene, S. C. (1956). Representation of events in nerve nets and finite automata. *In* "Automata Studies" (C. E. Shannon and J. McCarthy, eds.), pp. 3−41. Princeton Univ. Press, Princeton, New Jersey.

Kleinrock, L. (1975). "Queueing Systems," I. Wiley, New York.

Kleinrock, L. (1976). "Queueing Systems," II. Wiley, New York.

Kleitman, D., Leighton, F. T., Lepley, M., and Miller, G. L. (1981). New layouts for the shuffle-exchange graph. *Proc. 13th Ann. Symp. Theory Comput.*, pp. 278−292.

Klinger, A., and Dyer, C. R. (1976). Experiments on picture representations using regular decomposition. *Comput. Graphics Image Process.* **5**, 68−105.

Knuth, D. E., and Moore, R. W. (1975). An analysis of alpha-beta pruning. *Artificial Intelligence* **6**, 293−326.

Koren, I. (1981). A reconfigurable and fault-tolerant VLSI multiprocessor array. *Proc. 8th Ann. Symp. Comput. Archit.*, pp. 425−443.

Kornfeld, W. A., and Hewitt, C. (1981). The scientific community metaphor. *IEEE Trans. Systems Man Cybernet.* **11**, 24−33.

Kozdrowicki, E. W., and Thies, D. J. (1980). Second generation of vector supercomputers. *Computers* **13**, 71−83.

Kraft, C. D., and Toy, W. N. (1979). "Mini/Microcomputer Hardware Design." Prentice-Hall, Englewood Cliffs, New Jersey.

Kruse, B. (1973). A parallel picture processing machine. *IEEE Trans. Comput.* **22**, 1075−1087.

Kruse, B. (1976). The PICAP picture processing laboratory. *Proc. IJCPR-3* **4**, 875−881.

Kruse, B. (1978). Experience with a picture processor in pattern recognition processing. *Proc. AFIPS Natl. Comput. Conf.* **47**, 1015−1022.

Kruse, B. (1980). System architecture for image analysis. *In* "Structured Computer Vision: Machine Perception Through Hierarchical Computation Structures" (S. Tanimoto and A. Klinger, eds.), pp. 169−212. Academic Press, New York.

Kruse, B., Danielsson, P. E., and Gudmundsson, B. (1980). From PICAP I to PICAP II. *In* "Special Computer Architectures for Pattern Processing" (K. S. Fu and T. Ichikawa, eds.), CPR Press, New York.

Kuck, D. J. (1977). A survey of parallel machine organization and programming. *Comput. Surveys* **9**, 29−59.

Kuck, D. J. (1978). "The Structure of Computers and Computation," Vol. 1. Wiley, New York.

Kuck, D. J. and Stokes, R. A. (1982). The Burroughs scientific processor (BSP), *IEEE Trans. Computers.* **31**, 363−376.

Kuck, D. J., Lawrie, D. H., and Sameh, A. H., eds. (1977). "High Speed Computer and Algorithm Organization." Academic Press, New York.

Kung, H. T. (1980a). The structure of parallel algorithms. *Ad. Comput.* **19**, 293−326.

Kung, H. T. (1980b). Special-purpose devices for signal and image processing: An opportunity in VLSI. *Comput. Sci. Dept. Tech. Rep.* 80−132. Carnegie-Mellon Univ., Pittsburgh.

Kung, H. T., and Song, S. W. (1982). A systolic 2-d convolution chip. *In* "Multicomputer Algorithms for Image Processing" (K. Preston Jr., and L. Uhr, eds.). pp. 373–384. Academic Press, New York.

Kung, H. T., and Stevenson, D. (1977). A software technique for reducing the routing time on a parallel computer with a fixed interconnection network. *In* "High Speed Computer and Algorithm Organization" (D. J. Kuck, D. H. Lawrie, and A. H. Sameh, eds.), pp. 423–433. Academic Press, New York.

Kushner, T., Wu, A. Y., and Rosenfeld, A. (1980). Image processing on ZMOB. *Comput. Sci. Dept. Tech. Rep.* 987, Univ. of Maryland.

Lawler, E. L. (1976). "Combinatorial Optimization: Networks and Matroids." Holt, New York.

Lawrie, D. H. (1975). Access and alignment of data in an array processor. *IEEE Trans. Comput.* **24**, 1145–1155.

Lawrie, D. H., Layman, T., Baer, D., and Randal, J. M. (1975). GLYPNIR—a programming language for ILLIAC IV. *Comm. ACM* **18**, 157–164.

Ledley, R. S., Kulkarni, Y. G., Rotolo, L. S., Golab, T. J., and Wilson, J. B. (1977). TEXAC: A special purpose picture processing texture analysis computer. *Proc. IEEE Comput. Soc. Int. Conf., 15th*, pp. 66–71.

Lee, I., Yu, R. T., Smith, F. J.,. Wong, S., and Embrathiry, M. P. (1979a). A 64kb MOS dynamic RAM. *Dig. of Tech. Pap., ISSCC* **79**, 146–147.

Lee, J., Breivogel, M., Kunita, R., and Webb, C. (1979b). A 80ns 5v-only dynamic RAM. *Dig. Tech. Pap. ISSCC* **79**, 141–143.

Lee, R. B. (1980). Empirical results on the speed, efficiency, redundancy and quality of parallel computations. *Proc. Int. Conf. Parallel Process*, pp. 91-100.

Lehman, M. (1966). A survey of problems and preliminary results concerning parallel processing and parallel processors. *Proc. IEEE* **54**, 1889–1901.

Leiserson, C. E. (1980). Area-efficient graph layouts (for VLSI). *Comput. Sci. Dept. Tech. Rep.* 80–138, Carnegie-Mellon Univ., Pittsburgh.

Leland, W. (1982). Density and reliability of interconnection topologies for multicomputers. Unpublished PhD. Diss., University of Wisconsin.

Leland, W., Li, Q., and Uhr, L. (1980). Globally dense (d,k) graphs for computer network architectures, *Comput. Sci. Dept. Tech. Rep.* University of Wisconsin, Madison.

Lev, G. F., Pippenger, N., and Valiant, L. G. (1981). A fast parallel algorithm for routing in permutation networks. *IEEE Trans. Comput.* **30**, 93–100.

Levialdi, S., Isoldi, M., and Uccella, G. (1980). Programming in Pixal. *Pap., IEEE Workshop Picture Data Description Management*, 1980, Asilomar, Calif.

Levialdi, S., Maggiolo-Schettini, A., Napoli, M., and Uccella, G. (1981). PIXAL: A high level language for image processing. *In* "Real-Time Parallel Computing: Image Analysis" (M. Onoe, K. Preston, Jr., and A. Rosenfeld, eds.), pp. 131–143. Plenum, New York.

Levine, M. D. (1978). A knowledge-based computer vision system. *In* "Computer Vision Systems" (A. R. Hanson and E. M. Riseman, eds.), pp. 335–352. Academic Press, New York.

Lin, S., and Kernighan, B. W. (1973). An effective heuristic algorithm for the traveling salesman problem. *Oper. Res.* **21**, 498–516.

Lincoln, N. R. (1982). Technology and design tradeoffs in the creation of a moderr. supercomputer, *IEEE Trans. Comput.* **31**, 349–362.

Lipovski, J. (1977). On a varistructured array of microprocessors. *IEEE Trans. Comput.* **26**, 125–138.

Lipovski, J., and Tripathi, A. (1977). A reconfigurable varistructure array processor. *Proc. Int. Conf. Parallel Process.*, pp. 165−174.

Lipton, R. J., and Tarjan, R. E. (1977). A separator theorem for planar graphs. *Proc. Conf. Theor. Comput. Sci., 1977* pp. 1−10.

Lorin, H. (1972). "Parallelism in Hardware and Software: Real and Apparent Concurrency." Prentice-Hall, Englewood Cliffs, New Jersey.

Lougheed, R. L., and McCubbrey, D. L. (1980). The cytocomputer: A practical pipelined image processor. *Proc. 7th Ann. Symp. Comput. Archit.*, pp. 271−277.

Lundstrom, S. F., and Barnes, G. H. (1980). A controllable MIMD architecture. *Proc. Int. Conf. Parallel Process.*, pp. 19−27.

McCarthy, J., Abrahams, P. W., Edwards, D. J., Hart, T. P., and Levin, M. I. (1962). "LISP 1.5 Programmers Manual." MIT Press, Cambridge, Massachusetts.

McCormick, B. H. (1963). The Illinois pattern recognition computer ILLIAC III. *IEEE Trans. Comput.* **12**, 791−813.

McCulloch, W. S., and Pitts, W. (1943). A logical calculus of the ideas immanent in nervous activity. *Bull. Math. Biophys.* **5**, 115−133.

McGraw, J. (1980). Data flow computing: Software development. *IEEE Trans. Comput.* **29**, 1095−1103.

McQuillan, J. M. (1977). Graph theory applied to optimal connectivity in computer networks. *Comput. Commun. Rev.* **7**, 13−41.

McQuillan, J. M., and Walden, D. C. (1977). The ARPA network design decisions. *Comput. Networks* **1**, 243−289.

Mago, G. A. (1980). A cellular computer architecture for functional programming. *Proc. COMPCON Spring 80* pp. 179−187.

Manara, R., and Stringa, L. (1981). The EMMA system: An industrial experience on a multi-processor. *In* "Languages and Architectures for Image Processing" (M. J. B. Duff and S. Levialdi, eds.), pp. 215−227. Academic Press, New York.

Marcus, M. J. (1977). The theory of connecting networks and their complexity: A review. *Proc. IEEE* **65**, 1263−1271.

Marks, P. (1980). Low-level vision using an array processor. *Comput. Graphics Image Process.* **14**, 281−292.

Martin, H. G. (1977). A discourse on a new super computer, PEPE. *In* "High Speed Computer and Algorithm Organization" (D. J. Kuck, D. H. Lawrie, and A. H. Sameh, eds.), pp. 101−112. Academic Press, New York.

Masson, G. M. (1977). Binomial switching networks for concentration and distribution. *IEEE Trans. Commun.* **25**, 873−883.

Mead, C. A., and Conway, L. A. (1980). "Introduction to VLSI Systems." Addison-Wesley, Reading, Massachusetts.

Mead, C. A., and Rem, M. (1979). Cost and performance of VLSI computing structures. *IEEE J. Solid-State Circuits* **14**, 455−462.

Meilander, W. C. (1981). History of parallel processing at Goodyear Aerospace, *Proc. Int. Conf. Parallel Process.* pp. 5−15.

Mersereau, R. M. (1979). The processing of hexagonally sampled two-dimensional signals. *Proc. IEEE* **67**, 930−949.

Metcalfe, R. M., and Boggs, D. R. (1976). Ethernet: Distributed packet switching for local computer networks. *Comm. ACM* **19**, 395−404.

Miller, J. S., Lickley, D. J., Kosmala, A. L., and Saponaro, J. A. (1970). "Multiprocessor Computer System Study," Final report. Intermetrics, Inc., Cambridge, Massachusetts.

Minnick, R. C. (1966). Cellular arrays for logic and storage. *SRI Proj. 5087 Final Rep.* pp. 197−215.

Minnick, R. C. (1967). A survey of microcellular research. *J. Assoc. Comput. Mach.* **14**, 203−241.

Minsky, M. (1967). "Computation: Finite and Infinite Machines." Prentice-Hall, Englewood Cliffs, New Jersey.

Minsky, M., and Papert, S. (1971). On some associative parallel and analog computations. *In* "Associative Information Techniques" (E. J. Jacks, ed.), Am. Elsevier, New York.

Miranker, W. L. (1979). Hierarchical relaxation. *Computing* **23**, 267−285.

Moore, E. F. (1956). Gendanken-experiments on sequential machines. *In* "Automata Studies" (C. E. Shannon and J. McCarthy, eds.), pp. 129−153. Princeton Univ. Press, Princeton, New Jersey.

Moore, E. F. (1964). "Sequential Machines: Selected Papers." Addison-Wesley, Reading, Massachusetts.

Mori, K., Kidode, M., Shinoda, H., and Asada, H. (1978). Design of local parallel pattern processor for image processing. *Proc. AFIPS Natl. Comput. Conf.* **47**, 1025−1031.

Mukhopadhyay, A., and Stone, H. S. (1971). Cellular logic. *In* "Recent Developments in Switching Theory" (A. Mukhopadhyay, ed.), pp. 256−313. Academic Press, New York.

Murphy, B. T. (1964). Cost-size optima of monolithic integrated circuits. *Proc. IEEE* **52**, 1537−1545.

Narasimhan (1966). Syntax-directed interpretation of classes of pictures. *Communic. ACM* **9**, 166-173.

Newton, G. E., and Brenner, R. L. (1971). "Design of a Cellularly Organized APL Machine," Final Rep., NSF Grant GJ-723. University of Iowa, Iowa City.

Noe, J. D. (1971). A petri-net description of the CDC 6400. *Proc. ACM Workshop System Performance Evaluation,* pp. 362−378.

Noe, J. D. (1979). Nets in modelling and simulation. *In* "Net Theory and Applications" (W. Brauer, ed.), pp. 347−368. Springer-Verlag, Berlin and New York.

Onoe, M., Preston, K., Jr., and Rosenfeld, S., eds. (1981). "Real-Time Parallel Computing: Image Analysis." Plenum, New York.

Ore, O. (1968). Diameters in graphs. *J. Combin. Theory* **5**, 75−81.

Pashley, R. D., Liu, S. S., Owen, W. H., Owen, J. M., Shappir, J., Smith, R. J., and Jecmen, R. J. (1979). A 16k by 1b Static RAM. *Dig. Tech. Pap., ISSCC* **79**, pp. 106−107.

Patel, J. H. (1978). Design and analysis of processor-memory interconnections for multiprocessors, *Tech. Rep.* EE 78-40. School of Electrical Engineering, Purdue University, Lafayette, Indiana.

Paterson, H. S., Ruzzo, W. L., and Snyder, L. (1981). Bounds on minimax edge length for complete binary trees. *Proc. 13th Ann. Symp. Theory Comput.,* pp. 293−299.

Patterson, D. A., Fehr, E. S., and Sequin, C. H. (1979). Design considerations for the VLSI processor of X-tree. *Proc. 6th Ann. Symp. Comput. Archit.,* pp. 90−101.

Patton, P. C. (1972a). Data processing special studies final report. *AIC Tech. Rep.* HSU/72-006, Analysts International Corp.

Patton, P. C. (1972b). Estimated performance loss of 7600 multiprocessors due to LCS memory conflicts. *AIC Tech. Note* 1687-C-3, Analysts International Corp.

Pease, M. C. (1968). An adaptation of the fast fourier transform for parallel processing. *J. Assoc. Comput. Mach.* **15**, 252−264.

Pease. M. C. (1977). The indirect binary *n*-cube microprocessor array, *IEEE Trans. Comput.* **26**, 458−473.

Perrott, R., and Stevenson, D. (1978). ACTUS—a language for SIMD architectures. *Proc. LASL Workshop Vector Parallel Processors,* pp. 212–218.

Petersen. J. (1891). Die Theorie der regularen Graphen, *Acta Math.* **15**, 193–220.

Peterson, J. L. (1977). Petri nets. *Comput. Surveys* **9**, 223–252.

Petri, C. A. (1962). Fundamentals of a theory of asynchronous information flow. *Proc. IFIP Congr.* **62**.

Petri, C. A. (1979). Introduction to general net theory. *In* "Net Theory and Applications" (W. Brauer, ed.), pp. 1–19. Springer-Verlag, Berlin and New York.

Plas, A., Compte, D., Gelly, O., and Syre, J. C. (1976). LAU system architecture: A parallel data driven processor based on single assignment. *Proc. Int. Conf. Parallel Process.,* pp. 293–303.

Post, E. L. (1936). Finite combinatory processes. *J. Symbolic Logic* **1**, 103–105.

Post, E. L. (1943). Formal reductions of the general combinatorial decision problem. *Am. J. Math.* **65**, 197–268.

Potter, J. L. (1978). The STARAN architecture and its applications to image processing and pattern recognition algorithms. *Proc. AFIPS Natl. Comput. Conf.* **47**, 1041–1047.

Potter, J. L. (1982). MPP architecture and programming. *In* "Multicomputer Algorithms for Image Processing" (K. Preston, Jr., and L. Uhr, eds.), pp. 275–290. Academic Press, New York.

Preparata, F. P., and Vuillemin, J. (1979). The cube-connected-cycles: A versatile network for parallel computation. *Proc. 20th Ann. Symp. Found. Comput. Sci.,* pp. 140–147.

Preston, K., Jr. (1971). Feature extraction by Golay hexagonal pattern transformations. *IEEE Trans. Comput.* **20**, 1007–1014.

Preston, K., Jr. (1979). Image manipulative languages: A preliminary survey. *Pap., Workshop Conf. Higher-Level Lang. Image Process.* Windsor, U. K.

Preston, K., Jr., (1981). Languages for parallel processing of images. *In* "Real-Time Parallel Computing: Image Analysis" (M. Onoe, K. Preston, Jr., and A. Rosenfeld, eds.), pp. 145–158. Plenum, New York.

Preston, K., Jr., and Uhr, L., eds. (1981). "Multicomputer Algorithms and Image Processing." Academic Press, New York.

Radhakrishan, T., Barrera, R., Guzman, A., and Jinich, A. (1979). Design of a high level language for image processing, Insituto de Invesigaciones en Matematicas Aplicadas y en Sistemas, *Tech. Rep.* PR-78-22. Universidad Nacional Autonoma de Mexico.

Randell, B. (1973). "The Origins of Digital Computers: Selected Papers." Springer-Verlag, Berlin and New York.

Reddaway, S. F. (1978). DAP—a flexible number cruncher. *Proc. LASL Workshop Vector Parallel Processors,* pp. 233–234.

Reddaway, S. F. (1980). Revolutionary array processors. *In* "Electronics to Microelectronics" (W. A. Kaiser and W. E. Proebster, eds.), pp. 730–734. North-Holland Publ., Amsterdam.

Reeves, A. P. (1979). An array processing system with a Fortran-based realization. *Comput. Graphics Image Process.* **9**, 267–281.

Reeves, A. P. (1981). Parallel computer architectures for image processing, *Proc. Int. Conf. Parallel Process.* pp. 199–206.

Reeves, A. P., Bruner, J. D., and Poret, M. S. (1980). The programming language parallel PASCAL. *Proc. Int. Conf. Parallel Process.,* pp. 5–6.

Reeves, A. P., and Rindfuss, R. (1979). The base 8 binary array processor (BAP). *Proc. IEEE Conf. Pattern Recognition Image Process.,* pp. 246–249.

Rieger, C., Bane, J., and Trigg, R. (1980). A highly parallel multiprocessor. *Proc. IEEE Workshop Picture Data Description Manage.*, pp. 298−304.

Rohrbacker, D., and Potter, J. L. (1977). Image processing with the Staran parallel computer. *Computer* **10**, 54−59.

Rosenfeld, A. (1979). "Picture Languages." Academic Press, New York.

Rosenfeld, A. (1980). Quadtrees and pyramids for pattern recognition and image processing. *Proc. 5th Int. Conf. Pattern Recognition*, pp. 802−807.

Rosenfeld, A., and Kak, A. C. (1982). "Digital Picture Processing," 2nd ed. Academic Press, New York.

Rosenstiehl, P., Fiksel, J. R., and Holliger, A. (1972). Intelligent graphs: Networks of finite automata capable of solving graph problems. *In* "Graph Theory and Computing" (R. C. Read, ed.), pp. 219−265. Academic Press, New York.

Rosin, R. F. (1965). A special purpose computer for solution of partial differential equations and other iterative algorithms. *IEEE Trans. Electron. Comput.* **14**, 488−490.

Rumbaugh, J. (1977). A data flow multiprocessor. *IEEE Trans. Comput.* **26**, 138−146.

Russell, J., and Kime, C. (1975). System fault diagnosis, *IEEE Trans. Comput.* **24**, 1078−1089, 1155−1161.

Russell, R. M. (1978). The Cray-1 computer system. *Comm ACM* **21**, 63−72.

Samari, N. K., and Schneider, G. M. (1980). A queueing theory-based analytic model of a distributed computer network. *IEEE Trans. Comput.* **29**, 994−1001.

Sameh, A. H. (1977). Numerical parallel algorithms—a survey. *In* "High Speed Computer and Algorithm Organization" (D. J. Kuck, D. C. Lawrie, and A. H. Sameh, eds.), pp. 207−228. Academic Press, New York.

Sato, M., Matsuura, H., Ogawa, H., and Iijima, T. (1982). Multimicroprocessor system PX-1 for pattern information processing. *In* "Multicomputer Algorithms and Image Processing" (K. Preston, Jr., and L. Uhr, eds.), pp. 361−371. Academic Press, New York.

Schaeffer, D. H. (1979). Massively parallel information processing systems for space applications. *AIAA 2nd Conf. Comput. Aerosp.*, pp. 284−286.

Schaeffer, D. H., and Strong, J. P. (1977). TSE computers. *Proc. IEEE* **65**, 129−138.

Schmitt, L. (1979). Unpublished paper on parallel languages for general networks. Univ. of Wisconsin, Madison.

Schuster, S. A., Nguyen, H. B., Ozkarahan, E. A., and Smith, K. C. (1979). RAP.2−an associative processor for databases and its applications. *IEEE Trans. Comput.* **28**, 446−458.

Scott, A. (1977). "Neurophysics." Wiley, New York.

Sequin, C. H. (1981). Doubly twisted torus networks for VLSI processor arrays. *Proc. 8th Ann. Symp. Comput. Archit.*, pp. 471−480.

Shima, M. (1978). Two versions of 16-bit chip span microprocessor, minicomputer needs. *Electronics* **21**, 81−88.

Shoch, J. F., and Hupp, J. A. (1980). Performance of an Ethernet local network—a preliminary report. *Proc. Compcon 80* pp. 318−322.

Shooman, W. (1960). Parallel computing with vertical data. *Proc. AFIPS EJCC* **17**, 111−115.

Shore, J. E. (1973). Second thoughts on parallel processing. *Comput. Electron. Engrg.* **1**, 95−109.

Siegel, H. J. (1979). A model of SIMD machines and a comparison of various interconnection networks. *IEEE Trans. Comput.* **28**, 907−917.

Siegel, H. J. (1981). PASM: A reconfigurable multimicrocomputer system for image processing. *In* "Languages and Architectures for Image Processing" (M. J. B. Duff and S. Levialdi, eds.), pp. 252−265. Academic Press, New York.

Siegel, H. J., (1979). PASM: A partitionable multimicrocomputer SIMD/MIMD system for image processing and pattern recognition, *Tech. Rept.-EE* 79-40. School of Electrical Engineering, Purdue University, Lafayette, Indiana.

Slotnick, D. L. (1967). Unconventional systems. *Proc. AFIPS Spring Joint Comput. Conf.* pp. 477−481.

Slotnick, D. L., Borck, W. C., and McReynolds, R. C., (1962). The Solomon computer. *Proc. AFIPS Fall Joint Comput. Conf.* **20**, 97−107.

Spector, A. Z. (1980). Performing remote operations efficiently on a local network. *Comput. Sci. Dept. Tech. Rep.* 80−831. Stanford Univ.

Squire, J. S., and Palais, S. M. (1963). Programming and design considerations of a highly parallel computer. *Proc. AFIPS Spring Joint Comput. Conf.* **22**, 395−400.

Stanley, D. J., and Redstone, P. S. (1979). High throughput image preprocessing techniques for earth resources imagery. *Pap., Workshop Conf. Higher-Level Lang. Image Process. 1979,* Windsor, U. K.

Sternberg, S. R. (1978). Cytocomputer real-time pattern recognition. *Pap., 8th Pattern Recognition Symp.,* National Bureau of Standards, Washington, D. C.

Sternberg, S. R. (1980). Language and architecture for parallel image processing. *In* "Pattern Recognition in Practice" (E. S. Gelsema and L. N. Kanal, eds.), North-Holland Publ., Amsterdam.

Stone, H. S. (1971). Parallel processing with the perfect shuffle. *IEEE Trans. Comput.* **20**, 153−161.

Stone, H. S. (1973). An efficient parallel algorithm for the solution of a tridiagonal linear system of equations. *J. Assoc. Comput. Mach.* **20**, 27−38.

Stone, H. S. (1975). Parallel computers. *In* "Introduction to Computer Architecture" (H. S. Stone, ed.), pp. 318−374. SRA, Chicago, Illinois.

Stone, H. S. (1977). Multiprocessor scheduling with the aid of network flow analysis, *IEEE Trans. Software Engrg.* **3**, 315−321.

Stonebraker, M. (1981). Operating system support for database management, *Comm. ACM* **24**, 412−418.

Storer, J. A. (1980). The node cost measure of embedding graphs on the planar grid. *Proc. 12th Ann. Symp. Theory Comput.,* pp. 201−210.

Storwick, R. M. (1970). Improved construction for (d,k) graphs. *IEEE Trans. Comput.* **19**, 1214−1216.

Su, S. Y. W., Nguyen, H. B., Emam, A., and Lipovski, G. J. (1979). The architectural features and implementation techniques of the multicell CASSM. *IEEE Trans. Comput.* **28**, 430−445.

Sugarman, R. (1980). *IEEE Spectrum* **17**, 28−34.

Sullivan, H., Bashkov, T., and Klappholz, D. (1977). A large scale, homogenous, fully distributed parallel machine. *Proc. 4th Ann. Symp. Comput. Archit., 1977* pp. 105−124.

Surprise, J. M. (1981). Airborne associative memory processor (ASPRO), *Proc. Int. Conf. Parallel Process.* pp. 129−130.

Sutherland, I. E., Mead, C. A., and Everhart, T. E. (1976). "Basic Limitations in Microcircuit Fabrication Technology," Rep. R-1956-ARPA. RAND Corp., Santa Monica, California.

Swan, R. J., Fuller, S. H., and Siewiorek, D. P. (1977). Cm*−A modular, multimicroprocessor. *Proc. AFIPS NCC,* pp. 637−663.

Tanenbaum, A. S. (1981). "Computer Networks." Prentice-Hall, Englewood Cliffs, New Jersey.

Tanimoto, S. L. (1976). Pictorial feature distortion in a pyramid. *Comput. Graphics Image Process.* **5**, 333−352.

Tanimoto, S. L. (1978). Regular hierarchical image and processing structures in machine vision. *In* "Computer Vision Systems" (A. R. Hanson and E. M. Riseman, eds.), pp. 165–174. Academic Press, New York.

Tanimoto, S. L. (1982). Programming techniques for hierarchical image processors. *In* "Multicomputer Algorithms and Image Processing" (K. Preston, Jr., and L. Uhr, eds.), pp. 421–429. Academic Press, New York.

Tanimoto, S., and Klinger, A., eds. (1980). "Structured Computer Vision: Machine Perception Through Hierarchical Computation Structures." Academic Press, New York.

Thompson, C. D. (1979). Area-time complexity for VLSI. *Proc. 11th Ann. Symp. Theory Comput.*, pp. 81–88.

Thompson, C. D. (1980). A complexity theory for VLSI. Unpubl. Ph.D. Dissertation, Computer Science, Carnegie-Mellon University, Pittsburgh, Pennsylvania.

Thurber, K. J. (1976). "Large Scale Computer Architecture." Hayden, Rochelle Park, New Jersey.

Thurber, K. J. (1978). Circuit switching technology: A state of the art survey. *Proc. Compcon 78* pp. 116–124.

Thurber, K. J., and Freeman, H. A. (1979). Local computer network architectures. *Proc. Compcon 79* pp. 258–261.

Thurber, K. J., and Myrna, J. W. (1970). System design of a cellular APL computer. *IEEE Trans. Comput.* **19**, 291–303.

Thurber, K. J., and Wald, L. D. (1975). Associative parallel processors. *Comput. Surveys* **4**, 215–255.

Toda, M. (1962). The design of a fungus-eater: A model of human behavior in an unsophisticated environment. *Behav. Sci.* **7**, 164–183.

Toueg, S., and Steiglitz, K. (1979). The design of small-diameter networks by local search. *IEEE Trans. Comput.* pp. 537–542.

Travis, L., Honda, M., LeBlanc, R., and Zeigler, S. (1977). Design rationale for TELOS, a PASCAL based AI language. *ACM SIGPLAN* 12, No. 8, 67–76.

Trufanov, S. V. (1967). Some problems of distance on a graph. *Engrg. Cybernetics* May-June, pp. 60–66.

Turing, A. M. (1936). On computable numbers, with an application to the Entscheidungsproblem. *Proc. London Math. Soc.* (2) **42**, 230–265.

Tutte, W. T. (1966). "Connectivity in Graphs." Oxford Univ. Press, London and New York.

Uehara, T., and Van Cleemput, W. M. (1981). Optimal layout of CMOS functional arrays. *IEEE Trans. Comput.* **30**, 305–312.

Uhr, L. (1972). Layered "recognition cone" networks that preprocess, classify and describe. *IEEE Trans. Comput.* **21**, 758–768.

Uhr, L. (1973). "Pattern Recognition, Learning and Thought." Prentice-Hall, Englewood Cliffs, New Jersey.

Uhr, L. (1975). Toward integrated cognitive systems, which *must* make fuzzy decisions about fuzzy problems. *In* "Fuzzy Sets and Their Applications to Cognitive and Decision Processes" (L. A. Zadeh, K.-S. Fu, K. Tanaka, and M. Shimura, eds.), pp. 353–393. Academic Press, New York.

Uhr, L. (1976). "Recognition cones" that perceive and describe scenes that move and change over time. *Proc. 3rd Int. Joint Conf. Pattern Recognition*, **4**, 287–293.

Uhr, L. (1978). "Recognition cones" and some test results. *In* "Computer Vision Systems" (A. Hanson and E. Riseman, eds.), pp. 363–372. Academic Press, New York.

Uhr, L. (1979a). A higher-level language for a large parallel array computer. *Pap., Workshop Conf. Higher-Level Languages Image Process.*, (also *Univ. of Wisconsin Comput. Sci. Dept. Tech. Rep.* 354, 1979). Windsor, U. K.

Uhr, L. (1979b). A language for parallel processing of arrays, embedded in PASCAL. *In* "Languages and Architectures for Image Processing" (M. J. B. Duff and S. Levialdi, eds.) pp. 583−87. Academic Press, London and New York (also *Comput. Sci. Dept. Tech. Rep.* 365, Univ. of Wisconsin, Madison.)

Uhr, L. (1979c). Parallel-serial production systems with many working memories. *Proc. 5th, Int. Joint Conf. Artificial Intelligence.*

Uhr, L. (1980). Compounding denser (d,k) graph architectures for computer networks *Comput. Sci. Dept. Tech. Rept.* University of Wisconsin, Madison.

Uhr, L. (1981a). Computer analysis and perception of visual signals. *In* "Computer Analysis and Perception of Visual and Auditory Signals" (C. Y. Suen and R. De Mori, eds.), pp. 1−16. CRC Press, New York.

Uhr, L. (1981b). Network and array architectures for real-time perception. *Comput. Sci. Dept. Tech. Rep.* 424. Univ. of Wisconsin, Madison.

Uhr, L. (1981c). Comparing serial computers, arrays and networks using measures of "active resources." *Proc. CAPAIDM, 1981* pp. 188−194.

Uhr, L. (1981d). A language for parallel processing of arrays, embedded in PASCAL. *In* "Languages and Architectures for Image Processing" (M. J. B. Duff and S. Levialdi, eds.), pp. 53−87. Academic Press, New York.

Uhr, L., and Kochen, M. (1969). MIKROKOSMS and Robots. *Proc. 1st Int. Joint Conf. Artificial Intell.*, pp. 541−556.

Uhr, L., Lackey, J., and Thompson, M. (1980). A 2-layered SIMD/MIMD parallel pyramidal "array/net." *Comput. Sci. Dept. Tech. Rep.* 409. Univ. of Wisconsin, Madison.

Unger, S. H. (1958). A computer oriented toward spatial problems. *Proc. IRE* **46**, 1744−1750.

Unger, S. H. (1959). Pattern detection and recognition. *Proc. IRE* **47**, 1732−1752.

Valiant, L. G. (1981). Universality considerations in VLSI circuits. *IEEE Trans. Comput.* **30**, 135−140.

Valiant, L. G., and Brebner, G. J. (1981). Universal schemes for parallel communication. *Proc. 13th Ann. Symp. Theory Comput., 1981* pp. 263−277.

Vick, C. R., and Cornell, E. J. (1978). Pepe architecture−present and future. *Proc. AFIPS Natl. Comput. Conf.* **47**, 981−992.

Vick, C. R., Kartashev, S. P., and Kartashev, S. I. (1980). Adaptable architectures for supersystems. *Computer* **13**, 17−34.

von Neumann, J. (1945). "First Draft of a Report on the EDVAC." Moore School of Electrical Engineering, University of Pennsylvania Philadelphia (reprinted in part in Randell, 1973).

von Neumann, J. (1959). "The Computer and the Brain." Yale Univ. Press, New Haven, Connecticut.

von Neumann, J. (1966). A system of 29 states with a general transition rule. *In* "Theory of Self-Reproducing Automata" (A. Burks, ed.), pp. 132−156, 305−317. Univ. of Illinois Press, Urbana.

Vorgrimler, K. and Gemar, P. (1977). Structural programming of a multiprocessor system. *In* "Parallel Computers−Parallel Mathematics" (M. Feilmeier, Ed.), pp. 191−195) Int. Assoc. for Mathematics and Computers in Simulation.

Vuillemin, A. (1980). A combinatorial limit to the computing power of V.L.S.I. Circuits. *Proc. 21st Ann. Symp. Found Comput. Sci.*

Waterman, D. A. (1975). Adaptive production systems, *Proc. 4th Int. Joint Conf. Artificial Intelligence* pp. 296−303.

Widdoes, L. C. (1980). The S-1 project: Developing high-speed digital computers. *Proc. Compcon 80*, pp. 282−291.

Wilkov, R. S. (1970). Construction of maximally reliable communication networks with minimum transmission delay. *Proc. IEEE Int. Conf. Commun.*, **6**, 4210−4215.

Wilkov, R. S. (1972). Analysis and design of reliable computer networks. *IEEE Trans. Commun.* **20**, 660−678.

Wirth, N. (1971). Design of a PASCAL compiler. *Software Practice Experience* **1**, 309−333.

Wirth, N. (1977a). Toward a discipline of real-time programming. *Comm. ACM* **20**, 577−583.

Wirth, N. (1977b). Modula: A language for modular multi-programming. *Software Practice Experience* **7**, 3−35.

Wittie, L. D. (1976). Efficient message routing in mega-micro-computer networks. *Proc. 3rd Ann. Symp. Comput. Archit.*

Wittie, L. D. (1978). MICRONET: A reconfigurable microcomputer network for distributed systems research. *Simulation* **31**, 145−153.

Wittie, L. D., and van Tilborg, A. M. (1980). MICROS, a distributed operating system for MICRONET, a reconfigurable network computer. *IEEE Trans. Comput.* **29**, 1133−1144.

Wittie, L. D., Curtis, R. S., and Frank, A. J. (1982). MICRONET/MICROS − A network computer system for distributed applications. *In* "Multicomputer Algorithms and Image Processing" (K. Preston, Jr., and L. Uhr, eds.), pp. 307−318. Academic Press, New York.

Wood, A. (1977). "CAP4 Programmers Manual." Image Processing Group, University College, London.

Wu, C. L., and Feng, T. Y. (1978). Routing techniques for a class of multistage interconnection networks. *Proc. Int. Conf. Parallel Process.*, pp. 197−205.

Wu, C. L., and Feng, T. Y. (1981). The universality of the shuffle-exchange network. *IEEE Trans. Comput.* **30**, 324−332.

Wulf, W. A., and Bell, C. G. (1972). C.mmp−a multi-mini processor. *Proc. AFIPS Natl. Comput. Conf.* **41**, Part II, 765−777.

Wulf, W. A., Levin, R., and Harbison, S. P. (1981). "HYDRA/C.mmp: An Experimental Computer System." McGraw-Hill, New York.

Yao, A. C. (1981). The entropic limitations on VLSI computation. *Proc. 13th Ann. Symp. Theory Comput.*, pp. 308−311.

You, S. S., and Fung, H. S. (1977). Associative processor architecture−a survey. *Comput. Surveys* **9**, 3−27.

Zuse, K. (1958). Die Feldrechenmaschine. *Math., Tech., Wirtschaft-Mittelungen* **4**, 213−220.

Glossary

algorithm: A precise step-by-step effective procedure for solving some problem.

ALU: Arithmetic logic unit. The set of special-purpose processors in a conventional CPU that execute the standard set of arithmetic and logic instructions.

arc: A directed link in a digraph.

array: A two-dimensional structure over pieces of information such that each piece is stored in one cell of the array, and cells are linked to several nearest neighbors (usually to the four square, to the east, west, north, and south, or to the six hexagonal, or to the eight square and diagonal). This might be an array of computers, processors, memory, image pixels, or other objects.

ASPRO: A STARAN-like array built using VLSI chips with 32 processors on each chip, so that a 1024-processor system fits in a box substantially smaller than 1 cubic foot.

associative memory: A memory that is "content addressable"—that is, where every memory register that contains the specified string of symbols is accessed, rather than the single register whose location is specified.

binary: two-valued.

bipartite graph: A graph whose nodes can be partitioned into two subsets so that every link joins nodes in different subsets.

bit: BInary digiT. A symbol that can take on two values (i.e., is binary). Decimal numbers, letters of the alphabet, and any other symbol can be encoded as a string of binary symbols. This is usually done in today's computers, using switches to store and represent the binary encoded information.

boolean logic: Binary two-valued logic (named after George Boole, who algebraicized logic).

buffer: Intermediate memory store, often associated with input and output devices that must handle large bursts of data.

bus: A common connector (e.g., a coaxial cable connecting computers and other devices in different rooms, or a wire on a VLSI chip).

byte: An 8-bit string (enough to encode 256 different symbols).

cache: A small, very fast memory into which the computer loads information that it anticipates the processor will need to keep things moving as fast as possible.

CAM: Content addressable memory. See *associative memory*.

cellular array: An array (usually two-dimensional) of simple processors, each with its own memory, and with near-neighbor interconnections.

chip: A tiny (roughly 4-mm-square) piece of very pure flat (usually silicon) substrate on which thousands of electronic devices are printed, using LSI and VLSI technologies.

CLIP: Cellular Logic Information Processor. The series of parallel array computers being designed and constructed by M. J. B. Duff and his research group at University College, London. CLIP3 (1974) is a 12×16 array. CLIP4 (1979) is a 96×96 array of processors, each with 32 bits of memory. CLIP4 uses a custom-built CMOS chip with 8 processors plus their 3 registers and 32 bits of memory on each chip.

cluster: With respect to graphs and computer networks, a relatively small (sub)graph chosen for structural and/or formal properties appropriate to the problems for which it will be used, and compounded or tessellated or embossed into a larger graph structure.

CMOS: Complementary Metal−Oxide Semiconductor. The fastest of the metal oxide−semiconductor family of technologies, but usually slower than ECL or TTL. Low in power consumption and high in packing density.

code: A program for a computer is made up of lines of code, or basic instructions.

compiler: When one codes a program in a higher-level language, another program, the compiler for that language, is used to translate that program into its equivalent machine-language program.

complete bipartite graph: In graph theory, a graph with two sets of nodes, where each node in the first set is joined by one link to every node in the second set, and vice versa. For example, the graph $G_{m,n}$ is a complete bipartite graph with m nodes in the first set, each linked to all n nodes in the second set.

complete graph: In graph theory, a regular graph where every pair of nodes is connected by one link.

complete n-partite graph: In graph theory, a graph with n sets of nodes, where each node in each set is joined by one link to every node in every other set.

compounding operation: With respect to graphs, an operation that constructs a larger graph by compounding smaller graph clusters.

computer: A system built from switches (transistors, gates) that execute structures of logical operations to carry out the instructions given to it in a "program." The computer's major parts are the "CPU" ("central processing unit," usually containing one processor, or a set of specialized processors, plus a controller), "memory store," and "input−output" channels and devices. Today "computer" usually is taken to mean "stored-program" "serial" "digital" "general-purpose" computer. In this book the word is used a bit more broadly, to include "parallel" multicomputers, and occasionally a bit more narrowly, to include a processor-plus-memory, but possibly without its own controller.

content addressable memory: See *associative memory.*

controller: The hardware in the computer (usually considered a part of the CPU in traditional serial computers) that handles all the details of program execution (e.g., fetching, decoding, and executing the stream of instructions), input and output of data (from and to disks, tapes, etc.) and a variety of other control functions, as necessary.

CPU: Central processing unit. The single processor (including the controller) in a conventional 1-processor computer.

Cray-1: The fastest and most powerful single-CPU "supercomputer" built. It uses a pipeline and several special-purpose processors in parallel to speed up processing. (Named after Seymour Cray, who also designed the CDC 6600).

cross-point switch: An array of switches that can be thrown to connect any device linked to one of its rows to any device linked to one of its columns.

cycle: In graph theory, two different paths between the same pair of nodes.

Cytocomputer: A long pipeline of over 100 processors specialized for image processing, designed and built by Stanley Sternberg in 1978 at the Environmental Research

Institute of Michigan, in Ann Arbor. A new LSI version that can have pipelines of customized length is now being manufactured.

DAP: Distributed Array Processor. The parallel array computer designed and constructed by Stewart Reddaway and his group at ICL Research and Development, Stevenage, England. The Pilot DAP (1976) is a 32×32 array with 2000 bits of memory per processor. The production DAP (1980) is a 64×64 array with 4000 bits of external RAM memory. A custom TTL chip has been built, each chip with 4 processors.

data: Information (numbers, symbols, arrays, records, and other structures, usually represented as binary strings) that the program operates on.

data-flow graph: A directed graph used to represent a data-flow program, where nodes are instructions or processes whose outputs pass along links to subsequent processes. A node executes if all its inputs (that is, the nodes linking into it) have been enabled with data. Very similar to Petri nets.

degree: In graph theory, the number of links touching a node.

density: With respect to graphs, the number of nodes in a regular graph of given degree and diameter.

device: A basic component of a computer. The terms "device," "transistor-equivalent device," "gate," and "element" are often used as rough equivalents.

diameter: In graph theory, the longest shortest path between any two nodes.

digraph: A directed graph, where each link goes in only one of the two possible directions between the nodes it joins.

diode: A two-state silicon device used along with transistors as a basic component for modern electronic (including computer) circuits.

directed graph: See *digraph*.

distributed networks: Networks of computers, usually loosely coupled as in ARPANET and ETHERNET, but often used loosely to include all kinds of networks.

ECL: Emitter-coupled logic. The fastest of the widely used technologies for LSI and VLSI chips. Power consumption and dissipation are high; packing density is low.

edge: See *link*. In graph theory, the more commonly used term.

emboss: With respect to graphs, replacing a node of degree n by an n-node graph cluster, joining one link to each node.

finite-state automaton: A simple system (it might be embodied in a physical machine or represented by a directed graph) for executing processes. The "Moore machine" version moves from one of a finite set of states to a second, determined by the first state and an input; it then outputs, determined by this second state.

flip-flop: A simple circuit (built using $2-10$ transistors) that flips, or flops, into either of two stable states. The basic component of many high-speed memory stores.

gate: A device that computes a primitive logical function. Depending on the technology, a simple gate might need from 1 to 12 transistor-equivalent devices. For comparison purposes, this book usually equates gate and switch, assuming each needs roughly one transistor-equivalent device.

general-purpose computer: A computer that has the Turing machine capability of effective computability; able to execute any program that might be executed by a universal computer (Turing machine).

girth: In graph theory, the length of the shortest cycle from any node through other nodes and returning to that node, traversing no link more than once.

Goodyear-Aerospace: The division of Goodyear Tire, Akron, Ohio, that designs and builds parallel computers (as well as airplane and missile-related equipment, including tires, clutches, and blimps). The electronics group has probably done more work on parallel computers than any other group in the world, starting with STARAN (1974) through ASPRO (1981) and the MPP (1982).

graph: In graph theory, a set of nodes and links connecting them. (Usually refers to a connected graph, where an unbroken path exists between any two nodes.)

IC: Integrated circuit. A chip that has several devices. Continuing miniaturization turned the transistors of the 1950s into the ICs of the 1960s and the LSI chips of the 1970s and VLSI chips of the 1980s. ICs are still widely used. They give the greatest flexibility in designing and building a computer, and even when LSI chips (which are themselves ICs) are used, additional chips with only a few devices each are also needed.

ICL: International Computers Limited. The U.K. amalgamation of their largest computer manufacturers; until recently the world's fifth largest computer company, and the largest outside the U.S.

ILLIAC III: A system of three computers, including a 32×32 array, built (but never operational) by Bruce McCormick and his associates at the University of Illinois in the 1960s.

ILLIAC IV: The largest computer ever built. It was designed by Daniel Slotnick and his associates at the University of Illinois in the 1960s, and built by Burroughs. Planned to have 256 64-bit processors, it was actually completed with 64.

information: Any string of symbols used by the computer, whether as part of the system's programs, physical switch states, user programs, data input to programs, or information generated by programs (e.g., intermediate and final results, messages).

input (device): A piece of hardware that inputs information (e.g., a keyboard, tape reader, or television camera).

language: A program is coded in a programming language for which there is a compiler that translates that program into machine-language code for a particular computer, which can then execute that program.

lattice: An N-dimensional (often three-dimensional) structure with many nodes arranged in a very regular fashion, each node linked to a small number of near neighbors. A two-dimensional lattice is often called a (cellular) array.

link: A wire or cable connecting two nodes in a network. A connection (edge, line) between two nodes (vertices, points) in a graph.

LSI: Large-scale integration. Refers loosely to chip fabrication technologies where from several hundred to 50,000 or so devices are deposited on a single chip.

mega: Prefix meaning "million" or "very large."

memory: A store for information inside the computer.

milli: Prefix meaning "thousandth."

model: A representation inside the computer of some physical system that is being studied, by simulation, analysis, or some type of modeling. In this book, frequently used to refer to the modeling of three-dimensional masses of matter represented in three-dimensional arrays (e.g., wind tunnels, storms, missiles, turbulent water or air, earthquakes).

MPP: Massively Parallel Processor. The largest parallel array computer (preliminary design by David Schaeffer and his group at NASA-Goddard, the final design and construction by Kenneth Batcher and his group at Goodyear-Aerospace). The full system (1983) is to have 128×128 computers, each with 1000 bits of external RAM memory and a 32-bit high-speed register. All processors will execute the same instruction, in SIMD mode. A custom 40-MHz chip will have 8 processors and the registers on each chip.

multicomputer: A system with more than one computer (that is, processor plus that processor's main memory). It might have one controller for all processors (e.g., the ILLIAC IV as completed) or for a subset of processors (e.g., the planned ILLIAC IV and PASM); or it might have a controller for each processor.

multiprocessor: A system with more than one processor (often used loosely as synonymous with "multicomputer").

nano: Prefix meaning "billionths" (10^{-9}).

network: A structure of more than one computer, all connected together. There is a range of coupling, from loosely distributed to very tightly coupled networks.

network flow graphs: Directed graphs where nonnegative integer bandwidths are assigned to edges, and two disjoint sets of nodes are designated as "source" (into which material or information is input) and "sink" (from which finished products or solutions output). Used to investigate a variety of problems related to the overall channel capacity, bottlenecks, and productivity of a network (e.g., of machines in a factory, roads in a country, trunk lines in a telephone system).

node: A piece of hardware in a network. It might be processor, memory store, input device, output device, or register. A vertex or point in a graph.

NP: Nondeterministic polynomial. An NP problem can be solved in polynomial time only by a nondeterministic Turing machine, and therefore needs exponential time on a deterministic Turing machine. Any but a very small NP problem needs too much computing time, and is intractable. Many combinatorial problems, including many graph matching, network flow, and artificial intelligence problems, are NP. For example, the isomorphism of two graphs is NP (of a graph with a subgraph is NP-complete, which is even worse, as is $n \times n$ checkers).

$O(\cdots)$: Order $\cdots$ For example, "$O(\log n)$" means "of order $\log n$."

output (device): A piece of hardware that outputs information from the computer (e.g., a printer, plotter, or speech synthesizer).

parallel computer: A computer that executes several processes all at the same time, in parallel.

Petersen graph: A highly symmetric graph (see Appendix A for its picture) with 10 nodes and diameter 2, girth 5, degree 3.

Petri nets: Directed bipartite graphs of "places" and "transitions." Tokens are put in places. A transition "fires" if all places pointing to it have tokens. The result is the removal of one token from each of the transition's input places and the putting of one token in each of its output places. Used (in a number of variations) to model data flow and other types of languages and computers.

PICAP: A powerful picture processing system that combines a pipeline, several special-purpose processors, and a single-CPU computer host. Designed and built by Bjorn Kruse and his associates, at the University of Linköping, Sweden: PICAP I in 1978; PICAP II in 1980.

pico: Prefix meaning "trillionths" (10^{-12}).

pin: An IC, LSI, or VLSI chip has a small number of pins that serve to pass information to and from that chip.

pipeline: A set of processors, strung one after the other, through which data are pumped. Each processor repeatedly executes the same instruction on each successive piece of data, much as in an assembly line each worker effects the same process on each successive item.

pixel: Jargon for "picture element," or a single cell in an array containing an image (e.g., a spot in a TV picture).

planar: In graph theory, a graph is planar if it can be drawn on the plane without any lines crossing. (Any graph can be constructed in three dimensions.)

port: The place on a processor, memory, or other node where a bus or other link interfaces and connects to that node.

procedure: A self-contained set of code in a program that effects some subpart of the larger program (much like a "subroutine" or a "function"). We commonly talk of (hardware) processors executing (software) programs and procedures.

process: A program or some more or less self-contained transformation that is actually being executed by a processor.

processor: The set of gates that transforms data. The (physical) hardware that executes the (software) program's processes.

program: The set of instructions one writes in some programming language for some computer to execute some algorithm on that computer.

pyramid: A network in the general shape of a pyramid or a cone, often constructed from successively smaller arrays.

RAM: Random-access memory chip.

register: A small very high-speed memory store large enough to contain a word or two of information (used to make this information accessible to a processor as quickly as possible) is typically associated with each processor. The cache and main memory stores are also made of (slower) registers.

regular: In graph theory, a regular graph has the same degree (i.e., the same number of links touching) at every node.

ROM: Read only memory chip.

serial: One after another, one at a time. A serial computer carries out a sequential process by executing one instruction at a time.

software: Programs (i.e., information read into and stored in the computer's memory), which then tell the computer what sequence of instructions to carry out.

STARAN: A large one-dimensional array of processors connected to an associative memory through a "flip network" that serves to reconfigure that memory under program control. Designed and built by Kenneth Batcher and his associates at Goodyear-Aerospace.

store: A memory that stores and keeps a record of information.

supercomputer: A loose term for a very big, expensive, and, one hopes, powerful computer.

switch: A device that can change from one state to another. Today, binary two-state switches are almost always used.

systolic: A pipenet pipelinelike carefully designed structure of gates that computes some special-purpose function as efficiently as possible. As developed chiefly by H. T. Kung and his colleagues at Carnegie-Mellon University, systolic arrays are probably the best example we have today of algorithms designed in silicon.

tessellate: In graphs and networks, connect clusters (subgraphs) into a regular tilelike mosaic of repeated clusters.

threshold device: A threshold value must be exceeded for the device to fire (similar to a neuron that fires when enough neurons fire into it to exceed its threshold).

tightly coupled network: Loosely used term often applied when all processors share a common memory, are mastered by a single controller, or have good point-to-point links.

tile: In graphs and networks, tessellate.

toroidal structure: A structure whose opposite sides are connected, giving cycles. For example, a toroidal array connects the first and last columns and the top and bottom rows.

transistor: A solid-state amplifying device made from silicon that replaced the vacuum tube in the 1950s. In computer circuits it is usually used as a switch to build flip-flops and logic gates.

tree: In graph theory, a connected graph with no cycles. A network that has the form of a tree, usually but not always without cycles (see, e.g., the augmented X-trees with rings and shuffles).

TTL: Transistor–transistor logic. Fast, but slower than ECL. Power consumption and dissipation are high; packing density is low. The most commonly used technology for building large, relatively fast ICs and circuits on LSI and VLSI chips.

Turing machine: A very simple general-purpose logical construction of a computer that has been used to prove the generality and equivalence of all computers that have a Turing machine's capabilities (as do all modern computers), such that each can compute anything (i.e., carry out any algorithm) that any other can (given enough time). Essentially, a finite-state automaton with a potentially infinite memory (i.e., as much memory as is needed for whatever program it is asked to execute). (Named after Alan Turing.)

vacuum tube: A "radio tube" that can be used as a switch, and therefore as the basic component in a computer. (Obsolete.)

vertex: In graph theory, the elements (nodes) connected by edges (links), to form graphs.

VLSI: Very large-scale integration. Somewhat arbitrarily, when the number of transistor-equivalent devices on a single chip reaches 50,000 or more, it is characterized as a VLSI chip.

wafer: The large disk-shaped substrate on which several hundred chips are made. Typically, a 3-in-diameter wafer is diced into the individual 4−6-mm-square chips.

whorl: A very large three-dimensional network with near-neighbor connections and a general shape like a pyramid, lattice, or sphere, into which programs can be mapped so that their data swirls (much as in a three-dimensional pipeline) efficiently through processors.

window: CLIP, PICAP, and the Cytocomputer, along with other image processing systems, allow the programmer to write code that examines a local 3×3 window surrounding each cell. Pixal, PascalPL, and other languages for parallel processors offer a primitive window construct, to exploit and reflect this capability. The term is, however, a rather loose one, often meaning a local subarray of unspecified size.

word: A string of binary symbols that is input, processed, and output in parallel by the computer. A "1-bit," "8-bit," "16-bit," or "32-bit" computer refers to the size of the word with which it works.

Author Index

Subject Index

Computer Science and Applied Mathematics

A SERIES OF MONOGRAPHS AND TEXTBOOKS

Editor
Werner Rheinboldt
University of Pittsburgh

HANS P. KÜNZI, H. G. TZSCHACH, AND C. A. ZEHNDER. Numerical Methods of Mathematical Optimization: With ALGOL and FORTRAN Programs, Corrected and Augmented Edition

AZRIEL ROSENFELD. Picture Processing by Computer

JAMES ORTEGA AND WERNER RHEINBOLDT. Iterative Solution of Nonlinear Equations in Several Variables

AZARIA PAZ. Introduction to Probabilistic Automata

DAVID YOUNG. Iterative Solution of Large Linear Systems

ANN YASUHARA. Recursive Function Theory and Logic

JAMES M. ORTEGA. Numerical Analysis: A Second Course

G. W. STEWART. Introduction to Matrix Computations

CHIN-LIANG CHANG AND RICHARD CHAR-TUNG LEE. Symbolic Logic and Mechanical Theorem Proving

C. C. GOTLIEB AND A. BORODIN. Social Issues in Computing

ERWIN ENGELER. Introduction to the Theory of Computation

F. W. J. OLVER. Asymptotics and Special Functions

DIONYSIOS C. TSICHRITZIS AND PHILIP A. BERNSTEIN. Operating Systems

PHILIP J. DAVIS AND PHILIP RABINOWITZ. Methods of Numerical Integration

A. T. BERZTISS. Data Structures: Theory and Practice, Second Edition

N. CHRISTOPHIDES. Graph Theory: An Algorithmic Approach

SAKTI P. GHOSH. Data Base Organization for Data Management

DIONYSIOS C. TSICHRITZIS AND FREDERICK H. LOCHOVSKY. Data Base Management Systems

JAMES L. PETERSON. Computer Organization and Assembly Language Programming

WILLIAM F. AMES. Numerical Methods for Partial Differential Equations, Second Edition

ARNOLD O. ALLEN. Probability, Statistics, and Queueing Theory: With Computer Science Applications

ELLIOTT I. ORGANICK, ALEXANDRA I. FORSYTHE, AND ROBERT P. PLUMMER. Programming Language Structures

ALBERT NIJENHUIS AND HERBERT S. WILF. Combinatorial Algorithms, Second Edition

AZRIEL ROSENFELD. Picture Languages, Formal Models for Picture Recognition

ISAAC FRIED. Numerical Solution of Differential Equations

ABRAHAM BERMAN AND ROBERT J. PLEMMONS. Nonnegative Matrices in the Mathematical Sciences

BERNARD KOLMAN AND ROBERT E. BECK. Elementary Linear Programming with Applications

CLIVE L. DYM AND ELIZABETH S. IVEY. Principles of Mathematical Modeling

ERNEST L. HALL. Computer Image Processing and Recognition

ALLEN B. TUCKER, JR. Text Processing: Algorithms, Languages, and Applications

MARTIN CHARLES GOLUMBIC. Algorithmic Graph Theory and Perfect Graphs

GABOR T. HERMAN. Image Reconstruction from Projections: The Fundamentals of Computerized Tomography

WEBB MILLER AND CELIA WRATHALL. Software for Roundoff Analysis of Matrix Algorithms

ULRICH W. KULISCH AND WILLARD L. MIRANKER. Computer Arithmetic in Theory and Practice

LOUIS A. HAGEMAN AND DAVID M. YOUNG. Applied Iterative Methods

I. GOHBERG, P. LANCASTER, AND L. RODMAN. Matrix Polynomials

AZRIEL ROSENFELD AND AVINASH C. KAK. Digital Picture Processing, Second Edition, Vol. 1, Vol. 2

DIMITRI P. BERTSEKAS. Constrained Optimization and Lagrange Multiplier Methods

JAMES S. VANDERGRAFT. Introduction to Numerical Computations, Second Edition

FRANÇOISE CHATELIN. Spectral Approximation of Linear Operators

GÖTZ ALEFELD AND JÜRGEN HERZBERGER. Introduction to Interval Computations. Translated by Jon Rokne

ROBERT R. KORFHAGE. Discrete Computational Structures, Second Edition

MARTIN D. DAVIS AND ELAINE J. WEYUKER. Computability, Complexity, and Languages: Fundamentals of Theoretical Computer Science

LEONARD UHR. Algorithm-Structured Computer Arrays and Networks: Architectures and Processes for Images, Percepts, Models, Information

In preparation

O. AXELSSON AND V. A. BARKER. Finite Element Solution of Boundary Value Problems: Theory and Computation

PHILIP J. DAVIS AND PHILIP RABINOWITZ. Methods of Numerical Integration, Second Edition

ROBERT H. BONCZEK, CLYDE W. HOLSAPPLE, AND ANDREW B. WHINSTON. Micro Data Base Management: Practical Techniques for Application Development

NORMAN BLEISTEIN. Mathematical Methods for Wave Phenomena

PETER LANCASTER AND MIRON TISMENETSKY. The Theory of Matrices: With Applications, Second Edition